Jaap Wijker

Mechanical Vibrations in Spacecraft Design

Springer
Berlin
Heidelberg
New York
Hong Kong
London
Milan
Paris
Tokyo

Jaap Wijker

Mechanical Vibrations in Spacecraft Design

With 120 Figures and 29 Tables

Springer

Jaap Wijker

Dutch Space BV
P.O. Box 32070
2303 DB Leiden
The Netherlands

E-mail: j.wijker@dutchspace.nl

ISBN 3-540-40530-5 Springer-Verlag Berlin Heidelberg New York

Cataloging-in-Publication Data applied for.
Bibliographic information published by Die Deutsche Bibliothek. Die Deutsche Bibliothek lists this publication in the Deutsche Nationalbibliografie; detailed bibliographic data is available in the Internet at <http://dnb.ddb.de>.

Springer-Verlag Berlin Heidelberg New York
a member of BertelsmannSpringer Science+Business Media GmbH

http://www.springer.de

Printed in Germany

Typesetting: Dataconversion by author
Cover-design: medio, Berlin
Printed on acid-free paper 62 / 3020 hu – 5 4 3 2 1 0

Dedicated to my mother

Maartje Wijker-Gravemaker

and to the memory of my father

Job Wijker

Preface

This book about mechanical vibrations focuses on spacecraft structures design and reflects my experiences gained at Dutch Space B.V., formerly Fokker Space B.V., Fokker Space & Systems B.V. and the Space Division of Fokker Aircraft B.V., over a period of about 30 years.

Many books about mechanical vibrations have been published, however, in spacecraft structures design, many vibration topics are applied but can be read in different books. I have collected in this book most of the topics about mechanical vibrations techniques usually applied in spacecraft structures design.

I work as a part-time associate professor at the Chair Aerospace Structures & Computational Mechanics, Faculty of Aerospace Engineering, Delft University of Technology, and lecture "Spacecraft Structures" in the Master's program. The scientific environment at the university, in combination with my work in the aerospace industry, has amplified the wish to write a book about mechanical vibrations with focus on spacecraft structures design. To bring together most of the techniques of modal and dynamic response analysis is my greatest motivation to write this book.

I would like to express my admiration for the patient attitude of my wife Wil during the time I was preparing this book.

I would also like to acknowledge my colleagues at Dutch Space and the Delft University of Technology in general, but in particular I would like to thank my collegue John Tyrrell at Dutch Space, for all the discussions we had about vibration problems within the framework of spacecraft structures projects, and Gillian Saunders-Smits at the Delft University of Technology for reading the English text. Also, I would like to thank Bas Franssen for reading the sections on the Mode Acceleration Technique and Load Transformation Matrices.

Jaap Wijker
Velserbroek 2003

Table of Contents

1 Introduction

1.1 Why Another Book about Mechanical Vibrations?

Placing spacecraft (S/C) or satellites into an orbit around the Earth or in our Solar System is done by expendable and reusable launch vehicles, ELV (e.g. ARIANE 5, Atlas, Delta, etc.) and RLV (e.g. Shuttle) respectively. The spacecraft is placed on top of the ELV or in the cargo-bay of the RLV.

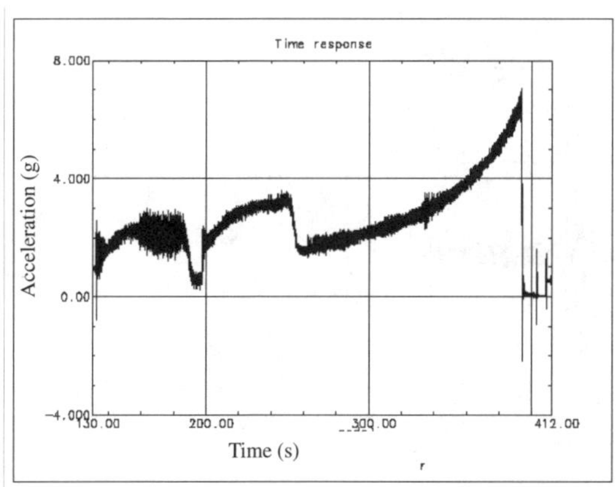

Fig. 1.1. Total acceleration measured during launch of the ACE S/C with Delta II launcher (Courtesy Goddard Space and Flight Centre, FEMCI[1] pages)

The propulsion forces, the aerodynamic forces, the acoustic and shock loads during launch of the S/C strongly interact with the low- and medium-

1. Finite Element Modelling Continuous Improvement

frequency dynamic characteristics of the launch vehicle (L/V) and will introduce mechanical vibrations throughout the L/V and also at the interface with the spacecraft.

The propulsion forces from the stage main engines and boosters will accelerate the L/V and will result in inertia forces. The total inertia force the S/C will encounter during launch is the summation of steady-state and vibrations loads and will help the engineer to dimension the S/C primary and secondary structures. This is illustrated in Fig. 1.1.

The primary structure is the backbone of the S/C structure and is considered to be the primary load path. The total inertia loads are frequently called the quasi static design loads when safety factors to cover uncertainties are introduced. The design of the S/C structures against the quasi static loads, considering minimum natural frequency considerations (stiffness), belongs to the area of strength of materials. If minimum natural frequency requirements are met, the S/C may be considered as a rigid body on top of or in the cargo-bay of the L/V.

The pure mechanical vibrations are illustrated in Fig. 1.2. The steady-state accelerations are subtracted. These mechanical vibrations are generally categorised as follows:

- Sinusoidal vibrations, 5–100 Hz;
- Random vibrations, 20–2000 Hz;
- Shock loads, accelerations, 100–5000 Hz.

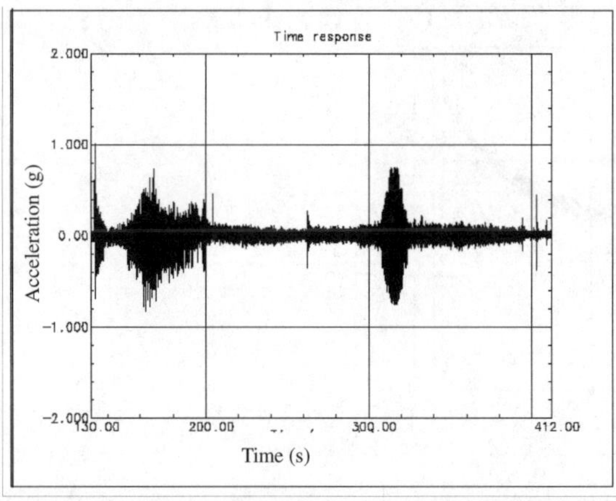

Fig. 1.2. Vibrations during launch of the ACE S/C with Delta II launcher (Courtesy Goddard Space and Flight Center, FEMCI pages)

The shock loads have very short time periods (0.5 ms) in comparison to the time cycles associated with the lowest and lower natural frequencies of the spacecraft (10–50 Hz). Shock loads are mostly represented in the form of a shock-response spectrum (SRS). The SRS reflects the maximum absolute acceleration of a single degree of freedom (sdof) excited at the base by the transient acceleration representing the short-period shock. An illustration of an SRS is given in Fig. 1.3. Shock loads are caused by the separation of stages and by the ignition and cutoff of engines. However, the separation of the spacecraft from the L/V will introduce, in general, the most severe shocks in the spacecraft. Firing of pyros and the latching of deployable systems, e.g. antennae, solar arrays, etc., cause shock loads in the spacecraft internally.

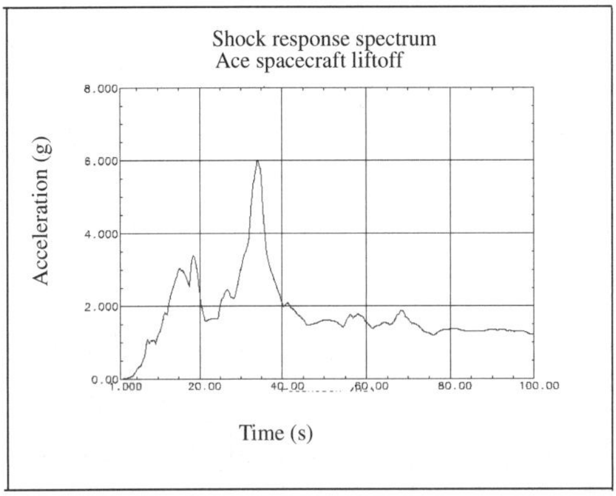

Fig. 1.3. Shock-response spectrum (SRS). (Courtesy Goddard Space and Flight Center, FEMCI pages)

In addition to the mechanical vibrations the outside of the spacecraft is also exposed to acoustic pressure (20–10000 Hz). Lightweight large-area structures, e.g. solar arrays, antenna dishes, are sensitive to acoustic loads. An illustration of the acoustic pressures is given Fig. 1.4. The acoustic pressure p has a random nature and is mostly denoted in sound pressure levels (SPL), specified in dB with a reference pressure $p_{ref} = 2x10^{-5}$ Pa. The SPL is defined as

$$SPL = 10\log\left(\frac{p^2}{p_{ref}^2}\right). \tag{1.1}$$

Acoustic excitation of the spacecraft causes mechanical vibrations of a random nature on top of the random mechanical vibrations enforced via the I/F spacecraft with the L/V.

Summarising, the spacecraft will encounter severe vibrations during launch and later less severe vibrations when the spacecraft is placed in orbit and the deployable appendices are released and latched.

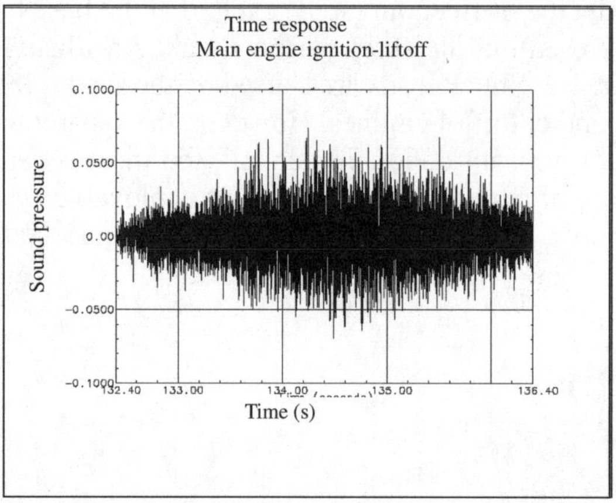

Fig. 1.4. Acoustic pressures (Pa). (Courtesy Goddard Space and Flight Center, FEMCI pages)

Structural engineers designing and analysing spacecraft structures, solar arrays, antennae, instruments, equipment, etc., have to investigate the structural responses (e.g. accelerations, forces, stress) of the spacecraft and its components to mechanical vibrations and acoustic loads. The exposure to mechanical vibration and acoustic loads is illustrated in Fig. 1.5.

The calculation of dynamic responses in the spacecraft, the primary structure, the secondary structures, the instruments, the equipment, etc., is called the dynamic analysis.

Besides the dynamic responses the modal properties of the complete spacecraft and subsystems, e.g. natural frequencies, mode shapes, effective masses, are of great interest. The modal and dynamic response analyses are mostly done using finite element methods [Cook 89, Petyt 90].

All typical and special modal and response analysis methods, applied within the frame of the design of spacecraft structures, are described in this book. To bring together most of the techniques of modal and dynamic response analysis is the greatest motivation to write this book.

The mechanical vibration topics discussed with focus on the design of spacecraft structures are:

- Single degree of freedom (sdof) systems
- Damping models
- Multi-degrees of freedom (mdof) systems
- Modal analysis, e.g. natural frequencies, mode shapes, modal effective masses
- Dynamic response analysis
    - Deterministic in frequency and time domain
    - Random vibration
    - Shock response spectrum

- Low-frequency acoustic loads, structural responses
    - Statistical energy analysis
    - Mode acceleration
    - Residual vectors
- Free-free dynamic systems
    - Inertia relief
- Reduced dynamic models
    - Dynamic model reduction
    - Component mode synthesis
    - Load transformation matrices

The above mentioned topics will be previewed in the next section.

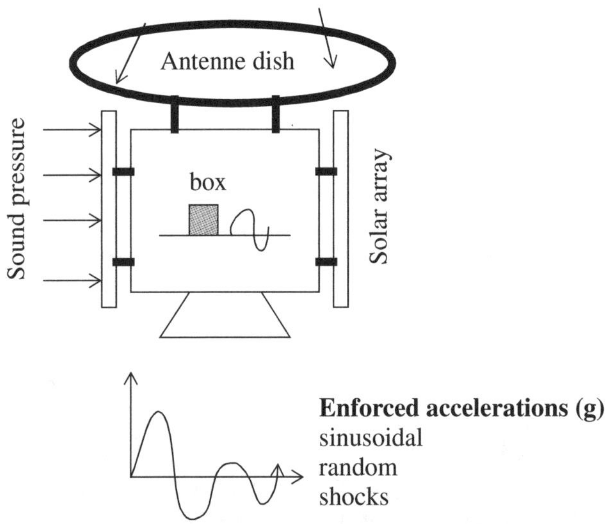

Fig. 1.5. Spacecraft exposed to mechanical and acoustic vibrations

1.2 A Short Overview of Theory

1.2.1 Single Degree of Freedom (sdof) Systems

Spacecraft are, mechanically speaking, dynamically excited via the interface between the L/V and the S/C. This is more or less illustrated in Fig. 1.6. Thus within the frame of sdof dynamic system base excitation (e.g. enforced acceleration) is of great importance!

An sdof system with a discrete mass m, a damper element c and a spring element k is placed on a moving base, which is accelerated with an acceleration $\ddot{u}(t)$. The natural frequency is given by $\omega_n = \sqrt{\frac{k}{m}}$ (rad/s) and the damping ratio by ζ. The resulting displacement of the mass is $x(t)$. We introduce a relative motion $z(t)$ which is the displacement of the mass with respect to the base. The relative displacement is

$$z(t) = x(t) - u(t). \tag{1.2}$$

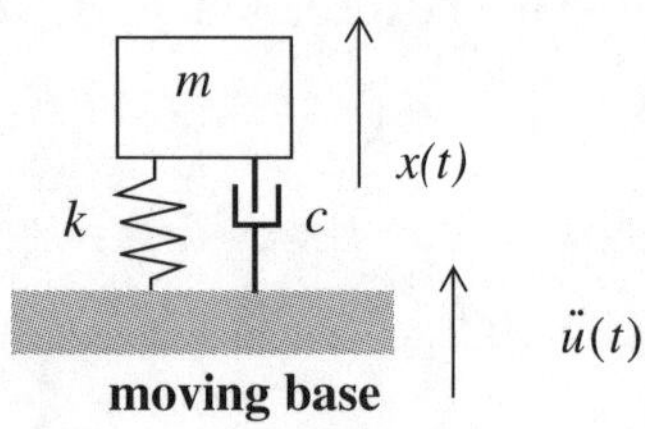

Fig. 1.6. Enforced acceleration on a damped sdof system

The equation of motion for the relative displacement $z(t)$ becomes

$$\ddot{z}(t) + 2\zeta\omega_n\dot{z}(t) + \omega_n^2 z(t) = -\ddot{u}(t). \tag{1.3}$$

with

- ζ damping ratio, $\zeta = \frac{c}{2\sqrt{km}}$.

The enforced acceleration of the sdof system is transformed into an external force. The absolute displacement $x(t)$ can be calculated with

$$\ddot{x}(t) = \ddot{z}(t) + \ddot{u}(t) = -2\zeta\omega_n\dot{z}(t) - \omega_n^2 z(t). \tag{1.4}$$

1.2.2 Damped Vibrations

For treating mechanical vibrations, the single degree of freedom (sdof) system is the simplest mechanical oscillator. In general a damping element (linear or nonlinear) may be active in parallel with the linear spring. The absolute displacement of the discrete mass is defined as $x(t)$. Applying an excitation force $F(t)$ (N) on the mass m (kg), the equation of motion of such an sdof system is [Beards 96]:

$$\ddot{x}(t) + \frac{f_d}{m} + \omega_n^2 x(t) = \frac{F(t)}{m} = f(t) \tag{1.5}$$

where

- $\omega_n = \sqrt{\frac{k}{m}}$ the natural circular frequency (Rad/s)
- f_d the damping force (N).

The damping force f_d may be caused by:

- Material damping
- Air damping
- Acoustic damping
- Joint damping (rivets, bolts, bonding)

For linear damping, the damping is either linearly proportional to the velocity $\dot{x}(t)$, the so-called viscous damping, or is linearly proportional to the elastic force, $kx(t)$, the so-called structural damping.
Hence the damping force can be defined as:

- Viscous damping $f_d = c\dot{x}$ (N)
- Structural damping $f_d = jgkx$ (N).

The mentioned damping models are applied throughout of this book.

1.2.3 Multi-Degrees of Freedom (mdof) Dynamic Systems

A number of coupled linear sdof systems lead to a linear dynamic system with more degrees of freedom, the so-called multi-degrees of freedom system (mdof). Such a dynamic system is often called a discrete dynamic system with *n* discrete masses coupled with springs and dampers.

Continuous dynamic systems may be transferred to an mdof system using generalised coordinates [Meirovitch 70].

The application of finite element methods to continuous structures will result in a discrete mdof system where node displacements and rotations are the discrete unknowns [Cook 89].

In general the equations of motion of a discrete dynamic system can be written as

$$[M]\{\ddot{x}(t)\} + [C]\{\dot{x}(t)\} + [K]\{x(t)\} = \{F(t)\}, \tag{1.6}$$

and consists of the following matrices and vectors:

- the mass matrix $[M]$

- the stiffness matrix $[K]$
- the damping matrix $[C]$
- the force vector $\{F(t)\}$
- the displacement, velocity and acceleration vectors $\{x(t)\}$, $\{\dot{x}(t)\}$ and $\{\ddot{x}(t)\}$

For linear mdof systems the mass matrix, stiffness and damping matrix do not vary in time, however, the displacement, velocity, acceleration and force vector do usually change with time.

The equations of motion of mdof dynamic systems can be derived by applying, amongst others, Newton's law and Lagrange's equations [Meirovitch 70]. Both methods will be used to derive the equations of motion.

1.2.4 Modal Analysis

For structures exposed to dynamic forces the knowledge of the dynamic characteristics of these structures is of great importance. The most important dynamic intrinsic (modal) characteristics of linear dynamic systems are:

- The natural frequency
- The associated mode shape
- Damping

The natural frequencies and associated mode shapes may be analysed for both undamped and damped linear dynamic systems. The main emphasis in this book is on undamped modal characteristics because the damping in spacecraft structures is, in general, low.

The modal characteristics of an undamped linear dynamic system are:

- The natural frequencies
- The associated mode shapes
- Orthogonality relations of modes (normal modes)
- Effective masses
- Rigid-body modes

To calculate the natural frequencies of complex dynamic linear systems, in general, the finite element analysis method is applied. However, it is good practice at first to apply a method to approximately calculate the natural frequency of that system to get a feel for of the value of the natural frequency. The system will be simplified as much as possible in order to use approximated methods, e.g.

- The static displacement method
- Rayleigh's Quotient

- Dunkerley's Equation

1.2.5 Modal Effective Mass

The modal effective mass is a modal dynamic property of a dynamic structure associated with the modal characteristics; natural frequencies, mode shapes, generalised masses, and participation vectors. The modal effective mass is a measure to classify the importance of a mode shape when a structure will be accelerated via its base (enforced acceleration). A high effective mass will lead to a high reaction force at the base, while mode shapes with low associated modal effective mass are nearly excited by the base acceleration and will give low reaction forces at the base. The effect of local modes is not well described with modal effective masses [Shunmugavel 95]. The modal effective mass matrix is a $6x6$ mass matrix. Within this matrix the coupling between translations and rotations, for a certain mode shape, can be traced. The summation over all modal effective masses will result in the mass matrix as a rigid body. We define the $6x6$ modal effective mass $[M_{\mathrm{em},k}]$ as follows

$$[M_{\mathrm{em},k}] = \frac{[L_k]^T[L_k]}{m_k}, \tag{1.7}$$

with

- $[L_k]$ the modal participation factors associated with mode shape k
- m_k the generalised mass.

1.2.6 Response Analysis

This section briefly outlines the response analysis of a mdof linear dynamic system due to dynamic forces or enforced motions. Displacements, velocities and acceleration will also be discussed. The general equations of motion are set up and a partitioning between internal and boundary dofs has been made in order to solve the internal dofs because boundary motions were applied in combination with forces. When solving the equations of motion a distinction has been made between relative motions, motions with respect to the base, and absolute motions. Also a distinction has been made between redundant and nonredundant boundaries. The equations are applicable to solve the responses both in the time and the frequency domain.

In general, the equation of motion, in particular the internal dofs, are solved using the mode displacement method (MDM), because the full damping characteristics, meaning a full damping matrix, are not readily available.

The enforced acceleration of a structure, can be treated using three different methods:

- Relative motions: The absolute motion $\{x(t)\}$ of the dynamic system is separated into a relative motion $\{z(t)\}$ with respect to the base and the motion at the base $\{x_b(t)\}$. The absolute motion $\{x(t)\}$ is the summation of the motion at the base and the relative motion $\{x(t)\} = \{z(t)\} + \{x_b(t)\}$. The dynamic system has fixed free boundary conditions, either determined or nondetermined.
- Absolute motions: The absolute motions $\{x(t)\}$ of the dynamic system are calculated in a direct manner as a result of forces $\{F(t)\}$ and enforced motion $\{x_b(t)\}$. The dynamic system has fixed free boundary conditions.
- Large-mass approach: The dynamic system is a free-free structure. However, attached to the interface is a very large mass to introduce a force that results in the enforced motion $\{x_b(t)\}$ to calculate the absolute motions $\{x(t)\}$.

The damped equations of motion of an mdof dynamic system can be obtained from (1.6)

$$[M]\{\ddot{x}(t)\} + [C]\{\dot{x}(t)\} + [K]\{x(t)\} = \{F(t)\}.$$

1.2.7 Transient Response Analysis

Transient response analysis is the solution of a linear sdof or linear mdof system in the time domain. For linear mdof dynamic systems, with the aid of the modal superposition, the mdof system can be broken down into a series of uncoupled sdof dynamic systems. For a very few cases the analytical solution of the second-order differential equation in the time domain may be obtained and numerical methods are needed to solve the sdof and the mdof dynamic systems. Often, the numerical solution schemes are the time-integration methods. The time-integration methods may have fixed or nonfixed (sliding) time increments per time integration step, and will solve the equations numerically for every time step, taking into account the initial values, the equation of motion, either for sdof or mdof dynamic systems. The sdof dynamic system may be written as

$$\ddot{x}(t) + 2\zeta\omega_n\dot{x}(t) + \omega_n^2 x(t) = \frac{F(t)}{m} = f(t). \tag{1.8}$$

Implicit and explicit numerical solution schemes will be discussed, such as the Runge–Kutta method, The Houbolt method, the Wilson-θ method, etc.

1.2.8 Random Vibrations

By random vibrations of linear dynamic systems we mean the vibration of deterministic linear systems exposed to random (stochastic) loads.

During the launch of a spacecraft with a launch vehicle the spacecraft will be exposed to random loads of both mechanical and acoustic nature. The mechanical random loads are the base acceleration excitation at the interface between the launch vehicle and the spacecraft. The random loads are caused by several sources, i.e. the interaction between the launch-vehicle structure and the engine exhaust noise, combustion. Also, turbulent boundary layers will introduce random loads.

The theory of random vibrations of linear systems will be briefly reviewed.

The following equation to calculate root mean square (rms) of acceleration response $\ddot{x}(t)$ is very useful and is called Miles' equation, [Miles 54]. It is mostly written as:

$$\ddot{x}_{\text{rms}} = \sqrt{\frac{\pi}{2} f_{\text{n}} Q W_{\ddot{u}}(f_{\text{n}})}\,, \tag{1.9}$$

with

- $\ddot{x}_{\text{rms}}$ the rms acceleration
- $Q = \dfrac{1}{2\zeta}$ the amplification factor
- f_{n} the natural frequency (Hz)
- $W_{\ddot{u}}(f_{\text{n}})$ the power spectral density (PSD) of the enforced acceleration at the natural frequency f_{n}

Other interesting properties of narrow-banded stationary processes will also be discussed in this chapter:

- Number of crossings per unit of time through a certain level
- Fatigue damage due to random excitation

The mentioned properties are important properties for further investigation of the strength characteristics.

1.2.9 Shock-Response Spectrum

Separation of stages, such as the separation of the spacecraft from the last stage of the launch vehicle will induce very short duration loads in the internal structure of the spacecraft, the so-called shock loads. The duration of the shock load is generally very short with respect to the duration associated with the fundamental natural frequencies of the loaded dynamic mechanical system.

The effects of the shock loads are generally depicted in a shock response spectrum (SRS). The SRS is essentially a plot that shows the responses of a number of sdof systems to an excitation. The excitation is usually an acceleration–time history. An SRS is generated by calculating the maximum response of an sdof system to a particular base transient excitation. Many sdof systems tuned to a range of natural frequencies are assessed using the same input time history. A damping value must be selected in the analysis. A damping ratio $\zeta = 0.05$, $Q = 10$, is commonly used. The final plot, the SRS, looks like a frequency-domain plot. It shows the largest response encountered for a particular sdof system anywhere within the analysed time. Thus the SRS provides an estimate of the response of an actual product and its various components to a given transient input (i.e. shock pulse) [Grygier 97]. An example of an SRS is given in Fig. 1.7.

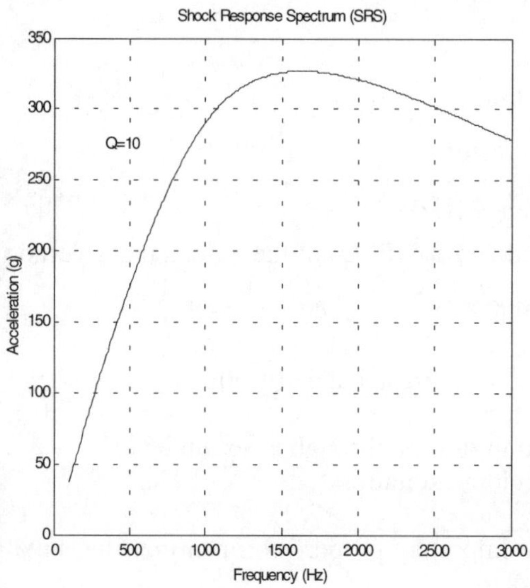

Fig. 1.7. SRS of half-sine pulse (HSP)

The response of an sdof system, due to enforced acceleration, will be reviewed.

Furthermore the calculation of SRSs will be discussed in detail. The maximum values occurring in time histories will be compared with the SRS approach and finally, it will be shown how an existing SRS can be matched (with synthesised decaying sinusoids).

1.2.10 Acoustic Loads, Structural Responses

By acoustic vibration we mean the structural responses of structures exposed to acoustic loads or sound pressures. In this chapter we discuss the low-frequency acoustic vibrations because the equations of motion are solved using the modal approach, namely mode superposition. In the higher-frequency bands the statistical energy analysis (SEA) is a good substitute for the classical modal approach.

In general, the modal characteristics of the dynamic system are calculated with the aid of the finite element method [Cook 89]. The accuracy is determined by the detail of the finite element model and the complexity of the structure. As stated above the equations of motion will be solved using the classical modal approach and therefore linear structural behaviour is assumed.

The structure is assumed to be deterministic, however, the acoustic loads have a random nature. In general, the sound field is assumed to be reverberant (diffuse). The sound intensity is the same in all directions.

Lightweight and large antenna structures and solar arrays of spacecraft (Fig. 1.5) are very sensitive to acoustic loads during the launch phase. Spacecraft external structures are subjected to severe acoustic loads.

1.2.11 Statistical Energy Analysis

The statistical energy analysis (SEA) is based on the principle of energy conservation. All the energy input in to a system, through mechanical or acoustic excitation, must leave the system through structural damping or acoustic radiation. The method is fast and is applicable over a wide frequency range. SEA is very good for problems that combine many different sources of excitation, whether mechanical or acoustic.

With the SEA a statistical description of the structural vibrational behaviour of elements (systems) is described. In the high-frequency band a deterministic modal description of the dynamic behaviour of structures is not very useful. The modes (oscillators) are grouped statistically and the energy transfer from one group of modes to another group of modes is statistically proportional with the difference in the subsystem total energies. Lyon and DeJong wrote a very interesting book about this topic [Lyon 95].

1.2.12 Inertia-Relief

Free-free systems can move as a rigid body through space, the structure is a so-called unconstrained structure. The stiffness matrix $[K]$ is singular and therefore the flexibility matrix $[G] = [K]^{-1}$ does not exist. Launch vehicles, aircraft and spacecraft are examples of free-free moving dynamic systems. In this chapter a method, the inertia-relief, will be derived to analyse free-free systems. The motion as a rigid body will be eliminated and a new set of applied loads (relative forces) will be used to analyse the elastic behaviour of the free-free system. The relative motion and relative forces will be introduced and a definition of the inertia-relief flexibility matrix $[G_f]$ will be given.

1.2.13 Mode Acceleration Method

The mode acceleration method (MAM) will improve accuracy of the responses; displacements and derivatives thereof such as element forces, stresses, etc., with respect to the mode displacement method (MDM) when a reduced set of mode shapes is used. The MDM is often called the mode superposition method. The MDM may only be used for linear dynamics systems. The MAM takes the truncated modes into account "statically". The MAM is, in fact, rearranging the matrix equations of motion in the following manner

$$\{x\} = [K]^{-1}(\{F(t)\} - [M]\{\ddot{x}\} - [C]\{\dot{x}\}) . \tag{1.10}$$

Using the MAM, less modes may be in taken into account compared to the MDM.

1.2.14 Residual Vectors

Residual vectors have been discussed by John Dickens and Ted Rose in [Dickens 00, Rose 91]. The modal base, when the modal displacement method [MDM] is applied, will be extended by residual vectors to account for the deleted modes. This method is quite similar to the mode acceleration method (MAM). Dickens proposed to construct a static mode (displacement) with respect to the boundaries based on the residual loads. Rose constructed a static mode, again with respect to the posed boundary conditions, however, based on the static part of the dynamic loads

Since the residual vectors are treated as modes, they will have associated modal mass, modal stiffness and damping. With the aid of artificial damping the responses due to the residual vectors will be minimised.

1.2.15 Dynamic Model Reduction

The combination of nonreduced finite element models (FEMs) of subsystems to a dynamic FEM of the complete system (spacecraft or launcher) will, in general, result in a finite element model with many degrees of freedom (dofs) and therefore will be difficult to handle. The responsible analyst will ask for a reduced dynamic FEM description of the subsystem to manipulate the total dynamic model and will prescribe the allowed number of left or analysis dofs of the reduced dynamic model. The reduced dynamic model is, in general, a modal description of the system involved.

The customer will prescribe the required accuracy of the reduced dynamic model, more specifically the natural frequencies, mode shapes in comparison to the complete finite element model or reference model. For example, the following requirements are prescribed:

- The natural frequencies of the reduced dynamic model must deviate less ± 3 % from the natural frequencies calculated using the reference model.
- The effective masses of the reduced dynamic model must be within ± 10 % of the effective masses calculated with the reference model.
- The diagonal terms at the cross orthogonality check [Ricks 91] must be greater than or equal to 0.95 and the off-diagonal terms must be less than or equal to 0.05. The cross orthogonality check is based upon the mass matrix.
- The diagonal terms at the modal assurance criteria (MAC) must be greater than or equal to 0.95 and the off-diagonal terms less than or equal to 0.10.

Sometimes the requirements concern the correlation of the response curves obtained with the reduced dynamic model and the reference model.

Reduced models are also used to support the modal survey, the experimental modal analysis. The reduced dynamic model will be used to calculate the orthogonality relations between measured and analysed modes. This reduced model is called the test-analysis model (TAM) [Kammer 87]. Also, various reduction methods will be discussed.

All mentioned reduction procedures are based upon the Ritz method [Michlin 62].

1.2.16 Component Model Synthesis

The component mode synthesis (CMS) or component modal synthesis [Hintz 75] or modal coupling technique [Maia 97] is used when components (substructures) are described by the mode displacement method (MDM) and coupled together (synthesis) via the common boundaries $\{x_b\}$ in order to perform a dynamic analysis, e.g. modal analysis, responses, of the complete structure (assembly of substructures). The CMS method can only be applied to linear structures. The component mode synthesis method can also be applied to components for which the modal characteristics are measured in combination with finite element reduced dynamic models.

In general, a component or substructure is a recognisable part of the structure, e.g. for a spacecraft; the primary structure, the solar arrays, the antennae, large instruments, etc.

In the past, the CMS method was applied to significantly reduce the number of dofs due to the imposed limitations on computers, however, nowadays, these limitations are more or less removed but the CMS method is still very popular. Subcontractors deliver their reduced FE dynamic models to the prime contractor who will combine (synthesise) all these models to the spacecraft dynamic FE model to perform the dynamic analysis on the complete spacecraft. The same applies to the coupled dynamic-load analysis when the reduced FE model of the complete spacecraft is placed on top of the launch vehicle. In general, the dynamic FE modal of the launch vehicle is a reduced dynamic FE model too.

Dynamic properties of substructures may be defined by experiment and may be coupled to other dynamic FE models of other substructures. Hence, there are many reasons to apply the CMS method.

For dynamic analysis the components may be obtained by reducing the number of dofs using the MDM. The physical dofs $\{x\}$ are generally depicted on a small number of kept modes (eigenvectors), the modal base.

$$\{x\} = [\Phi]\{\eta\}, \tag{1.11}$$

- $[\Phi]$ The modal base consisting of the kept mode
- $\{\eta\}$ The generalised or principal coordinates.

The number of generalised coordinates $\{\eta\}$ is, in general, much less than the number of physical dofs $\{x\}$.

There are many CMS techniques described in the literature. The modal description of the components strongly depends on the boundary conditions applied by building the reduced FE model of the component. The discussion of the CMS method will focus on:

- Components with fixed-interface dofs $\{x_b\}$

- Components with free interfaces
- Components with loaded interfaces

However, we will only consider undamped components.

1.2.17 Load Transformation Matrices

The mathematical reduced (condensed) dynamic model consists of the reduced mass and stiffness matrices. The damping matrix is generally not delivered in a reduced form because the damping characteristics will be introduced later in the dynamic response analyses.

Because the reduced dynamic model only consists of reduced matrices during the dynamic response analysis no direct information about physical responses; e.g. forces, stresses, can be made available. The reduced dynamic model will only produce response characteristics of physical (e.g. I/F dofs) and generalised degrees of freedom such as displacements, velocities and accelerations.

To be able to produce responses, stresses and forces in selected structural elements during the dynamic response analyses using (coupled) reduced dynamic models the so-called load transformation matrix (LTM) can be used. The LTM defines a relation between forces and stresses in certain structural elements and the degrees of freedom of the reduced dynamic model. In general, the transformation matrix is called the output transformation matrix (OTM) [Fransen 02]. Besides LTMs displacement transformation matrices (DTM), acceleration transformation matrices (ATM) can also be defined, however, only LTM's will be discussed. The creation of DTMs and ATMs is the same as the generation of LTMs.

Two methods, fixed-free and free-free systems, of obtaining LTMs are discussed. Both methods are based upon the mode displacement method (MDM) in combination with the mode acceleration method (MAM).

1.3 Problems

1.3.1 Problem 1

Scan an available L/V users manual with respect to the dynamic load specifications that must be applied in the design of spacecraft structure. Users manuals of L/Vs are available on the website of launcher authorities such as ARIANESPACE, www.arianespace.com!

2 *Single Degree of Freedom System*

2.1 Introduction

The single degree of freedom (sdof) system consists, in general, of a mass, a spring and a damper and is a very important dynamic element within the framework of mechanical vibration. The motion of the mass represents the degree of freedom. A degree of freedom for a system is analogous to an independent variable for a mathematical function. The mass represents the kinetic energy, the spring the potential energy and the damper introduces the dissipation of energy. The sdof element is illustrated in Fig. 2.1.

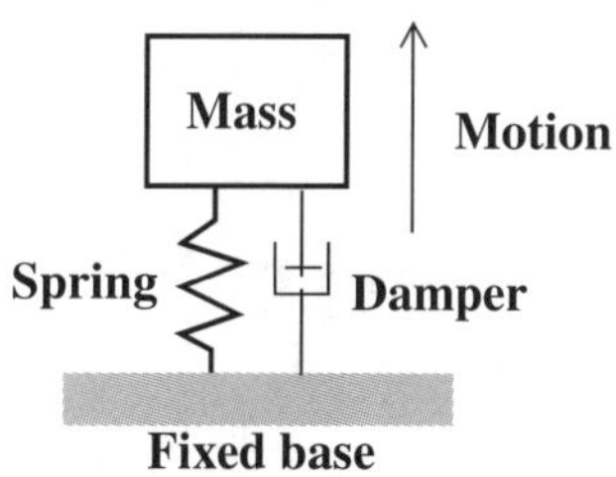

Fig. 2.1. Simple degree of freedom (sdof) element

The sdof element without a damper is called the undamped sdof system and the sdof element with a damper the damped sdof element. In general, the damper is a dashpot, in which the damping force is proportional to the relative velocity of the mass. The relative motion of the mass is defined with respect to the fixed base. In the case of linear mechanical vibration the dynamic properties of continuous dynamic systems, such as rods, beams, plates, etc., can be related to the sdof dynamic system.

Continuous dynamic systems can be discretised in many connected sdof systems, the so-called multi-degrees of freedom (mdof) dynamic systems. Many techniques can be applied to discretise continuous systems (see Table 2.1). The finite element method is one such frequently used technique.

Linear mdof dynamic systems can be transformed in uncoupled sdof systems using properties from linear algebra [Strang 88, Zurmuehl 64]. Therefore much attention will be given to the sdof system.

In this chapter both the dynamics of undamped and damped sdof systems will be discussed.

The equation of motion of the sdof system will be solved both with the Laplace transform and the Fourier transform. The Laplace transform is used to solve the equation of motion in the time domain (transient vibration) and the Fourier transform in the frequency domain (steady-state vibration, harmonic vibration or frequency responses).

2.2 Undamped Sdof System

The undamped sdof consists of a mass m (kg) and a spring k (N/m) and is illustrated in Fig. 2.2. The properties of the mass element and the spring element do not vary in time. An external force $F(t)$ is applied to the mass element. The force is dependent on time.

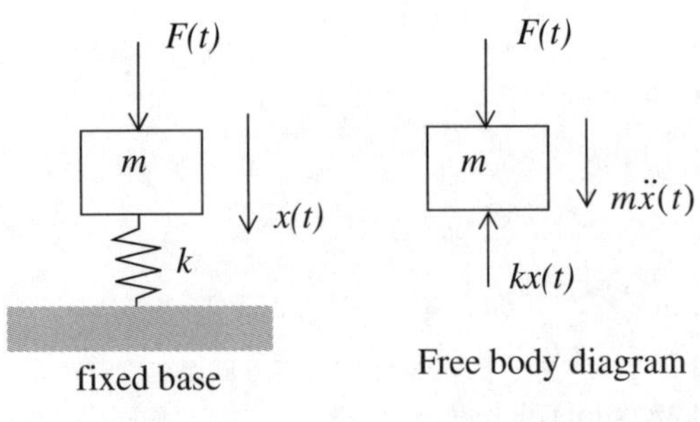

Fig. 2.2. Undamped sdof and free-body diagram

If the mass has a small displacement $x(t)$ (m) and has an acceleration $\ddot{x}(t)$ (m/s^2) and the mass element is exposed to several forces; an external applied force $F(t)$ and two internal forces; the spring forces $kx(t)$ and the inertia force $m\ddot{x}(t)$. In the sense of the D'Alembert principle (Jean Le Rond d'Alembert 1717–1783) [Meirovitch 70] a dynamic problem is treated as a static problem. The equilibrium of the mass element will result in an equation of motion of the sdof dynamical system.

$$m\ddot{x}(t) + kx(t) = F(t). \tag{2.1}$$

We will now introduce the radian (circular) natural frequency ω_n (rad/s) which is defined as

$$\omega_n = \sqrt{\frac{k}{m}}. \tag{2.2}$$

The dimensions of the spring stiffness k are (N/m) or (kgm/s^{2}m=kg/s^2) and the dimension of the natural frequency now becomes (rad/s). The undamped natural period T_n (s) is

$$T_n = \frac{2\pi}{\omega_n}. \tag{2.3}$$

The frequency is denoted by f_n (Hz)

$$f_n = 2\pi\omega_n = \frac{1}{T_n}. \tag{2.4}$$

The unit (Hz), Hertz, was named after the German physisist Heinrich Hertz (1857–1894) [Nahin 98]. Quite often instead of the unit (Hz) the unit cycles per second (cps) is used.

Dividing (2.1) by m will result in the following equation of motion for the sdof system

$$\ddot{x}(t) + \omega_n^2 x(t) = \frac{F(t)}{m} = f(t). \tag{2.5}$$

Equation (2.5) is a linear second-order nonhomogeneuos differential equation. The coefficients are constants.

In Table 2.1 a few examples of sdof systems are illustrated. The mass of the beams is neglected and no shear effects are taken into account. The bending stiffness of the beam is EI (Nm^2).

Equation (2.5) and corresponding initial and boundary conditions will be solved using the Laplace transform method and the Inverse Laplace Transform [Kreyszig 93].

Table 2.1. Examples

Example of single degree of freedom dynamic system	Undamped equation of motion $\ddot{x} + \omega_n^2 x = 0$
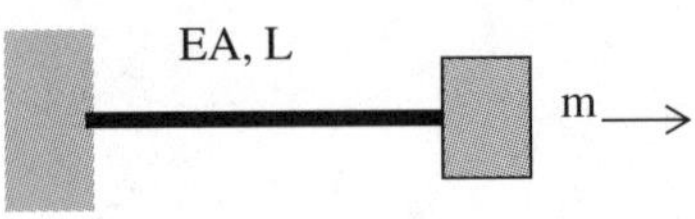	$\omega_n^2 = \dfrac{AE}{mL}$
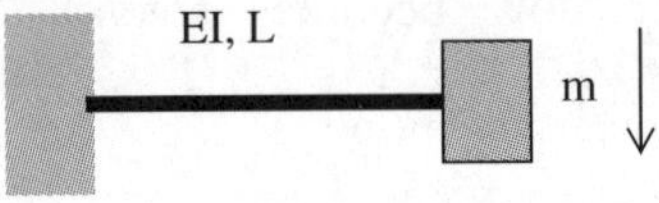	$\omega_n^2 = \dfrac{3EI}{mL^3}$
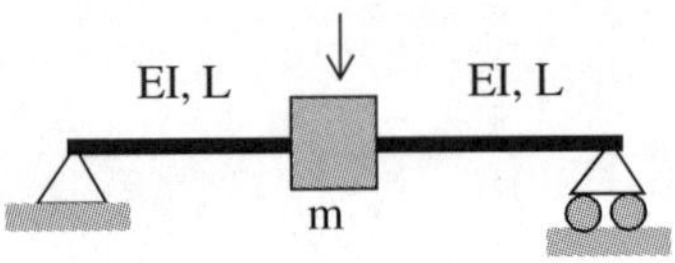	$\omega_n^2 = \dfrac{6EI}{mL^3}$
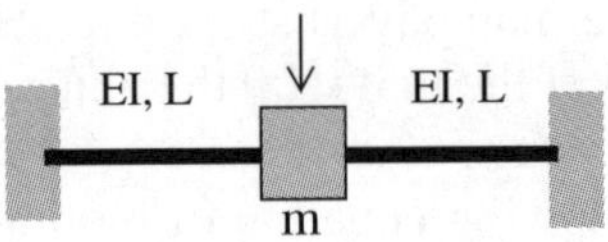	$\omega_n^2 = \dfrac{24EI}{mL^3}$
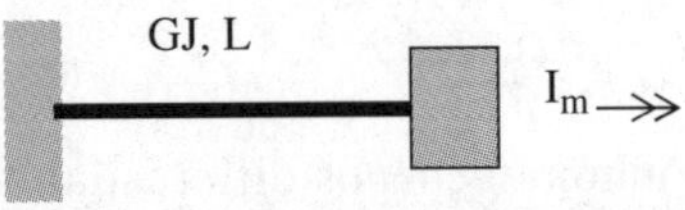	$\omega_n^2 = \dfrac{GJ}{I_m L}$

where

- E Modulus of elasticity (Young's modulus of elasticity) (Pa)
- I Second moment of area (m^4)
- J Torsion constant (m^4)
- G Shear modulus (Pa)
- L Length (m)
- m Discrete mass element (kg)
- I_m Second moment of inertia (kgm^2)

The standard Laplace transform (Pierre Simon de Laplace 1749–1827 [Nahin 98]) of the functions $x(t)$, $\dot{x}(t)$, $\ddot{x}(t)$ and $f(t)$, with $t \geq 0$, are [Kreyszig 93]:

$$L(x(t)) = \int_0^\infty e^{-st} x(t) ds = X(s), \tag{2.6}$$

and

$$L(\dot{x}(t)) = \int_0^\infty e^{-st} \dot{x}(t) ds = sX(s) - x(0), \tag{2.7}$$

and

$$L(\ddot{x}(t)) = \int_0^\infty e^{-st} \ddot{x}(t) ds = s^2 X(s) - sx(0) - \dot{x}(0), \tag{2.8}$$

and

$$L(f(t)) = \int_0^\infty e^{-st} f(t) ds = F(s). \tag{2.9}$$

2.2.1 Solution of an Sdof System with Initial Conditions

We will solve (2.5) with the aid of the Laplace transform with a zero external force, $f(t) = 0$, however, with the initial conditions, namely $x(0) = x_o$ and $\dot{x}(0) = 0$. The Laplace Transform of the homogeneous equation of motion (2.5) is

$$L[\ddot{x}(t)] + \omega_n^2 L[x(t)] = L[0],$$

or

$$s^2 X(s) - sx(0) - \dot{x}(0) + \omega_n^2 X(s) = 0.$$

After the introduction of the initial conditions we have

$$s^2 X(s) - sx_o + \omega_n^2 X(s) = 0.$$

We will find the following relation for $X(s)$

$$X(s) = \frac{sx_o}{s^2 + \omega_n^2}. \tag{2.10}$$

The inverse Laplace transform is defined as

$$x(t) = L^{-1}[X(s)]. \tag{2.11}$$

The Laplace transform of the function $g(t) = \cos\omega_n t$ is

$$L[ag(t)] = \frac{as}{s^2 + \omega_n^2}. \tag{2.12}$$

Finally the solution of the homogeneous equation (2.5) with initial conditions $x(0) = x_o$ and $\dot{x}(0) = 0$ is

$$x(t) = L^{-1}[X(s)] = L^{-1}\left[\frac{sx_o}{s^2 + \omega_n^2}\right] = x_o\cos(\omega_n t). \tag{2.13}$$

Assume the initial conditions are specified for both the velocity and the displacement, $\dot{x}(0) = v_o$ and $x(0) = x_o$.

The expression for $X(s)$ becomes

$$X(s) = \frac{v_o}{s^2 + \omega_n^2} + \frac{sx_o}{s^2 + \omega_n^2}. \tag{2.14}$$

The Laplace transform of the function $g(t) = \sin\omega_n t$ is

$$L[ag(t)] = \frac{a\omega_n}{s^2 + \omega_n^2}. \tag{2.15}$$

The final solution of the homogeneous equation (2.5) with initial conditions $x(0) = x_o$ and $\dot{x}(0) = v_o$ is

$$x(t) = L^{-1}[X(s)] = L^{-1}\left[\frac{v_o}{s^2 + \omega_n^2} + \frac{sx_o}{s^2 + \omega_n^2}\right] = x_o\cos(\omega_n t) + v_o\frac{\sin(\omega_n t)}{\omega_n}. \tag{2.16}$$

This solution for $x(t)$ is called the complementary function and is associated with the initial conditions.

2.2.2 Solution of an Sdof System with Applied Forces

The solution of an sdof system with applied forces can be divided into two categories:

- Enforced motion; displacement, velocity and acceleration or combinations. In this book we will only consider sdof systems with enforced accelerations or base excitation. Enforced motion can be translated into applied forces.
- External forces applied to the sdof system

Enforced Accelerations

An sdof system with a discrete mass m and a spring element k is placed on a moving base, which is accelerated with an acceleration $\ddot{u}(t)$. The resulting displacement of the mass is $x(t)$. We introduce a relative motion $z(t)$, which is the displacement of the mass with respect to the base. The relative displacement is

$$z(t) = x(t) - u(t)\,. \tag{2.17}$$

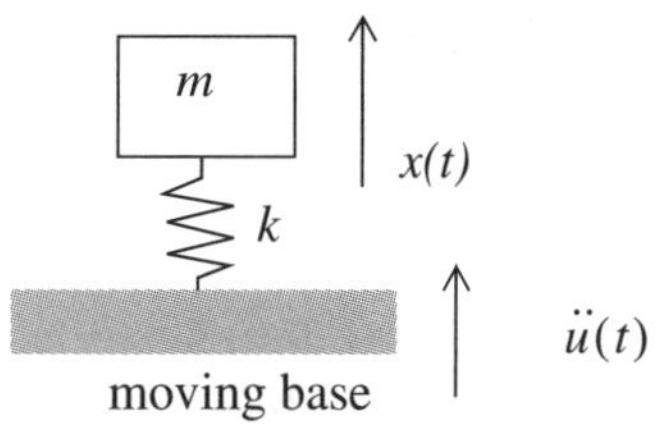

Fig. 2.3. Enforced acceleration on an sdof system

The spring force $F_{sf}(t)$ acting on the mass element is

$$F_{sf}(t) = k\{z(t)\} = k\{x(t) - u(t)\}\ . \tag{2.18}$$

With reference to Fig. 2.2 and (2.1) the equation of motion of the enforced sdof system in Fig. 2.3 is

$$m\ddot{x}(t) + k\{x(t) - u(t)\} = 0\,. \tag{2.19}$$

Introducing the relative motion $z(t)$ and $\ddot{z}(t)$ (2.19) becomes

$$m\ddot{z}(t) + kz(t) = -m\ddot{u}(t)\,, \tag{2.20}$$

or

$$\ddot{z}(t) + \omega_n^2 z(t) = -\ddot{u}(t)\,. \tag{2.21}$$

The enforced acceleration of the sdof system is transformed into an external force. The absolute displacement $x(t)$ can be calculated with (2.17) or

$$\ddot{x}(t) = \ddot{z}(t) + \ddot{u}(t) = -\omega_n^2 z(t)\,. \tag{2.22}$$

A Sinusoidal Enforced Acceleration

We start with the zero initial conditions $z(0) = 0$ and $\dot{z}(0) = 0$ the sdof system will be enforced with sinusoidal acceleration, assuming $\omega \neq \omega_n$

$$\ddot{u}(t) = A\sin\omega t\,. \tag{2.23}$$

The Laplace transform of (2.21) is now

$$s^2 Z(s) + \omega_n^2 Z(s) = -A\frac{\omega}{s^2 + \omega^2}\,. \tag{2.24}$$

The Laplace transform $Z(s)$ becomes

$$Z(s) = -A\frac{1}{s^2 + \omega_n^2}\frac{\omega}{s^2 + \omega^2}\,. \tag{2.25}$$

The inverse Laplace transform of (2.25) is from [Kreyszig 93]

$$z(t) = L^{-1}\{Z(s)\} = \frac{-A}{\omega^2 - \omega_n^2}\left(\frac{\omega}{\omega_n}\sin\omega_n t - \sin\omega t\right). \tag{2.26}$$

The solution for $z(t)$ is called the particular integral and is associated with the forcing function. When $\omega = \omega_n$ the solution of (2.26) does not exist. Applying the l'Hôpital rule, differentiating the numerator and the denominator and let $\omega \to \omega_n$, we find

$$z(t) = \frac{-A}{2\omega_n^2}(\sin\omega_n t - \omega_n t\cos\omega_n t)\,. \tag{2.27}$$

The last term of (2.27) grows out of bounds, indicating resonance. The relative acceleration $\ddot{z}(t)$ for $\omega \neq \omega_n$ is found after differentiation of (2.26) twice with respect to time.

$$\ddot{z}(t) = -\frac{\omega^2 A}{\omega^2 - \omega_n^2}\left(\frac{\omega_n}{\omega}\sin\omega_n t - \sin\omega t\right). \tag{2.28}$$

External Forces

We recall here (2.5) of the sdof system illustated in Fig. 2.2

$$\ddot{x}(t) + \omega_n^2 x(t) = \frac{F(t)}{m} = f(t)\,.$$

Assume zero initial conditions with respect to the initial displacement and velocity; $x(0) = \dot{x}(0) = 0$. The Laplace tranform of (2.5) becomes

$$s^2 X(s) - sx(0) - \dot{x}(0) + \omega_n^2 X(s) = F(s)\,.$$

After the introduction of the initial conditions the expression of $X(s)$ will be

$$X(s) = \frac{F(s)}{s^2 + \omega_n^2} = F(s)\frac{1}{s^2 + \omega_n^2}. \tag{2.29}$$

The convolution product is defined as [Stephenson 70]:

$$f(t)*g(t) = \int_0^t f(t-\tau)g(\tau)d\tau = \int_0^t f(t)g(t-\tau)d\tau, \tag{2.30}$$

and the Laplace transform of the convolution product is [Stephenson 70]:

$$L[f(t)*g(t)] = F(s)G(s). \tag{2.31}$$

Applying the convolution theorem to (2.29) we obtain

$$x(t) = \int_0^t \frac{\sin\omega_n(t-\tau)}{\omega_n}f(\tau)d\tau = \int_0^t \frac{\sin\omega_n\tau}{\omega_n}f(t-\tau)d\tau, \tag{2.32}$$

or

$$x(t) = \int_0^t \frac{\sin\omega_n\tau}{\omega_n}f(t-\tau)d\tau = \int_0^t h(\tau)f(t-\tau)d\tau \tag{2.33}$$

where $h(t) = \dfrac{\sin\omega_n t}{\omega_n}$ is the undamped impulse response function.

The total solution of (2.5) is the superposition of the solutions of (2.16), with the initial conditions $x(0) = x_o$ and $\dot{x}(0) = v_o$, and the particular solution (2.32)

$$x(t) = x_o\cos(\omega_n t) + v_o\frac{\sin(\omega_n t)}{\omega_n} + \int_0^t \frac{\sin\omega_n\tau}{\omega_n}f(t-\tau)d\tau. \tag{2.34}$$

2.3 Damped Vibration and the Damping Ratio

A viscous damper element will now be added to the sdof system as illustrated in Fig. 2.4. The damping force is proportional to the relative velocity of the mass element. The mass m, spring stiffness k and the damping constant c are constant with time. The damped sdof has linear properties.

In the sense of d'Alembert the mass is in equilibrium, the internal forces are in equilibrium with the external force, with $F_{\text{inertia}} = -m\ddot{x}(t)$, the equation of equilibrium can be written as follows

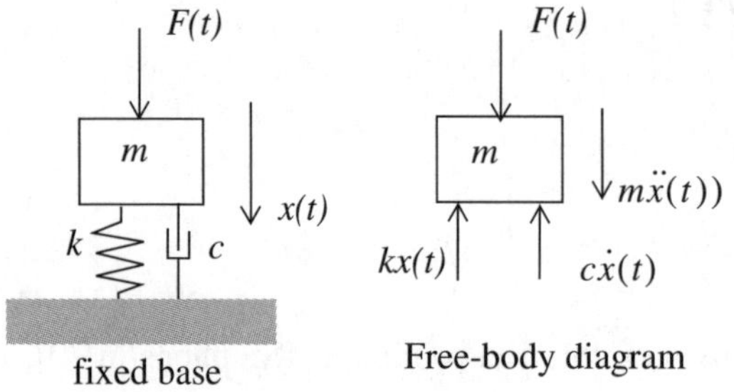

Fig. 2.4. Sdof system with damper

$$F_{\text{inertia}} + F_{\text{damping}} + F_{\text{stiffness}} = F_{\text{external}} . \tag{2.35}$$

In terms of the motion of the mass this means

$$m\ddot{x}(t) + c\dot{x}(t) + kx(t) = F(t) , \tag{2.36}$$

where

- c the viscous damping constant (Ns/m)

With the introduction of the natural radian frequency $\omega_n = \sqrt{\frac{k}{m}}$ and the critical damping constant $c_{\text{crit}} = 2\sqrt{km}$ and the damping ratio $\zeta = \frac{c}{c_{\text{crit}}}$ the damping constant c can be expressed as follows

$$\frac{c}{m} = \zeta\frac{c_{\text{crit}}}{m} = 2\zeta\sqrt{\frac{k}{m}} = 2\zeta\omega_n . \tag{2.37}$$

The critical damping is the minimum damping that results in non-periodic motion of the sdof system under free vibration. The damping ratio is the ratio of the sdof system's actual viscous damping to its critical damping.

Dividing (2.36) by the mass m the equation of motion of the damped sdof system becomes

$$\ddot{x}(t) + 2\zeta\omega_n\dot{x}(t) + \omega_n^2 x(t) = \frac{F(t)}{m} = f(t) . \tag{2.38}$$

We will investigate the influence of the damping ratio on the solution of the second-order linear homogeneous equation of motion, with $f(t) = 0$.

$$\ddot{x}(t) + 2\zeta\omega_n\dot{x}(t) + \omega_n^2 x(t) = 0 .$$

We try the following solution for $x(t)$

$$x(t) = e^{\lambda t} , \tag{2.39}$$

which results in the characteristic equation

$$\lambda^2 + 2\zeta\omega_n\lambda + \omega_n^2 = 0. \tag{2.40}$$

The solution of (2.40) is

$$\lambda_{1,2} = \omega_n(-\zeta \pm \sqrt{\zeta^2 - 1}). \tag{2.41}$$

We consider three cases for the damping:

1. The system is underdamped, $\zeta < 1$; the damping ratio is less than the critical damping ratio, the sdof system will show oscillatory (periodic) behaviour (two conjugate complex roots).
2. The system is critically damped, $\zeta = 1$; the damping ratio is the critical damping ratio, the sdof system shows aperodic (nonoscillatory) behaviour (repeated real roots).
3. The system is overdamped, $\zeta > 1$; the damping ratio is greater than the critical damping ratio, the sdof system shows nonoscillatory and exponentiallly decaying response behaviour (real roots).

2.3.1 Solution of the Sdof System in the Time Domain

The Laplace transform of (2.38) is

$$s^2X(s) - sx(0) - \dot{x}(0) + 2\zeta\omega_n\{sX(s) - x(0)\} + \omega_n^2X(s) = F(s). \tag{2.42}$$

With initial conditions at $t = 0$, $x(0) = x_o$ and $\dot{x}(0) = v_o$, $X(s)$ becomes

$$X(s) = \frac{1}{s^2 + s(2\zeta\omega_n) + \omega_n^2}\{x_o(s + \zeta\omega_n) + x_o\zeta\omega_n + v_o + F(s)\}. \tag{2.43}$$

Below critical damping $\zeta < 1$
Equation (2.43) becomes

$$X(s) = \frac{1}{(s + \zeta\omega_n)^2 + \omega_n^2(1 - \zeta^2)}\{x_o(s + \zeta\omega_n) + x_o\zeta\omega_n + v_o + F(s)\}. \tag{2.44}$$

The damped natural (circular) frequency ω_d (rad/s) is defined as

$$\omega_d = \omega_n\sqrt{1 - \zeta^2}. \tag{2.45}$$

The solution of (2.44), the inverse Laplace transform, is

$$x(t) = x_o e^{-\zeta\omega_n t}\left(\cos\omega_d t + \frac{\zeta}{\sqrt{1-\zeta^2}}\sin\omega_d t\right)$$

$$+ v_o e^{-\zeta\omega_n t}\frac{\sin\omega_d t}{\omega_d} + \int_0^t \left[e^{-\zeta\omega_n \tau} \cdot \frac{\sin\omega_d \tau}{\omega_d} \cdot f \cdot (t-\tau) \cdot d\tau\right]. \tag{2.46}$$

The function

$$h(t) = e^{-\zeta\omega_n t}\frac{\sin\omega_d t}{\omega_d}, \tag{2.47}$$

is called the damped impulse response function, which is, in fact the response of the damped sdof system, with zero initial conditions, and as external force the Dirac delta function $\delta(t)$ (P.A.M. Dirac, 1902–1984). The Dirac delta function has the following properties:

- $\delta(t) = 0$ for $t \neq 0$
- $\int_{-\infty}^{\infty} \delta(t)dt = 1$
- $\int_{-\infty}^{\infty} f(t))\delta(t-a)dt = f(a)$

In (2.47) $e^{-\zeta\omega_n t}$ represents the exponentially decay and $\frac{\sin\omega_d t}{\omega_d}$ the periodic motion.

Equal to critical damping $\zeta = 1$
Equation (2.43) becomes

$$X(s) = \frac{1}{(s+\omega_n)^2}\{x_o(s+\omega_n) + x_o\omega_n + v_o + F(s)\}. \tag{2.48}$$

The solution of (2.48) is

$$x(t) = x_o e^{-\zeta\omega_n t}(1+t) + \frac{v_o t}{\omega_n}e^{-\zeta\omega_n t} + \int_0^t \tau e^{-\zeta\omega_n \tau} f(t-\tau)d\tau. \tag{2.49}$$

Above critical damping $\zeta > 1$
Equation (2.43) becomes

$$X(s) = \frac{1}{(s+\zeta\omega_n)^2 - \omega_n^2(\zeta^2-1)}\{x_o(s+\zeta\omega_n) + x_o\zeta\omega_n + v_o + F(s)\}. \tag{2.50}$$

The solution of (2.50) is, with

$$\Omega_{\mathrm{d}} = \omega_{\mathrm{n}}\sqrt{\zeta^2 - 1} \tag{2.51}$$

$$x(t) = x_{\mathrm{o}}e^{-\zeta\omega_{\mathrm{n}}t}\left(\cos h\Omega_{\mathrm{d}}t + \frac{\zeta}{\sqrt{1-\zeta^2}}\sin h\Omega_{\mathrm{d}}t\right)$$

$$+ v_{\mathrm{o}}e^{-\zeta\omega_{\mathrm{n}}t}\frac{\sin h\Omega_{\mathrm{d}}t}{\Omega_{\mathrm{d}}} + \int_0^t e^{-\zeta\omega_{\mathrm{n}}\tau}\frac{\sin h\Omega_{\mathrm{d}}\tau}{\Omega_{\mathrm{d}}}f(t-\tau)d\tau\,. \tag{2.52}$$

In this book no further attention will be paid to damped dynamical systems with damping ratios equal to or higher than the critical damping ratio $(\zeta = 1)$. In general the damping ratios $0 \le \zeta \le 0{,}1$ are very common in spacecraft primary structure design, appendices and electronic boxes. In most cases the damping in structures is represented by the modal viscous damping ratio. Later in this book another model of damping idealisation, the structural damping (hysteresis), will be discussed.

An sdof dynamic system is given by [Kreyszig 93]

$$9.082\ddot{x}(t) + c\dot{x}(t) + 890x(t) = 0$$

When it is subjected to $x(0) = 0.15$ m and $\dot{x}(0) = 0$ m/s. How does the motion change if the sdof has damping given by

1. $c = 200.0$ Ns/m
2. $c = 179.8$ Ns/m
3. $c = 100.0$ Ns/m

Solution 1

$$9.082\ddot{x}(t) + 200.0\dot{x}(t) + 890x(t) = 0\,.$$

The damping ratio $\zeta = \dfrac{c}{2\sqrt{km}} = 1.11 > 1$.

The solution can be written as $x(t) = Ae^{\lambda t}$, thus with $\lambda_{1,2} = -11.01 \pm 4.822$ we can solve the constants A_1 and A_2 of equation $x(t) = A_1e^{\lambda_1 t} + A_2e^{\lambda_2 t}$. With $A_1 + A_2 = 0.1500$ and $\lambda_1 A_1 + \lambda_2 A_2 = 0$ the solution becomes

$$x(t) = 0.246\mathrm{e}^{-6.190t} - 0.096\mathrm{e}^{-15.83t}\,.$$

Solution 2

$$9.082\ddot{x}(t) + 179.8\dot{x}(t) + 890x(t) = 0\,.$$

The damping ratio $\zeta = \frac{c}{2\sqrt{km}} = 1$.

The solution can be written as $x(t) = Ae^{\lambda t}$, thus with $\lambda = -9.899$ we can solve the constants A_1 and A_2 of the equation $x(t) = A_1 e^{\lambda t} + A_2 t e^{\lambda t}$. With $A_1 = 0.1500$ and $\lambda A_1 + A_2 = 0$ the solution becomes

$$x(t) = (0.1500 + 1.485t)e^{-9.899t}.$$

Solution 3

$$9.082\ddot{x}(t) + 100.0\dot{x}(t) + 890x(t) = 0.$$

The damping ratio $\zeta = \frac{c}{2\sqrt{km}} = 0.56 < 1$.

The solution can be written as $x(t) = Ae^{\lambda t}$, thus with $\lambda_{1,2} = -5.506 \pm j8.227$ we can solve the constants A_1 and A_2 of the equation $x(t) = A_1 e^{\lambda_1 t} + A_2 e^{\lambda_2 t}$. With $A_1 + A_2 = 0.1500$ and $\lambda_1 A_1 + \lambda_2 A_2 = 0$ the solution becomes

$$x(t) = \{0.1500\cos(8.227t) + 0.1004\sin(8.227t)\}e^{-5.506t}.$$

2.3.2 Solution of the Damped Sdof System with Applied Forces

The solution of a damped sdof system with applied forces can be divided into two categories:

- Enforced motion; displacement, velocity and acceleration or combinations. In this book we will only consider sdof systems with enforced accelerations or base excitation. Enforced motion can be translated into applied forces. The sdof system is illustrated in Fig. .
- External forces applied to the sdof system.

Enforced Accelerations

An sdof system with a discrete mass m, a damper element c and a spring element k is placed on a moving base which is accelerated with an acceleration $\ddot{u}(t)$. The resulting displacement of the mass is $x(t)$. We introduce a relative motion $z(t)$ which is the displacement of the mass with respect to the base. The relative displacement is

$$z(t) = x(t) - u(t) \tag{2.53}$$

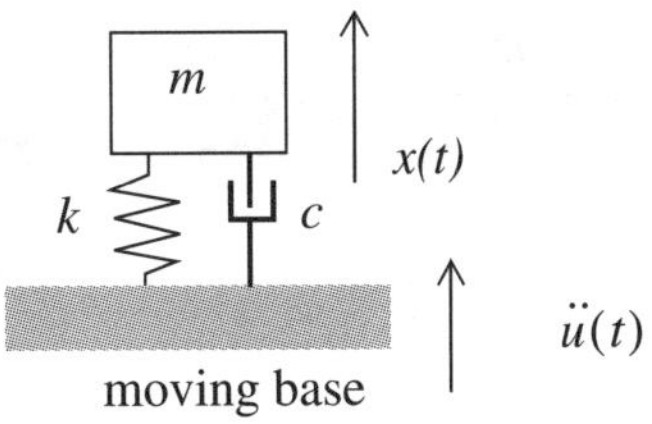

Fig. 2.5. Enforced acceleration on a damped sdof system

Analogue to (2.21) we can write

$$\ddot{z}(t) + 2\zeta\omega_n\dot{z}(t) + \omega_n^2 z(t) = -\ddot{u}(t) \tag{2.54}$$

The enforced acceleration of the sdof system is transformed into an external force. The absolute displacement $x(t)$ can be calculated with (2.54) or

$$\ddot{x}(t) = \ddot{z}(t) + \ddot{u}(t) = -2\zeta\omega_n\dot{z}(t) - \omega_n^2 z(t) \tag{2.55}$$

We start with zero initial conditions $z(0) = 0$ and $\dot{z}(0) = 0$ and the sdof system will be enforced with a sinusoidal acceleration

$$\ddot{u}(t) = A\sin\omega t \tag{2.56}$$

Equation (2.54) may be solved with the aid of and with the initial conditions, the displacement $z(0)$ and thevelocity $\dot{z}(0)$, the solution for $z(t)$ is

$$z(t) = z(0)e^{-\zeta\omega_n t}\left(\cos\omega_d t + \frac{\zeta}{\sqrt{1-\zeta^2}}\sin\omega_d t\right)$$

$$+ \dot{z}(0)e^{-\zeta\omega_n t}\frac{\sin\omega_d t}{\omega_d} - A\int_0^t e^{-\zeta\omega_n\tau}\frac{\sin\omega_d\tau}{\omega_d}\sin\omega(t-\tau)d\tau \tag{2.57}$$

The solution of (2.57) , the relative displacement $z(t)$ is

$$z(t) = e^{-\zeta\omega_n t}(B\cos\omega_d t + C\sin\omega_d t) - \frac{A}{\omega_n^2}\left[\frac{\left\{1-\left(\frac{\omega}{\omega_n}\right)^2\right\}\sin\omega t - 2\zeta\left(\frac{\omega}{\omega_n}\right)\cos\omega t}{\left\{1-\left(\frac{\omega}{\omega_n}\right)^2\right\}^2 + \left(2\zeta\frac{\omega}{\omega_n}\right)^2}\right]. \tag{2.58}$$

The relative velocity $\dot{z}(t)$ becomes

$$\dot{z}(t) = e^{-\zeta\omega_n t}[(C\omega_d - B\omega_n\zeta)\cos\omega_d t - (C\omega\zeta + B\omega_d)\sin\omega_d t]$$

$$-\frac{A}{\omega_n}\left(\frac{\omega}{\omega_n}\right)\left[\frac{\left\{1-\left(\frac{\omega}{\omega_n}\right)^2\right\}\cos\omega t+2\zeta\left(\frac{\omega}{\omega_n}\right)\sin\omega t}{\left\{1-\left(\frac{\omega}{\omega_n}\right)^2\right\}^2+\left(2\zeta\frac{\omega}{\omega_n}\right)^2}\right]. \tag{2.59}$$

The relative acceleration $\ddot{z}(t)$ becomes

$$\ddot{z}(t) = e^{-\zeta\omega_n t}[(B\omega_d^2\zeta^2 - 2C\omega_d\omega_n\zeta - B\omega_d^2)\cos\omega_d t)]$$

$$+e^{-\zeta\omega_n t}[(C\omega_n^2\zeta^2 + 2B\omega_d\omega_n\zeta - C\omega_d^2)\sin\omega_d t]$$

$$+A\left(\frac{\omega}{\omega_n}\right)^2\left[\frac{\left\{1-\left(\frac{\omega}{\omega_n}\right)^2\right\}\sin\omega t-2\zeta\left(\frac{\omega}{\omega_n}\right)\cos\omega t}{\left\{1-\left(\frac{\omega}{\omega_n}\right)^2\right\}^2+\left(2\zeta\frac{\omega}{\omega_n}\right)^2}\right] \tag{2.60}$$

The constants B and C can be determined by evaluating (2.58) and (2.59) for the boundary conditions $z(0) = \dot{z}(0) = 0$:

$$B = -\frac{A}{\omega_n^2}\left[\frac{2\zeta\left(\frac{\omega}{\omega_n}\right)}{\left[\left\{1-\left(\frac{\omega}{\omega_n}\right)^2\right\}^2+\left(2\zeta\frac{\omega}{\omega_n}\right)^2\right]}\right], \tag{2.61}$$

and

$$C = \frac{1}{\omega_d}\left[B\omega_n\zeta + \frac{A}{\omega_n}\left(\frac{\omega}{\omega_n}\right)\frac{1-\left(\frac{\omega}{\omega_n}\right)^2}{\left\{1-\left(\frac{\omega}{\omega_n}\right)^2\right\}^2+\left(2\zeta\frac{\omega}{\omega_n}\right)^2}\right]. \tag{2.62}$$

2.3.3 Solution in the Frequency Domain

Solving (2.38) in the frequency domain means forced-response characteristics (particular integral) of the sdof system are taken into account. No initial conditions are taken into account. They are assumed to be damped out. This type of response analysis is often called forced responses, frequency responses or steady-state responses. We will solve (2.38) with the use of the

Fourier transform technique (Jean Baptiste Joseph Fourier 1768–1830). The Fourier-transform of a function $x(t)$ is defined by [James 93]

$$F\{x(t)\} = X(\omega) = \int_{-\infty}^{\infty} x(t)e^{-j\omega t}dt, \tag{2.63}$$

and the inverse of the Fourier transform is

$$F^{-1}\{X(t)\} = x(t) = \frac{1}{2\pi}\int_{-\infty}^{\infty} X(\omega)e^{j\omega t}d\omega. \tag{2.64}$$

Some authors introduce the factor $\frac{1}{2\pi}$ not in (2.64) but in (2.63), or $\frac{1}{\sqrt{2\pi}}$ in both equations for reasons of symmetry. We have adopted the above because it is commonly used in the engineering literature.

The Fourier-transform of the first time derivative of $x(t)$, the velocity $\dot{x}(t)$

$$F\{\dot{x}(t)\} = j\omega X(\omega) = \int_{-\infty}^{\infty} \dot{x}(t)e^{-j\omega t}dt \tag{2.65}$$

The Fourier transform of the second time derivative of $x(t)$, the acceleration $\ddot{x}(t)$ is

$$F\{\ddot{x}(t)\} = (j\omega)^2X(\omega) = -\omega^2X(\omega) = \int_{-\infty}^{\infty} \ddot{x}(t)e^{-j\omega t}dt \tag{2.66}$$

The Fourier-transform of (2.38) is

$$-\omega^2X(\omega) + 2j\zeta\omega_n\omega X(\omega) + \omega_n^2X(\omega) = F(\omega). \tag{2.67}$$

We make $X(\omega)$ now explicit, thus

$$X(\omega) = \frac{F(\omega)}{(\omega_n^2 - \omega^2) + 2j\zeta\omega_n\omega} \tag{2.68}$$

The receptance $\alpha(\omega)$ becomes

$$\alpha(\omega) = \frac{X(\omega)}{F(\omega)} = \frac{(\omega_n^2 - \omega^2) - 2j\zeta\omega_n\omega}{(\omega_n^2 - \omega^2)^2 + (2\zeta\omega_n\omega)^2}. \tag{2.69}$$

The absolute value of the receptance $|\alpha(\omega)|$ is

$$|\alpha(\omega)| = \left|\frac{X(\omega)}{F(\omega)}\right| = \frac{1}{\sqrt{(\omega_n^2-\omega^2)^2+(2\zeta\omega_n\omega)^2}}. \tag{2.70}$$

The absolute value of the receptance $|\alpha(\omega)|$ reaches a maximum value at the resonance frequency ω_{res} (rad/s)] if , [Friswell 98],

$$\omega_{res} = \omega_n\sqrt{1-2\zeta^2}. \tag{2.71}$$

The damped natural frequency is $\omega_d = \omega_n\sqrt{1-\zeta^2}$.

Enforced Accelerations

We will solve (2.54)

$$\ddot{z}(t) + 2\zeta\omega_n\dot{z}(t) + \omega_n^2 z(t) = -\ddot{u}(t)$$

in the frequency domain assuming a base excitation like (2.56)

$$\ddot{u}(t) = A\sin(\omega t)$$

The Fourier-transform of (2.54) and equation (2.56) is

$$-\omega^2 Z(\omega) + 2j\zeta\omega_n\omega Z(\omega) + \omega_n^2 Z(\omega) = -A\frac{\pi}{j}[\delta(\omega-\omega)-\delta(\omega+\omega)]. \tag{2.72}$$

The inverse Fourier-transform of (2.72) is

$$z(t) = F^{-1}\{Z(\omega)\} = -A\frac{j\pi}{2\pi}\int_{-\infty}^{\infty}\frac{[\delta(\omega+\omega)-\delta(\omega-\omega)]}{(\omega_n^2-\omega^2)+2j\zeta\omega_n\omega}e^{j\omega t}d\omega. \tag{2.73}$$

Finally, the solution of the relative displacement $z(t)$ becomes

$$z(t) = -A\frac{\sin(\omega-\varphi)t}{\sqrt{(\omega_n^2-\omega^2)^2+(2\zeta\omega_n\omega)^2}}, \tag{2.74}$$

where the phase angle φ can be calculated with

$$\tan\varphi = \frac{2\zeta\omega_n\omega}{(\omega_n^2-\omega^2)}. \tag{2.75}$$

The relative velocity $\dot{z}(t)$

$$\dot{z}(t) = -A\omega\frac{\cos(\omega-\varphi)t}{\sqrt{(\omega_n^2-\omega^2)^2+(2\zeta\omega_n\omega)^2}}, \tag{2.76}$$

and the relative acceleration $\ddot{z}(t)$

$$\ddot{z}(t) = A\omega^2 \frac{\sin(\omega-\varphi)t}{\sqrt{(\omega_n^2-\omega^2)^2+(2\zeta\omega_n\omega)^2}} = -\omega^2 z(t)\,. \tag{2.77}$$

We can rewrite (2.74), (2.76) and (2.77) as follows

$$z(t) = |Z(\omega)|\sin\left((\omega-\varphi)t+1\frac{1}{2}\pi\right), \tag{2.78}$$

and

$$\dot{z}(t) = \omega|Z(\omega)|\sin((\omega-\varphi)t+2\pi)\,, \tag{2.79}$$

and

$$\ddot{z}(t) = \omega^2|Z(\omega)|\sin\left\{(\omega-\varphi)t+2\frac{1}{2}\pi\right\}. \tag{2.80}$$

This means that the relative velocity $\dot{z}(t)$ has a positive phase shift of $\theta = \frac{1}{2}\pi$ rad with respect to the relative displacement $z(t)$ and the relative acceleration $\ddot{z}(t)$ has a positive phase shift of $\theta = \pi$ rad with respect to the relative displacement $z(t)$. $|Z(\omega)|$ is the modulus or vector length of the harmonic displacement of $z(t)$ with a forced excitation frequency ω.

$$|Z(\omega)| = \frac{A}{\sqrt{(\omega_n^2-\omega^2)^2+(2\zeta\omega_n\omega)^2}}\,. \tag{2.81}$$

In fact, the moduli of the displacement $z(t)$. the velocity $\dot{z}(t)$ and the acceleration $\ddot{z}(t)$ are the frequency responses due to the enforced harmonic acceleration $\ddot{u}(t) = A\sin(\omega t)$.

In the complex plane (Wessel[1] or Argand diagram) means the multipica-tion with j a positive phase shift of $\theta = \frac{1}{2}\pi$ rad. The multiplication of the displacement $z(t)$ with $j\omega$ is differentiating with respect to time of with $z(t)$. Instead of using the harmonic function $\sin\omega t$ we will use the harmonic function $e^{j\omega t}$, with $j = \sqrt{-1}$ (rotation operator). Euler's formula, published in 1748 by Euler[2], [Nahin 98], tells us that

1. Casper Wessel 1745–1818, [Nahin 98]
2. Euler's identity

$$e^{j\omega t} = \sin(\omega t) + j\cos(\omega t). \tag{2.82}$$

The series of e^{jx} can be written as

$$e^{jx} = 1 + jx + \frac{(jx)^2}{2!} + \frac{(jx)^3}{3!} + \frac{(jx)^4}{4!} + \frac{(jx)^5}{5!} + ...$$

$$e^{jx} = \left(1 - \frac{x^2}{2!} + \frac{x^4}{4!} - ...\right) + j\left(x - \frac{x^3}{3!} + \frac{x^5}{5!} - ...\right) = \cos(x) + j\sin(x).$$

Equation (2.82) is illustrated in Fig. 2.6.

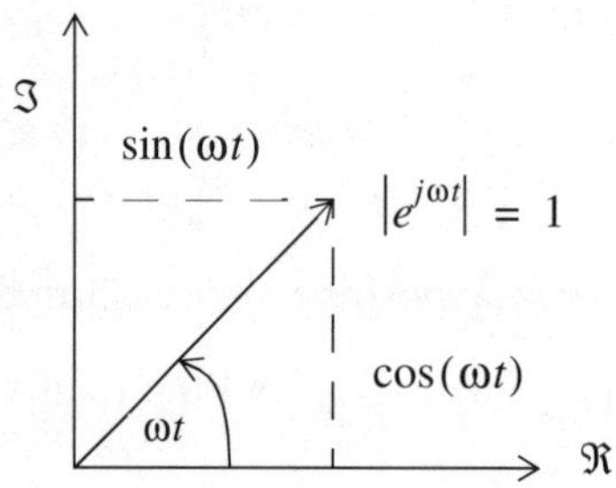

Fig. 2.6. Complex plane (Wessel or Argand diagram)

We will solve (2.54)

$$\ddot{z}(t) + 2\zeta\omega_n\dot{z}(t) + \omega_n^2 z(t) = -\ddot{u}(t)$$

in the frequency domain assuming a base excitation like (2.56)

$$\ddot{u} = Ae^{j\omega t}. \tag{2.83}$$

We assume now that the displacement is

$$z(t) = Z(\omega)e^{j\omega t}. \tag{2.84}$$

Substituting (2.84) into (2.54) we find for $Z(j\omega)$

$$Z(\omega) = \frac{-A}{\omega_n^2 - \omega^2 + 2j\zeta\omega_n\omega}. \tag{2.85}$$

We rewrite (2.85) as follows

$$Z(\omega) = \frac{-A(\omega_n^2 - \omega^2 - 2j\zeta\omega_n\omega)}{(\omega_n^2 - \omega^2)^2 + (2\zeta\omega_n\omega)^2}. \tag{2.86}$$

The modulus of $Z(\omega)$ is

$$|Z(\omega)| = \sqrt{\{\Re(Z)\}^2 + \{\Im(Z)\}^2} = \frac{A}{\sqrt{(\omega_n^2 - \omega^2)^2 + (2\zeta\omega_n\omega)^2}}, \tag{2.87}$$

and the associated argument $\mathrm{Arg}(Z(\omega))$

$$\mathrm{Arg}(Z(\omega)) = \mathrm{atan}\left\{\frac{\Im(Z)}{\Im(Z)}\right\} = \mathrm{atan}\left\{\frac{-2\zeta\omega_n\omega}{\omega_n^2 - \omega^2}\right\}. \tag{2.88}$$

The velocity $\dot{z}(t)$ now becomes

$$\dot{z}(t) = j\omega Z(\omega)e^{j\omega t} = \dot{Z}(\omega)e^{j\omega t}, \tag{2.89}$$

and the acceleration $\ddot{z}(t)$

$$\ddot{z}(t) = (j\omega)^2 Z(\omega)e^{j\omega t} = -\omega^2 Z(\omega)e^{j\omega_{ex}t} = \ddot{Z}(\omega)e^{j\omega t}. \tag{2.90}$$

The associated arguments for $\dot{z}(t)$ and $\ddot{z}(t)$ are respectively,

$$\mathrm{Arg}(\dot{z}(t)) = \frac{\pi}{2} + \mathrm{Arg}(Z(\omega)) \tag{2.91}$$

$$\mathrm{Arg}(\ddot{z}(t)) = \pm\pi + \mathrm{Arg}(Z(\omega)). \tag{2.92}$$

The absolute acceleration becomes

$$\ddot{X}(\omega) = \ddot{Z}(\omega) + A = \frac{\omega^2 A}{\omega_n^2 - \omega^2 + 2j\zeta\omega_n\omega} + A \tag{2.93}$$

or

$$\ddot{X}(\omega) = \ddot{Z}(\omega) + A = \frac{A\left(1 + 2j\zeta\frac{\omega}{\omega_n}\right)}{1 - \frac{\omega^2}{\omega_n^2} + 2j\zeta\frac{\omega}{\omega_n}}. \tag{2.94}$$

The result of (2.94) is similar when (2.56) has been used.

Instead of using the formal Fourier-transform for solving the frequency responses of the sdof system exposed forced vibration we use $e^{j\omega t}$ to achieve solutions of response functions in the frequency domain.

Frequency Response Function

The frequency response function (FRF) $H(\omega)$ is defined as the quotient of the frequency response of the accelerations with respect to the base acceleration (2.83), however, both in the frequency domain

$$H_z(\omega)A = \ddot{Z}(\omega). \tag{2.95}$$

Thus the FRF becomes

$$H_z(\omega) = \frac{\ddot{Z}(\omega)}{A} = \frac{\omega^2}{\omega_n^2 - \omega^2 + 2j\zeta\omega_n\omega} \tag{2.96}$$

$$H_z(\omega) = \frac{\omega^2}{\omega_n^2} \frac{1}{1 - \frac{\omega^2}{\omega_n^2} + 2j\zeta\frac{\omega}{\omega_n}}. \tag{2.97}$$

The FRF with respect to the absolute accelerations $H_x(\omega)A = \ddot{X}(\omega)$

$$H_x(\omega) = \frac{1 + 2j\zeta\frac{\omega}{\omega_n}}{1 - \frac{\omega^2}{\omega_n^2} + 2j\zeta\frac{\omega}{\omega_n}}. \tag{2.98}$$

Forcing Function $f(t)$

We will solve (2.38):

$$\ddot{x}(t) + 2\zeta\omega_n\dot{x}(t) + \omega_n^2 x(t) = \frac{F(t)}{m} = f(t)$$

in the frequency domain assuming the forcing function

$$f(t) = \frac{F_o}{m}e^{j\omega t} \tag{2.99}$$

We now assume that the displacement is

$$x(t) = X(\omega)e^{j\omega t} \tag{2.100}$$

The frequency response function $X(\omega)$ becomes

$$X(\omega) = \frac{F_o}{m(\omega_n^2 - \omega^2 + 2j\zeta\omega_n\omega)} = \frac{F_o}{m\omega_n^2\left(1 - \frac{\omega^2}{\omega_n^2} + 2j\zeta\frac{\omega}{\omega_n}\right)}, \tag{2.101}$$

or expressed in the static displacement $x_{\text{stat}} = \frac{F_o}{k}$ (m), using (2.2).

$$X(\omega) = \frac{F_o}{k\left(1 - \frac{\omega^2}{\omega_n^2} + 2j\zeta\frac{\omega}{\omega_n}\right)} = \frac{x_{stat}}{\left(1 - \frac{\omega^2}{\omega_n^2} + 2j\zeta\frac{\omega}{\omega_n}\right)} \tag{2.102}$$

The absolute velocity $\dot{x}(t)$ now becomes

$$\dot{x}(t) = j\omega X(\omega)e^{j\omega t} = \dot{X}(\omega)e^{j\omega t}, \tag{2.103}$$

and the absolute acceleration $\ddot{x}(t)$

$$\ddot{x}(t) = (j\omega)^2 X(\omega)e^{j\omega t} = -\omega^2 X(\omega)e^{j\omega t} = \ddot{X}(\omega)e^{j\omega t}. \tag{2.104}$$

Frequency Response Function
The following FRF can be derived:

$$\frac{X(\omega)}{x_{stat}} = \frac{1}{\left(1 - \frac{\omega^2}{\omega_n^2} + 2j\zeta\frac{\omega}{\omega_n}\right)}, \tag{2.105}$$

and the modulus is

$$\left|\frac{X(\omega)}{x_{stat}}\right| = \frac{1}{\sqrt{\left(1 - \frac{\omega^2}{\omega_n^2}\right)^2 + \left(2\zeta\frac{\omega^2}{\omega_n^2}\right)^2}}. \tag{2.106}$$

Alternative forms (names) of the frequency response function [McConnel 95, Maia 97] are:

- Receptance $\alpha(\omega) = H(\omega) = \frac{\text{displacement response}}{\text{force excitation}}$
- Admittance $\alpha(\omega) = \frac{\text{displacement response}}{\text{force excitation}}$
- Dynamic compliance $\alpha(\omega) = \frac{\text{displacement response}}{\text{force excitation}}$
- Mobility $Y(\omega) = j\omega H(\omega) = \frac{\text{velocity response}}{\text{force excitation}}$
- Accelerance $A(\omega) = j\omega Y(\omega) = \frac{\text{acceleration responsee}}{\text{force excitation}}$
- Inertance $A(\omega) = \frac{\text{acceleration response}}{\text{force excitation}}$
- Dynamic stiffness $\frac{1}{\alpha(\omega)} = \frac{\text{force excitation}}{\text{displacement response}}$

- Mechanical impedance $\dfrac{1}{Y(\omega)} = \dfrac{\text{force excitation}}{\text{velocity response}}$
- Apparent mass $\dfrac{1}{A(\omega)} = \dfrac{\text{force excitation}}{\text{acceleration response}}$
- Dynamic mass $\dfrac{1}{A(\omega)} = \dfrac{\text{force excitation}}{\text{acceleration response}}$

2.3.4 State Space Representation of the Sdof System

In (2.38) the equation of motion of a sdof dynamic system has been derived. We assume underdamped damping ratio $(\zeta < 1)$ characteristics.

$$\ddot{x}(t) + 2\zeta\omega_n\dot{x}(t) + \omega_n^2 x(t) = \frac{F(t)}{m} = f(t)\,.$$

We will now introduce the state space variables $y(t)$

$$y(t) = \begin{Bmatrix} y_1(t) \\ y_2(t) \end{Bmatrix} = \begin{Bmatrix} x(t) \\ \dot{x}(t) \end{Bmatrix}. \tag{2.107}$$

Rearranging (2.38) we get the space state equations of motion

$$\begin{Bmatrix} \dot{y}_1(t) \\ \dot{y}_2(t) \end{Bmatrix} = \begin{bmatrix} 0 & 1 \\ -\omega_n^2 & -2\zeta\omega_n \end{bmatrix} \begin{Bmatrix} y_1(t) \\ y_2(t) \end{Bmatrix} + \begin{Bmatrix} 0 \\ f(t) \end{Bmatrix}. \tag{2.108}$$

Response Analysis in Time Domain

With initial condition s at $t = 0$, $y_1(0) = x_o$ and $y_2(0) = v_o$, the equation to solve the Laplace-transform, becomes

$$\begin{bmatrix} s & -1 \\ 0 & s + \omega_n^2 + 2\zeta\omega_n \end{bmatrix} \begin{Bmatrix} Y_1(s) \\ Y_2(s) \end{Bmatrix} = \begin{Bmatrix} x_n \\ v_o + F(s) \end{Bmatrix} \tag{2.109}$$

The solution for $y_1(t) = x(t)$ becomes (see also (2.52))

$$y_1(t) = x(t) = x_o e^{-\zeta\omega_n t}\left(\cos\omega_d t + \frac{\zeta}{\sqrt{1-\zeta^2}}\sin\omega_d t\right)$$

$$+ v_o e^{-\zeta\omega_n t}\frac{\sin\omega_d t}{\omega_d} + \int_0^t e^{-\zeta\omega_n\tau}\frac{\sin\omega_d\tau}{\omega_d} f(t-\tau)d\tau\,, \tag{2.110}$$

and the solution for $y_2(t) = \dot{x}(t)$ is

$$y_2(t)= \dot{x}(t) = v_o e^{-\zeta\omega_n t}\left(\cos\omega_d t - \frac{\zeta}{\sqrt{1-\zeta^2}}\sin\omega_d t\right)$$

$$-x_o e^{-\zeta\omega_n t}\frac{\sin\omega_d t}{\omega_d} + \int_0^t e^{-\zeta\omega_n \tau}\left(\cos\omega_d \tau - \frac{\zeta}{\sqrt{1-\zeta^2}}\sin\omega_d \tau\right) f(t-\tau)d\tau. \quad (2.111)$$

Frequency Response Analysis

The forcing function in the frequency domain is assumed to be

$$f(t) = \frac{F_o}{m}e^{j\omega t}. \quad (2.112)$$

We assume now that the state vector is

$$\{y(t)\} = \{Y(\omega)\}e^{j\omega t}. \quad (2.113)$$

The equations written in the frequency domain are

$$\begin{bmatrix} j\omega & -1 \\ 0 & j\omega + \omega_n^2 + 2\zeta\omega_n \end{bmatrix}\begin{Bmatrix} Y_1(\omega) \\ Y_2(\omega) \end{Bmatrix} = \begin{Bmatrix} 0 \\ \frac{F_o}{m} \end{Bmatrix} \quad (2.114)$$

The solution of (2.114), $Y_1(\omega) = X(\omega)$ becomes

$$Y_1(\omega) = X(\omega) = \frac{F_o}{m}\frac{1}{\omega_n^2\left(1 - \frac{\omega^2}{\omega_n^2} + 2j\zeta\frac{\omega}{\omega_n}\right)}, \quad (2.115)$$

and the solution for $Y_2(\omega) = \dot{X}(\omega)$ is

$$Y_2(\omega) = \dot{X}(\omega) = \frac{F_o}{m}\frac{j\omega}{\omega_n^2\left(1 - \frac{\omega^2}{\omega_n^2} + 2j\zeta\frac{\omega}{\omega_n}\right)} = j\omega X(\omega). \quad (2.116)$$

The damped natural frequencies can be obtained by calculating the poles of $X(\omega)$ or $\dot{X}(\omega)$ with

$$\left(1-\frac{\omega^2}{\omega_n^2}+2j\zeta\frac{\omega}{\omega_n}\right) = 0\,. \tag{2.117}$$

This means that the damped natural frequences become

1. $\dfrac{\omega_1}{\omega_n} = -\zeta - j\sqrt{1-\zeta^2}$

2. $\dfrac{\omega_2}{\omega_n} = -\zeta + j\sqrt{1-\zeta^2}$

The natural frequency ω_n is obtained with

$$\sqrt{\frac{\omega_1\omega_2}{\omega_n^2}} = 1\,, \tag{2.118}$$

and the damping ratio ζ is calculated with

$$\zeta = -\frac{(\omega_1+\omega_2)}{2\omega_n}\,. \tag{2.119}$$

A certain sdof dynamic system, enforced with a Dirac delta force, can be described with the following equation of motion.

$$\ddot{x}(t)+\dot{x}(t)+100x(t) = \delta(t-1)\,.$$

The characteristics of the damped sdof system, the displacements, the velocities and the accelerations in the time domain will be calculated.

The space state equation of motion becomes

$$\begin{Bmatrix} \dot{y}_1(t) \\ \dot{y}_2(t) \end{Bmatrix} = \begin{bmatrix} 0 & 1 \\ -100 & -1 \end{bmatrix} \begin{Bmatrix} y_1(t) \\ y_2(t) \end{Bmatrix} + \begin{Bmatrix} 0 \\ \delta(t-1) \end{Bmatrix},$$

with initial condition at $t = 0$, $y_1(0) = 0$ and $y_2(0) = 0$.

The eigenvalue problem is obtained by substituting for $\{y\} = \{y(\lambda)\}e^{\lambda t}$ in the homogeneous state space equations of motion

$$\begin{bmatrix} \lambda & -1 \\ 100 & 1+\lambda \end{bmatrix} \begin{Bmatrix} y_1(\lambda) \\ y_2(\lambda) \end{Bmatrix} = \begin{Bmatrix} 0 \\ 0 \end{Bmatrix}.$$

The eigenvalues λ_1 and λ_2 are

- $\lambda_1 = -0.5 - 9.9875j$
- $\lambda_1 = -0.5 + 9.9875j$

From (2.118) we obtained the natural circular frequency of the sdof dynamic system

$$\omega_n = \sqrt{\lambda_1 \lambda_1} = \sqrt{(0.5)^2 + (9.9875)^2} = 10,$$

and from (2.119) the viscous damping ratio ζ

$$\zeta = -\frac{(\lambda_1 + \lambda_2)}{2\omega_n} = 0.05.$$

The solution for the state variables is $y_1(t) = x(t)$ and $y_2(t) = \dot{x}(t)$.

With $\omega_n = 10$, $\zeta = 0.05$ and $\omega_d = \omega_n\sqrt{1-\zeta^2} = 9.98749$

$$y_1(t) = x(t) = 100\int_0^t e^{-\zeta\omega_n\tau}\frac{\sin\omega_d\tau}{\omega_d}\delta(t-\tau-1)d\tau,$$

and

$$y_2(t) = \dot{x}(t) = 100\int_0^t e^{-\zeta\omega_n\tau}\left(\cos\omega_d\tau - \frac{\zeta}{\sqrt{1-\zeta^2}}\sin\omega_d\tau\right)\delta(t-\tau-1)d\tau.$$

It is a good excercise to evaluate the two previous relations.

2.4 Problems

2.4.1 Problem 1

A 1000 kg mass is attached to a spring such that the natural frequency is 25 Hz. What is the spring constant k (N/m)?

Answer: $k = 24.67\ 10^6$ N/m

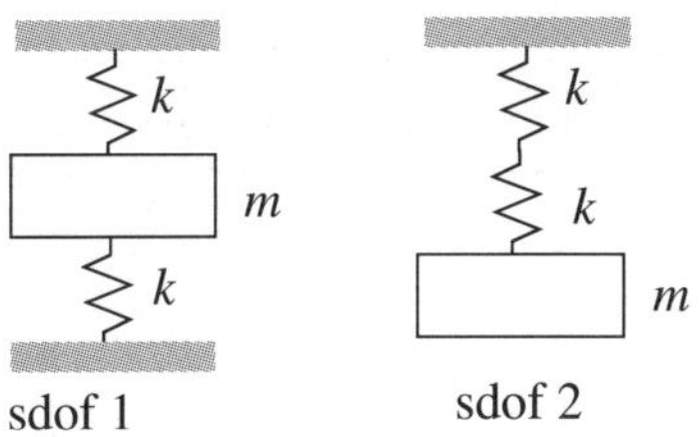

Fig. 2.7. sdof systems (spring–mass systems)

2.4.2 Problem 2

Derive the radian natural frequencies for both sdof dynamic systems as shown Fig. 2.7, [Norton 89]

Answers: $\omega_1 = \sqrt{\frac{2k}{m}}$, $\omega_2 = \sqrt{\frac{k}{2m}}$.

2.4.3 Problem 3

Assuming a clamped-free massless elastic beam with bending stiffness EI, derive an expression for the natural frequency of the dynamic system.

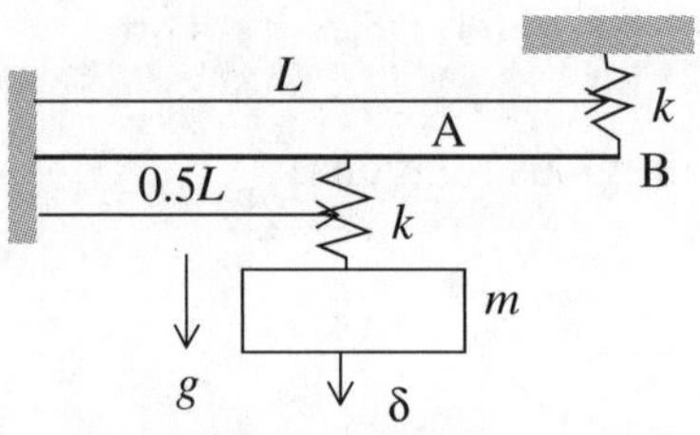

Fig. 2.8. Beam–springs–mass system

Hint: Calculate the displacement δ caused by the inertia force mg applied to the mass. The spring stiffness is given by $k = \frac{3EI}{2l^3}$. The natural frequency f (Hz) can be calculated with $f = \frac{1}{2\pi}\sqrt{\frac{g}{\delta}}$.

Answer: $f = 0.191\sqrt{\frac{EI}{ml^3}}$ Hz.

2.4.4 Problem 4

A mass oscillates harmonically with an amplitude of 0.002 m and a frequency of 35 Hz. What are the values of the maximum velocity v in (m/s) and the maximum acceleration a in (g) (9.81m/s^2)?

Answers: $v = 0.44$ m/s, $a = 9.86$ g.

2.4.5 Problem 5

Derive the expressions for the mobility and impedance of a mass element, a spring element, viscous-damping element and a mass–spring–damper system. For the answers see Table 2.2.

Table 2.2. Element mobility and impedance

Element	Mobility	Impedance
Mass	$\frac{1}{jm\omega}$	$jm\omega$
Spring	$\frac{j\omega}{k}$	$\frac{k}{j\omega}$
Damper	$\frac{1}{c}$	c
Mass–spring–damper	$\frac{1}{c+j\left(m\omega-\frac{k}{\omega}\right)}$	$c+j\left(m\omega-\frac{k}{\omega}\right)$

2.4.6 Problem 6

An sdof dynamic system is desribed by

$$\ddot{x}(t)+4\dot{x}(t)+5x(t) = 8\cos t.$$

Calculate the radian natural frequency ω_n and the damping ratio ζ.
Solve the forced equation of motion using the Laplace-transform subject to the initial conditions $x(0) = \dot{x}(0) = 0$.

Answers: $\omega_n = \sqrt{5}$, $\zeta = 0.89$, and $x(t) = \cos t + \sin t - e^{-2t}(\cos t + 3\sin t)$ [James 93].

2.4.7 Problem 7

The mass m of the mechanical system of Fig. 2.9 is subjected to a harmonic forcing $\sin(\omega t)$. Define the steady-state response of the sdof system.

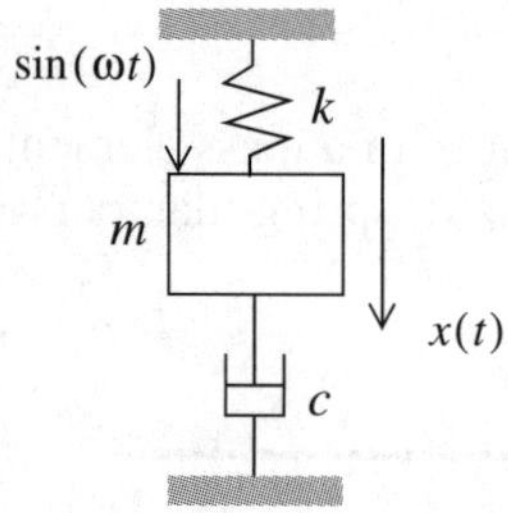

Fig. 2.9. Mechanical system

2.4.8 Problem 8

Show that the maxima and minima of an underdamped motion occur at equidistant values of t, the distance between two consecutive maxima being $\frac{2\pi}{\omega_d}$.

3 Damping Models

3.1 Introduction

For transient vibration, damping is less important than for harmonic (sinusoidal) vibration. In the case of harmonic vibration the damping in structures plays a very important role to reduce the responses. At the resonance frequencies the inertia loads are in equilibrium with the elastic loads; therefore the applied external loads are balanced by the induced damping loads. The lower the damping, the higher the responses in the structure, because less energy is dissipated. When the damping is very low the responses can be so large that damage is caused to the structure.

Damping may be categorised as follows [Beards 96]:

- Passive damping
- Inherent damping
- Added damping
- Active damping

Inherent damping in a structure is largely determined by the materials used, the type of construction and the assembly. In general the damping in pure materials is very low. The friction in structural joints, cable booms, etc., will give a much higher damping than the material damping only. In lightweight and large structures; solar arrays, antenna dishes, etc., the acoustic radiation will also contribute to the damping due to the dissipated energy of the occurring sound.

Damping from other (external) sources may be added to the inherent damping in the structure. The application of passive dampers and damping layers are examples of extra damping introduced in a design. In the case of prescribed response amplitudes (acceleration, etc.), responses will be controlled by use of active damping.

Only the inherent damping in structures is discussed in this chapter. The phenomena of increased damping due to increases in the input levels and frequencies is not considered here.

3.2 Damped Vibration

For treating mechanical vibration the single degree of freedom (sdof) system is the simplest mechanical oscillator. The sdof system consists of a discrete mass and a massless linear spring connected to the mass and the base. In general a damping element (linear or nonlinear) may be active in parallel with the linear spring. The relative displacement of the discrete mass with respect to the base is defined as $x(t)$ (m). Applying an excitation force $F(t)$ (N) to the mass m (kg), the equation of motion of such an sdof system is [Beards 96]:

$$\ddot{x}(t) + \frac{f_d}{m} + \omega_n^2 x(t) = \frac{F(t)}{m} = f(t), \tag{3.1}$$

where

- $\omega_n = \sqrt{\frac{k}{m}}$ is the natural circular frequency (rad/s).

The damping force f_d (N) may be caused by:

- Material damping
- Air damping
- Acoustic damping
- Joint damping (rivets, bolts, bonding)

3.2.1 Linear Damping

For linear damping, the damping is either linearly proportional to the velocity $\dot{x}(t)$, the so-called viscous damping, or is linearly proportional to the elastic force, $kx(t)$, the so-called structural damping. Hence the damping force can be defined as:

- Viscous damping $f_d = c\dot{x}$ (N)
- Structural damping $f_d = jgkx$ (N)

At the resonance frequency of a harmonic vibration, both viscous damping and structural damping forces exhibit a positive phase shift of $\frac{\pi}{2}$ rad with respect to the elastic force. Structural damping is proportional to the stiffness. The phase shift of $\frac{\pi}{2}$ rad for structural damping is obtained by multiplying the proportional part of the elastic force with the complex number $j = \sqrt{-1}$ [James 93].

The introduction of complex numbers in the mathematical model limits the application of the structural damping only for harmonic vibration. At the resonance frequency the harmonic inertia loads have a $\pm\pi$ rad phase shift with respect to the elastic forces.

In general, the mathematical models of viscous damping and structural damping are applied to linear mechanical vibration. They are easy to understand and to use in dynamic response analysis of harmonic vibration.

The concept of viscous damping may also be used in transient vibration.

3.2.2 Viscous Damping

The equation of motion of an sdof system, assuming linear viscous damping and a mass excited by an external load is:

$$m\ddot{x}(t) + c\dot{x}(t) + kx(t) = F(t), \tag{3.2}$$

where

- c is the viscous damping constant (Ns/m).

Dividing (3.2) by m

$$\ddot{x}(t) + 2\zeta\omega_n\dot{x}(t) + \omega_n^2 x(t) = \frac{F(t)}{m} = f(t), \tag{3.3}$$

where

- $c_{\text{crit}} = 2\sqrt{km}$ is the critical damping (Ns/m)
- $\zeta = \frac{c}{c_{\text{crit}}}$ is the damping ratio
- $\frac{c}{m} = \zeta\frac{c_{\text{crit}}}{m} = 2\zeta\sqrt{\frac{k}{m}} = 2\zeta\omega_{\text{n}}$

3.2.3 Structural Damping

The equation of motion of an sdof system, assuming linear structural damping and the mass excited by an external load is:

$$m\ddot{x}(t) + k(1 + jg)x(t) = F(t), \tag{3.4}$$

where

- g is the structural damping ratio (hysteric damping) (-)

Dividing (3.4) by the mass m gives

$$\ddot{x}(t) + \omega_n^2(1 + jg)x(t) = \frac{F(t)}{m} = f(t). \tag{3.5}$$

3.2.4 Loss Factor

When damping is general it is commonly quantified by the loss factor. The definition of the loss factor is [Thomson 98]:

$$\eta = \frac{\Pi_{\text{diss}}}{2\pi E_{\max}}, \tag{3.6}$$

where

- $\Pi_{\text{diss}} = \oint f_D dx$ is the dissipated energy per cycle
- $\dfrac{\Pi_{\text{diss}}}{2\pi}$ is the damping energy per radian
- $E_{\max}$ is the peak potential energy.

For the viscous damping c and with $x(t) = A\sin(\omega t - \varphi)$, the damping energy loss is given by

$$\Pi_{\text{diss}} = \oint f_D dx = \oint c\dot{x}dx = \oint c\dot{x}^2 dt = \pi c\omega A^2 = 2\zeta\pi A^2\omega\sqrt{km}, \tag{3.7}$$

and with

$$E_{\max} = \frac{1}{2}kA^2, \tag{3.8}$$

we obtain

$$\eta = \frac{\Pi_{\text{diss}}}{2\pi E_{\max}} = \left(\frac{\omega}{\omega_n}\right)2\zeta. \tag{3.9}$$

At the natural frequency ω_n the loss factor $\eta = 2\zeta$.

In the case of harmonic vibration the structural damping

$$\ddot{x}(t) + \omega_n^2(1 + jg)x(t) = \frac{F(t)}{m} = f(t)$$

can be expressed as the following equation $x(t) = A\sin(\omega t - \varphi)$:

$$\ddot{x}(t) + \left(\frac{\omega_n^2}{\omega}\right)g\dot{x}(t) + \omega_n^2 x(t) = \frac{F(t)}{m} = f(t) .$$

The dissipated energy per cycle is

$$\Pi_{diss} = \oint f_D dx = \oint \frac{\omega_n^2}{\omega} g\dot{x}dx = \oint \frac{\omega_n^2}{\omega} g\dot{x}^2 dt = \pi g\omega_n^2 A^2 , \tag{3.10}$$

and with the peak potential energy given by

$$E_{max} = \frac{1}{2}\omega_n^2 A^2 ,$$

we obtain

$$\eta = \frac{\Pi_{diss}}{2\pi E_{max}} = g . \tag{3.11}$$

3.3 Amplification Factor

The maximum absolute response of the sdof system, either with viscous or structural damping is very close to, or the same as, the natural frequency. (With viscous damping the maximum amplification occurs at $\omega = \omega_n\sqrt{1 - 2\zeta^2}$, but the damped natural frequency is given by $\omega_d = \omega_n\sqrt{1 - \zeta^2}$). The maximum absolute value of the frequency response function (FRF) at the resonance frequency is called the amplification factor (quality factor, magnification factor) Q. The FRF can be obtained by applying a unit forcing function $f(t) = e^{j\omega t}$. The response of the discrete mass can then be assumed to be $x(t) = X(j\omega)e^{j\omega t}$. This solution for $x(t)$ will be substituted in the equations of motion of a simple mass–spring system with either viscous or structural damping.

3.3.1 Modal Viscous Damping

The equation of motion of the sdof system with modal viscous damping is

$$\ddot{x}(t) + 2\zeta\omega_n\dot{x}(t) + \omega_n^2 x(t) = f(t) .$$

The frequency response of $x(t)$ is $X(j\omega)$

$$X(\omega) = \frac{1}{\omega_n^2} H(\omega), \tag{3.12}$$

where

- $H(\omega) = \dfrac{1}{\left[1 - \left(\frac{\omega}{\omega_n}\right)^2 + 2j\zeta\left(\frac{\omega}{\omega_n}\right)\right]}$ is the frequency response function (FRF)

The modulus of the FRF, $H(j\omega)$ is

$$|H(\omega)| = \frac{1}{\sqrt{\left[\left\{1 - \left(\frac{\omega}{\omega_n}\right)^2\right\}^2 + \left(2\zeta\frac{\omega}{\omega_n}\right)^2\right]}}. \tag{3.13}$$

The maximum value of $|H(\omega)|$ is

$$|H(\omega)| = \frac{1}{2\zeta\sqrt{1-\zeta^2}},$$

and will be reached at an excitation frequency

$$\omega = \omega_n\sqrt{1 - 2\zeta^2}. \tag{3.14}$$

In general, the natural frequency ω_n is:

$$\left|H(\omega_n\sqrt{1-2\zeta^2})\right| = \frac{1}{2\zeta\sqrt{1-\zeta^2}} = Q \approx |H(\omega_n)| = \frac{1}{2\zeta}. \tag{3.15}$$

At the natural frequency:

$$|H(\omega_n)| = \frac{1}{2\zeta} \approx Q, \tag{3.16}$$

where Q is called the amplification factor.

3.3.2 Modal Structural Damping

The equation of motion of the sdof system with modal structural damping is

$$\ddot{x}(t) + \omega_n^2(1 + jg)x(t) = f(t).$$

The frequency response of $x(t)$ is $X(j\omega)$,

$$X(\omega) = \frac{1}{\omega_n^2} H(\omega), \tag{3.17}$$

where

- $H(\omega) = \dfrac{1}{\left[1 - \left(\dfrac{\omega}{\omega_n}\right)^2 + jg\right]}$ is the frequency response function (FRF)

The modulus of the FRF $H(j\omega)$ is

$$|H(\omega)| = \frac{1}{\sqrt{\left[\left\{1 - \left(\dfrac{\omega}{\omega_n}\right)^2\right\}^2 + g^2\right]}} . \quad (3.18)$$

The maximum value of $|H(\omega)|$ is reached at the natural frequency ω_n,

$$|H(\omega_n)| = \frac{1}{g} .$$

At the natural frequency the absolute value of the FRF becomes

$$|H(\omega_n)| = \frac{1}{g} = Q . \quad (3.19)$$

3.3.3 Discussion of Modal Damping

For harmonic oscillations the damping force f_D is

- for modal viscous damping $\left|\dfrac{f_D}{\omega_n^2 X(\omega)}\right| = 2\zeta\left(\dfrac{\omega}{\omega_n}\right)$
- for modal structural damping $\left|\dfrac{f_D}{\omega_n^2 X(\omega)}\right| = g$

At the natural frequency ω_n the damping forces are equal to each other. Below the natural frequency the modal viscous damping force is lower than the damping force due to structural damping, however, beyond the natural frequency the viscous damping force is greater than the structural damping force.

3.4 Method of Determining Damping from Measurements

3.4.1 The Half-Power Point Method

For the half-power method the modal viscous damping is estimated by measuring the frequency increment $\Delta\omega$ at the half-power points (i.e. where the response R is equal to $R = \frac{R_{max}}{\sqrt{2}} \approx \frac{Q}{\sqrt{2}}$). The modal viscous damping ζ is

$$\zeta = \frac{\Delta\omega}{2\omega_n}. \tag{3.20}$$

In (3.20) the value of the modal viscous damping must be $\zeta \leq 0.1$, [Carrington 75] (see Fig. 3.1).

Equation (3.20) can be very easily proved because

$$|H(j\omega)| = \frac{1}{\sqrt{\left[\left\{1-\left(\frac{\omega}{\omega_n}\right)^2\right\}^2+\left(2\zeta\frac{\omega}{\omega_n}\right)^2\right]}} = \frac{Q}{\sqrt{2}} = \frac{1}{\sqrt{2}}\frac{1}{2\zeta}. \tag{3.21}$$

Then with

$$\left(\frac{\omega_1}{\omega_n}\right)^2 \approx 1-2\zeta \text{ and } \left(\frac{\omega_2}{\omega_n}\right)^2 \approx 1+2\zeta, \tag{3.22}$$

we get

$$\left(\frac{\omega_2}{\omega_n}\right)^2-\left(\frac{\omega_1}{\omega_n}\right)^2 \approx \frac{\Delta\omega}{\omega_n} = 2\zeta, \tag{3.23}$$

with

$$\Delta\omega = \omega_2-\omega_1. \tag{3.24}$$

To measure the damping ratio ζ with the aid of the half-power method the response peaks at the resonance frequency must be well separated. Damping may be measured or derived from a modal survey test[1], however, modal

1. A modal survey test is a test that identifies a set of modal properties of a mechanical system.

survey is beyond the scope of this book. Readers who are interested in modal survey are advised to read [Ewins 00, Maia 97].

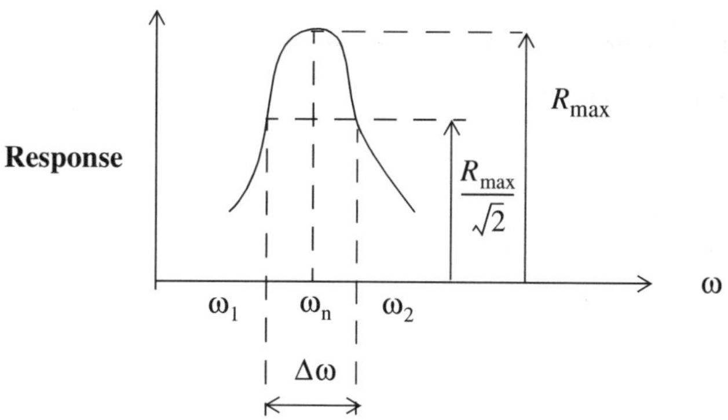

Fig. 3.1. Half-power method

3.5 Problems

3.5.1 Problem 1

Define the critical damping and damping ratio of the single degree of freedom dynamic system as illustrated in Fig. 3.2. The bar rotates about the support at point A. The only degree of freedom is the angle of rotation φ.

1. Set up the equation of motion $\ddot{\varphi}(k, m, c, L) = 0$.
2. Define the natural frequency ω_n.
3. Define the critical damping c_{crit}.
4. Define the damping ratio ζ, the equation of motion is then $\ddot{\varphi} + 2\zeta\omega_n\dot{\varphi} + \omega_n^2\varphi = 0$.

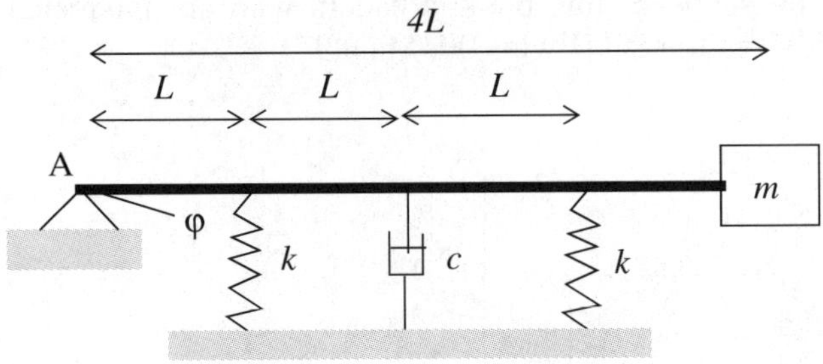

Fig. 3.2. Single degree of freedom dynamic system

Answers:

1. $\ddot{\varphi} + \frac{c}{4m}\dot{\varphi} + \frac{5}{8}\frac{k}{m}\varphi = 0$

2. $\omega_n^2 = \frac{5}{8}\frac{k}{m}$

3. $c_{crit} = 2\sqrt{10km}$

4. $\zeta = \frac{c}{c_{crit}}$

3.5.2 Problem 2

The damper in the single degree of freedom dynamic system, as illustrated Fig. 3.2, is deleted. The structural damping g is proportional to the stiffness k.

1. Define the undamped equation of motion $\ddot{\varphi}(k, m, L) = 0$.
2. Introduce the structural damping in the equation of motion.
3. Assume a periodic motion $e^{j\omega t}$. Derive the expression for the structural damping in the damper c, $g(c, m, k, \omega)$.

Answers:

1. $\ddot{\varphi} + \frac{5}{8}\frac{k}{m}\varphi = 0$

2. $\ddot{\varphi} + \omega_n^2(1 + jg)\varphi = 0$

3. $g = \frac{2}{5}\frac{c}{k}\omega$

4 Multi-Degrees of Freedom Linear Dynamic Systems

4.1 Introduction

A number of coupled linear sdof systems form a linear dynamic system with more degrees of freedom, the so-called multi-degrees of freedom system (mdof). Such a dynamic system is often called a discrete dynamic system with n discrete masses coupled with springs and dampers. We talk about a dynamic system with n degrees of freedom (dofs).

Continuous dynamic systems may be transferred to an mdof discrete system using generalised coordinates.

The application of the finite element method to continuous structures will result in a discrete mdof system where node displacements and rotations are the discrete unknowns.

In this chapter the derivation of the equations of motion of an mdof dynamic system will be discussed. In general, the equations of motion of a discrete dynamic system can be written as

$$[M]\{\ddot{x}(t)\} + [C]\{\dot{x}(t)\} + [K]\{x(t)\} = \{F(t)\}, \tag{4.1}$$

and consists of the following matrices and vectors:

- the mass matrix $[M]$
- the stiffness matrix $[K]$
- the damping matrix $[C]$
- the force vector $\{F(t)\}$
- the displacement, velocity and acceleration vectors $\{x(t)\}$, $\{\dot{x}(t)\}$ and $\{\ddot{x}(t)\}$

For linear mdof systems the mass matrix, stiffness and damping matrix do not vary with time, however, the displacement, velocity, acceleration and force vector do usually change with time.

4.2 Derivation of the Equations of Motion

The equations of motion of mdof dynamic systems can be derived applying, amongst others, Newton's law, differentiation of energies and work and the Lagrange's equations [Meirovitch 70]. The mentioned methods will be used to derive the equations of motion for a simple 3-dof discrete dynamic system (Fig. 4.1). No damping is taken into account.

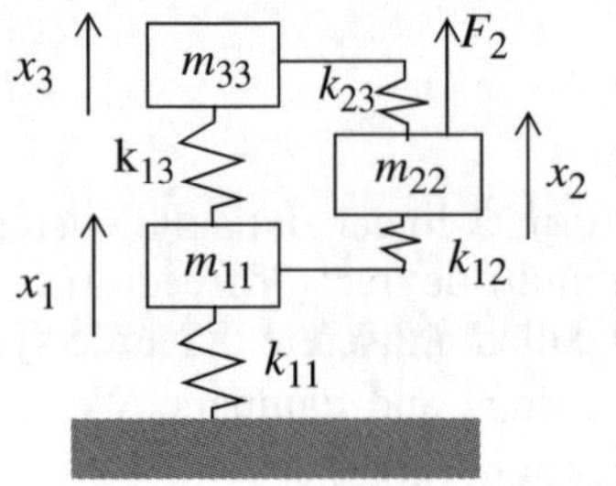

Fig. 4.1. 3-dof discrete linear dynamic system

4.2.1 Undamped Equations of Motion with Newton's Law

We will assume for the displacements, the following relations:

- $(x_1 - x_2) \geq 0$
- $(x_1 - x_3) \geq 0$
- $(x_2 - x_3) \geq 0$

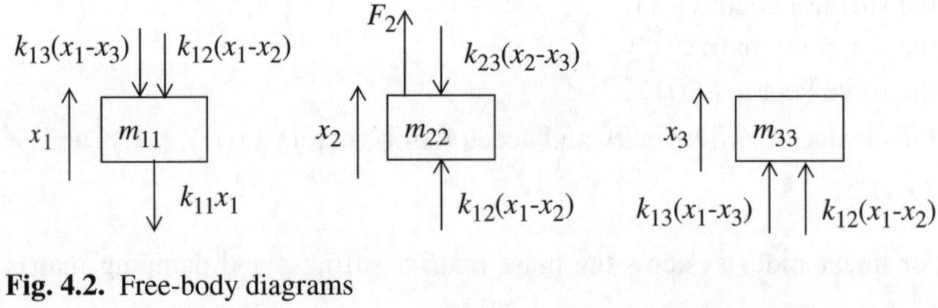

Fig. 4.2. Free-body diagrams

The spring force F (N) is equal to the spring constant k (N/m) multiplied by the difference in displacement Δx (m) between the discrete masses, $F = k\Delta x$.

The free-body diagrams are illustrated in Fig. 4.2.

The equations of motion per discrete mass (Newton's law) are then as follows:

$$m_{11}\ddot{x}_1 = -k_{13}(x_1 - x_3) - k_{12}(x_1 - x_2) - k_{11}x_1 \tag{4.2}$$

$$m_{22}\ddot{x}_2 = F_2 + k_{12}(x_1 - x_2) - k_{23}(x_2 - x_3) \tag{4.3}$$

$$m_{33}\ddot{x}_3 = k_{13}(x_1 - x_3) + k_{23}(x_2 - x_3). \tag{4.4}$$

Equations (4.2), (4.3) and (4.4) may be transferred into matrix notation

$$\begin{bmatrix} m_{11} & 0 & 0 \\ 0 & m_{22} & 0 \\ 0 & 0 & m_{33} \end{bmatrix} \begin{Bmatrix} \ddot{x}_1 \\ \ddot{x}_2 \\ \ddot{x}_3 \end{Bmatrix} + \begin{bmatrix} k_{11}+k_{12}+k_{13} & -k_{12} & -k_{13} \\ -k_{12} & k_{12}+k_{23} & -k_{23} \\ -k_{13} & -k_{23} & k_{13}+k_{23} \end{bmatrix} \begin{Bmatrix} x_1 \\ x_2 \\ x_3 \end{Bmatrix} = \begin{Bmatrix} 0 \\ F_2 \\ 0 \end{Bmatrix} \tag{4.5}$$

or

$$[M]\{\ddot{x}\} + [K]\{x\} = \{F\}. \tag{4.6}$$

The mass matrix $[M]$ and the stiffness matrix $[K]$ are symmetric. This can be easily proven. The general expressions for the kinetic energy T and the potential (strain) energy U of a linear system are

$$T = \frac{1}{2}m_{ij}\dot{x}_i\dot{x}_j, \text{ summed over } i,j=1,2,\ldots \text{ (Einstein convention)} \tag{4.7}$$

$$U = \frac{1}{2}k_{ij}x_ix_j, \text{ summed over } i,j=1,2,\ldots\ . \tag{4.8}$$

Equations (4.7) and (4.8) may be written in matrix notation, the kinetic energy in the mdof system is

$$T = \frac{1}{2}\{\dot{x}\}^T[M]\{\dot{x}\}, \tag{4.9}$$

and the potential energy is

$$U = \frac{1}{2}\{x\}^T[K]\{x\}. \tag{4.10}$$

The second derivative of a function has a symmetric property

$$\frac{\partial^2 T}{\partial \dot{x}_i \partial \dot{x}_j} = m_{ij} = \frac{\partial^2 T}{\partial \dot{x}_j \partial \dot{x}_i} = m_{ji}, \tag{4.11}$$

and

$$\frac{\partial^2 U}{\partial x_i \partial x_j} = k_{ij} = \frac{\partial^2 U}{\partial x_j \partial x_i} = k_{ji}, \tag{4.12}$$

thus

$$[M] = [M]^T, [K] = [K]^T. \tag{4.13}$$

The reciprocity properties can also be proved using (4.9) and (4.10). The transposition of a scalar quantity is equal to the scalar quantity.

$$T = \frac{1}{2}(\{\dot{x}\}^T[M]\{\dot{x}\})^T = \frac{1}{2}\{\dot{x}\}^T[M]^T\{\dot{x}\} = \frac{1}{2}\{\dot{x}\}^T[M]\{\dot{x}\}.$$

The same applies for the potential energy.

The work done by the external forces can be expressed as

$$W = F_i x_i \text{, summed over } i=1,2,\ldots., \tag{4.14}$$

hence

$$F_i = \frac{\partial W}{\partial x_i}. \tag{4.15}$$

4.2.2 Undamped Equations of Motion using Energies

The undamped equations of motion for the linear dynamic system, as shown in Fig. 4.1 will be derived using the kinetic and potential energies and the work done by the forces.

The kinetic energy T of the discrete linear dynamic system is

$$T = \frac{1}{2}[m_{11}\dot{x}_1^2 + m_{22}\dot{x}_2^2 + m_{33}\dot{x}_3^2], \tag{4.16}$$

and the potential energy U of the linear system is

$$U = \frac{1}{2}[k_{11}(x_1)^2 + k_{12}(x_1 - x_2)^2 + k_{13}(x_1 - x_3)^2 + k_{23}(x_2 - x_3)^2], \tag{4.17}$$

and the work W done by the forces becomes

$$W = F_2 x_2. \tag{4.18}$$

For the mass matrix $[M]$, the elements $\tilde{m}_{ij}$ become

- $\tilde{m}_{11} = \dfrac{\partial^2 T}{\partial \dot{x}_1 \partial \dot{x}_1} = m_{11}$, $\tilde{m}_{12} = \dfrac{\partial^2 T}{\partial \dot{x}_1 \partial \dot{x}_2} = 0$ and $\tilde{m}_{13} = \dfrac{\partial^2 T}{\partial \dot{x}_1 \partial \dot{x}_3} = 0$
- $\tilde{m}_{21} = \dfrac{\partial^2 T}{\partial \dot{x}_2 \partial \dot{x}_1} = 0$, $\tilde{m}_{22} = \dfrac{\partial^2 T}{\partial \dot{x}_2 \partial \dot{x}_2} = m_{22}$ and $\tilde{m}_{23} = \dfrac{\partial^2 T}{\partial \dot{x}_2 \partial \dot{x}_3} = 0$
- $\tilde{m}_{31} = \dfrac{\partial^2 T}{\partial \dot{x}_3 \partial \dot{x}_1} = 0$, $\tilde{m}_{32} = \dfrac{\partial^2 T}{\partial \dot{x}_3 \partial \dot{x}_2} = 0$ and $\tilde{m}_{33} = \dfrac{\partial^2 T}{\partial \dot{x}_3 \partial \dot{x}_3} = m_{33}$,

for the stiffness matrix $[K]$, the elements $\tilde{k}_{ij}$ become

- $\tilde{k}_{11} = \dfrac{\partial^2 U}{\partial x_1 \partial x_1} = k_{11} + k_{12} + k_{13}$, $\tilde{k}_{12} = \dfrac{\partial^2 U}{\partial x_1 \partial x_2} = -k_{12}$ and

  $\tilde{k}_{13} = \dfrac{\partial^2 U}{\partial x_1 \partial x_3} = -k_{13}$
- $\tilde{k}_{21} = \dfrac{\partial^2 U}{\partial x_2 \partial x_1} = -k_{12}$, $\tilde{k}_{22} = \dfrac{\partial^2 U}{\partial x_2 \partial x_2} = k_{12} + k_{23}$ and

  $\tilde{k}_{23} = \dfrac{\partial^2 U}{\partial x_2 \partial x_3} = -k_{13}$
- $\tilde{k}_{31} = \dfrac{\partial^2 U}{\partial x_3 \partial x_1} = -k_{13}$, $\tilde{k}_{32} = \dfrac{\partial^2 U}{\partial x_3 \partial x_2} = -k_{23}$ and

  $\tilde{k}_{33} = \dfrac{\partial^2 U}{\partial x_3 \partial x_3} = k_{13} + k_{23}$,

and the force vector $[F]$, the elements $\tilde{f}_i$ become

- $\tilde{f}_1 = \dfrac{\partial W}{\partial x_1} = 0$
- $\tilde{f}_2 = \dfrac{\partial W}{\partial x_2} = F_2$
- $\tilde{f}_3 = \dfrac{\partial W}{\partial x_3} = 0$

We have achieved equations of motion similar to (4.5).

4.2.3 Undamped Equations of Motion using Lagrange's Equations

The equation of motion of the mdof dynamic system, as illustrated in Fig. 4.1, can also be derived by Lagrange's equations. Lagrange's equations, without damping, are as follows [Gatti 99, Meirovitch 70]

$$\frac{d}{dt}\frac{\partial L}{\partial \dot{\eta}_i} - \frac{\partial L}{\partial \eta_i} = Q_i,\ i=1,2,......, \tag{4.19}$$

with

- η_i the i-th (generalised) coordinate, degree of freedom
- $L(\eta, \dot{\eta})$ Lagrangian of the dynamic system, $L = T - U$
- $T(\eta, \dot{\eta})$ kinetic energy of dynamic system
- $U(\eta, \dot{\eta})$ potential or strain energy of the dynamic system
- Q_i the i-th (generalised) force, $\delta W = Q_i \delta\eta_i$, $Q_i = \dfrac{\delta W}{\delta \eta_i}$
- δW virtual work of external forces

The scalar function $L = T - U$ determines the entire dynamics of the given system. The kinetic energy T of the mdof dynamic system is given by (4.16)

$$T = \frac{1}{2}[m_{11}\dot{x}_1^2 + m_{22}\dot{x}_2^2 + m_{33}\dot{x}_3^2],$$

and the potential energy U by (4.17)

$$U = \frac{1}{2}[k_{11}(x_1)^2 + k_{12}(x_1 - x_2)^2 + k_{13}(x_1 - x_3)^2 + k_{23}(x_2 - x_3)^2].$$

The derivation of the undamped equations of motion is as follows

- $\dfrac{d}{dt}\dfrac{\partial L}{\partial \dot{x}_1} = m_{11}\ddot{x}_1$, $\dfrac{\partial L}{\partial x_1} = -\{k_{11}x_1 + k_{12}(x_1 - x_2) + k_{13}(x_1 - x_3)\}$
- $\dfrac{d}{dt}\dfrac{\partial L}{\partial \dot{x}_2} = m_{22}\ddot{x}_2$, $\dfrac{\partial L}{\partial x_2} = -\{-k_{12}(x_1 - x_2) + k_{23}(x_2 - x_3)\}$
- $\dfrac{d}{dt}\dfrac{\partial L}{\partial \dot{x}_3} = m_{33}\ddot{x}_3$, $\dfrac{\partial L}{\partial x_3} = -\{-k_{13}(x_1 - x_3) - k_{23}(x_2 - x_3)\}$
- $\delta W = F_2 \delta x_2$, $Q_1 = Q_3 = 0$ and $Q_2 = F_2$

Finally the derived equations of motion are equal to (4.5).

$$\begin{bmatrix} m_{11} & 0 & 0 \\ 0 & m_{22} & 0 \\ 0 & 0 & m_{33} \end{bmatrix} \begin{Bmatrix} \ddot{x}_1 \\ \ddot{x}_2 \\ \ddot{x}_3 \end{Bmatrix} + \begin{bmatrix} k_{11}+k_{12}+k_{13} & -k_{12} & -k_{13} \\ -k_{12} & k_{12}+k_{23} & -k_{23} \\ -k_{13} & -k_{23} & k_{13}+k_{23} \end{bmatrix} \begin{Bmatrix} x_1 \\ x_2 \\ x_3 \end{Bmatrix} = \begin{Bmatrix} 0 \\ F_2 \\ 0 \end{Bmatrix},$$

or in the general matrix notation

$$[M]\{\ddot{x}\} + [K]\{x\} = \{F\}.$$

4.2.4 Damped Equations of Motion using Lagrange's Equations

The damped equations of motion of the mdof dynamic system may also be derived from Lagrange's equations. Lagrange's equation with damping is the same as (4.19), [Gatti 99, Meirovitch 70]

$$\frac{d}{dt}\frac{\partial L}{\partial \dot{\eta}_i} - \frac{\partial L}{\partial \eta_i} + \frac{\partial D}{\partial \dot{\eta}_i} = Q_i,\ i=1,2,\ldots\ldots, \tag{4.20}$$

with

- $D(\dot{\eta}_i)$ Viscous damping energy, $D = \frac{1}{2}c_{ij}\dot{\eta}_i\dot{\eta}_j$, summed indices. The damping terms are symmetric $c_{ij} = c_{ji}$, because $\frac{\partial^2 D}{\partial \dot{x}_i \partial \dot{x}_j} = c_{ij} = \frac{\partial^2 D}{\partial \dot{x}_j \partial \dot{x}_i} = c_{ji}$.

The damping force is proportional to the (relative) velocity. The damping force F_D (N) is equal to the damping constant c (Ns/m) multiplied by the difference in velocity $\Delta\dot{x}$ (m/s) between the discrete masses, $F_D = c\Delta\dot{x}$.

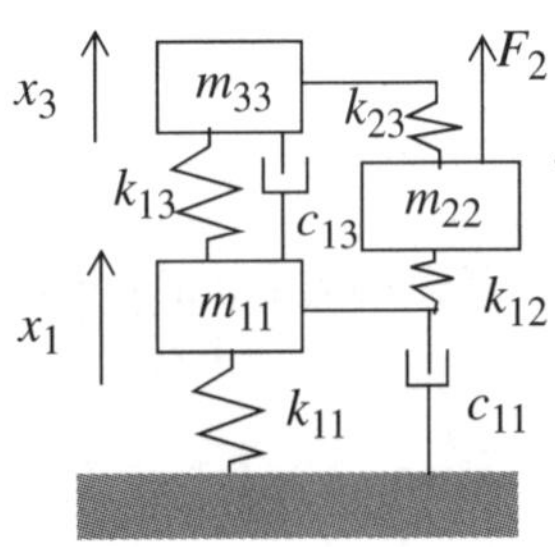

Fig. 4.3. 3-dof dynamic system with 2 discrete dampers

The damping energy of the dynamic mdof system, as shown in Fig. 4.3, becomes

$$D = \frac{1}{2}[c_{11}(\dot{x}_1)^2 + c_{13}(\dot{x}_1 - \dot{x}_3)^2]. \tag{4.21}$$

The derivation of the damping terms in the damped equations of motion is as follows

- $\frac{\partial D}{\partial \dot{x}_1} = c_{11}\dot{x}_1 + c_{13}(\dot{x}_1 - \dot{x}_3)$, $\frac{\partial D}{\partial \dot{x}_2} = 0$ and $\frac{\partial D}{\partial \dot{x}_3} = -c_{13}(\dot{x}_1 - \dot{x}_3)$

Thus the equation of damped motion becomes

$$\begin{bmatrix} m_{11} & 0 & 0 \\ 0 & m_{22} & 0 \\ 0 & 0 & m_{33} \end{bmatrix} \begin{Bmatrix} \ddot{x}_1 \\ \ddot{x}_2 \\ \ddot{x}_3 \end{Bmatrix} + \begin{bmatrix} c_{11}+c_{13} & 0 & -c_{13} \\ 0 & 0 & 0 \\ -c_{13} & 0 & c_{13} \end{bmatrix} \begin{Bmatrix} \dot{x}_1 \\ \dot{x}_2 \\ \dot{x}_3 \end{Bmatrix}$$

$$+ \begin{bmatrix} k_{11}+k_{12}+k_{13} & -k_{12} & -k_{13} \\ -k_{12} & k_{12}+k_{23} & -k_{23} \\ -k_{13} & -k_{23} & k_{13}+k_{23} \end{bmatrix} \begin{Bmatrix} x_1 \\ x_2 \\ x_3 \end{Bmatrix} = \begin{Bmatrix} 0 \\ F_2 \\ 0 \end{Bmatrix},$$

or in the general matrix notation

$$[M]\{\ddot{x}\} + [C]\{\dot{x}\} + [K]\{x\} = \{F\}$$

In the following example we will generate the equations of motion of an mdof dynamic system composed of a continuous elastic structure – a cantilevered bending beam- and two discrete elements; a mass–spring system and a viscous damper. The elastic behaviour of the beam will be translated with the aid of an "assumed mode" multiplied by a generalised coordinate.

The dynamic system is shown in Fig. 4.4. The bending beam is clamped at point A and has a bending stiffness EI (Nm^2), a length L (m) and a mass per unit length m (kg/m). The damper has a damping constant c (Ns/m) and is connected between the base and point B of the beam. The spring with a spring stiffness k (N/m) is connected between the discrete mass M (kg) and point B of the beam.

We split the deflection $w(x, t)$ in an assumed mode $\phi(x)$ and a generalised coordinate $\eta(t)$, hence

$$w(x, t) = \phi(x)\eta(t). \tag{4.22}$$

The whole system will be excited by enforced base acceleration $\ddot{u}(t)$ (m/s^2). The displacement $\delta(t)$ of the discrete mass M with respect to the base is (relative motion).

The assumed mode $\phi(x)$ will be calculated from a clamped beam loaded with a unit-point load at point B (see Fig. 4.5).
The bending moment at x is

$$M(x) = -(L-x). \tag{4.23}$$

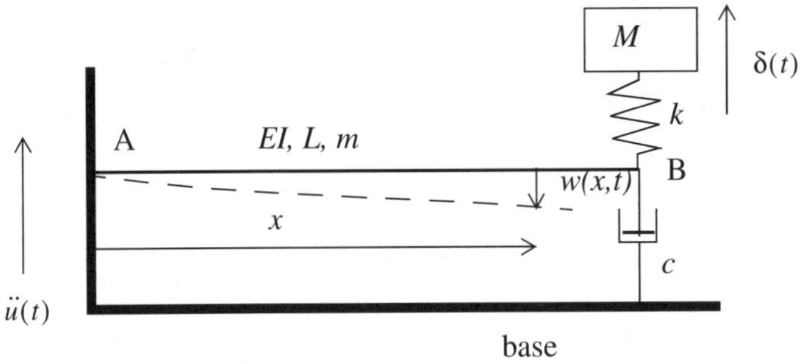

Fig. 4.4. Mdof dynamic system with bending beam, mass–spring and damper

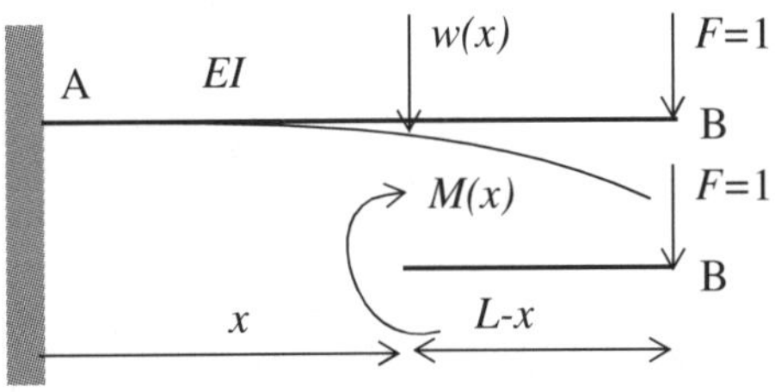

Fig. 4.5. Clamped beam loaded with unit load

The deflection $w(x)$ is related to the bending moment $M(x)$ [Wempner 95]

$$w''(x) = -\frac{M(x)}{EI}, \tag{4.24}$$

with

- E the modulus of elasticity (Young's modulus) (Pa, N/m^2)
- I the second moment of area (m^4)

The boundary conditions at point A are

$$w'(0) = 0 \text{ and } w(0) = 0. \tag{4.25}$$

The solution of (4.24) with (4.25) is

$$w(x) = \frac{1}{EI}\left(\frac{1}{2}Lx^2 - \frac{1}{6}x^3\right). \tag{4.26}$$

For the dimensionless assumed mode $\phi(x)$ we select

$$\phi(x) = 3\left(\frac{x}{L}\right)^2 - \left(\frac{x}{L}\right)^3. \tag{4.27}$$

The strain energy per unit of volume is given by

$$dU = \frac{1}{2}\sigma_{\text{bending}}\varepsilon dV \tag{4.28}$$

For a bending beam the strain energy U becomes

$$U = \frac{1}{2}\int_V \sigma_{\text{bending}}\varepsilon dV = \frac{1}{2}\int_V \frac{Mz}{I}\frac{Mz}{EI}dV = \frac{1}{2}\int_0^L\int_A \frac{M^2}{EI^2}z^2 dA dx = \frac{1}{2}\int_0^L \frac{M^2}{EI}dx, \tag{4.29}$$

where

- z the fibre distance (m)
- x the coordinate in the longitudinal direction of the beam (m)
- L the length of the beam (m)
- A the cross section of the beam (m^2)

Equation (4.29) can be written as

$$U = \frac{1}{2}\int_0^L \frac{M^2}{EI}dx = \frac{1}{2}\int_0^L EIw''(x)^2 dx. \tag{4.30}$$

The displacement $w(x, t)$ of the beam at point B ($x = L$) is

$$w(L, t) = \phi(L)\eta(t) = 2\eta(t). \tag{4.31}$$

We assume that $w(L, t) - \delta(t) > 0$. The damped equations of motion will be generated using (4.20).

The total strain U or potential energy of dynamic system, as shown in Fig. 4.4, becomes

$$U = \frac{1}{2}EI\int_0^L w''^2(x, t)dx + \frac{1}{2}k[w(L, t) - \delta(t)]^2, \tag{4.32}$$

or, using (4.1), the strain energy will be

$$U = \frac{1}{2}EI\eta^2(t)\int_0^L \phi''^2(x)dx + \frac{1}{2}k[\phi(L)\eta(t) - \delta(t)]^2. \tag{4.33}$$

The kinetic energy T is similar to the strain energy, however, we must take into account the base velocity $\dot{u}$, thus

$$T = \frac{1}{2}m\int_0^L [\dot{w}^2(x,t) + \dot{u}]^2 dx + \frac{1}{2}M[\dot{\delta}(t) + \dot{u}]^2, \tag{4.34}$$

or

$$T = \frac{1}{2}m\int_0^L [\phi(x)\dot{\eta}(t) + \dot{u}]^2 dx + \frac{1}{2}M[\dot{\delta}(t) + \dot{u}]^2. \tag{4.35}$$

The damping energy D is

$$D = \frac{1}{2}c\dot{w}^2(L,t) = \frac{1}{2}c[\phi(L)\dot{\eta}(t)]^2. \tag{4.36}$$

We are now able to set up the damped equations of motion for the unknowns $\eta(t)$ and $\delta(t)$. The procedure is as follows:

- $\frac{d}{dt}\frac{\partial L}{\partial \dot{\eta}} = m\int_0^L [\phi(x)\ddot{\eta}(t) + \ddot{u}]\phi(x)dx = \ddot{\eta}(t)m\int_0^L \phi^2(x)dx + m\ddot{u}\int_0^L \phi(x)dx$
- $\frac{d}{dt}\frac{\partial L}{\partial \dot{\delta}} = M\ddot{\delta}(t) + M\ddot{u}$
- $-\frac{\partial L}{\partial \eta} = -\frac{\partial(T-U)}{\partial \eta} = \eta(t)EI\int_0^L \phi'^2(x)dx + \eta(t)k\phi^2(L) - k\phi(L)\delta(t)$
- $-\frac{\partial L}{\partial \delta} = -\frac{\partial(T-U)}{\partial \delta} = \delta(t)k - k\phi(L)\eta(t)$
- $\frac{\partial D}{\partial \dot{\delta}} = 0, \frac{\partial D}{\partial \dot{\eta}} = c\phi(L)\dot{\eta}(t)$

The damped equations of motion are

$$\begin{bmatrix} m\int_0^L \phi^2(x)dx & 0 \\ 0 & M \end{bmatrix} \begin{Bmatrix} \ddot{\eta}(t) \\ \ddot{\delta}(t) \end{Bmatrix} + \begin{bmatrix} c\phi(L) & 0 \\ 0 & 0 \end{bmatrix} \begin{Bmatrix} \dot{\eta}(t) \\ \dot{\delta}(t) \end{Bmatrix}$$

$$+ \begin{bmatrix} EI\int_0^L \phi'^2(x)dx + k\phi^2(L) & -k\phi(L) \\ -k\phi(L) & k \end{bmatrix} \begin{Bmatrix} \eta(t) \\ \delta(t) \end{Bmatrix} = -\begin{Bmatrix} m\int_0^L \phi(x)dx \\ M \end{Bmatrix} \ddot{u}. \tag{4.37}$$

When substituting (4.27) into (4.37) the equations of motion become

$$\begin{bmatrix} \frac{33}{35}mL & 0 \\ 0 & M \end{bmatrix} \begin{Bmatrix} \ddot{\eta}(t) \\ \ddot{\delta}(t) \end{Bmatrix} + \begin{bmatrix} 2c & 0 \\ 0 & 0 \end{bmatrix} \begin{Bmatrix} \dot{\eta}(t) \\ \dot{\delta}(t) \end{Bmatrix} + \begin{bmatrix} \frac{12EI}{L^2} + 4k & -2k \\ -2k & k \end{bmatrix} \begin{Bmatrix} \eta(t) \\ \delta(t) \end{Bmatrix} = -\begin{Bmatrix} \frac{3}{4}mL \\ M \end{Bmatrix} \ddot{u} .$$

This equation is similar to (4.1).

$$[M]\{\ddot{x}(t)\} + [C]\{\dot{x}(t)\} + [K]\{x(t)\} = \{F(t)\}$$

4.3 Finite Element Method

The finite element method will give us the damped or undamped equations of motion [amongst many textbooks see Cook 89, Przemieniecki 85]. The structural system will be divided into nodes (grids) connected with each other with the so-called finite elements; springs, rods, beams, membranes, plate bending, volumes, etc. In general, a node represents 6 degrees of freedom; 3 translations and 3 rotations. For every finite element a mass, damping and stiffness matrix and external load vector will be generated. All the matrices are assembled in the overall mass matrix $[M]$, the overall damping matrix $[C]$, the overall stiffness matrix $[K]$ and overall load vector $\{F(t)\}$. The dofs of the nodes form, in fact, the displacement vector $\{x(t)\}$, the velocity vector $\{\dot{x}(t)\}$ and the acceleration vector $\{\ddot{x}(t)\}$. Together we will get (4.1)

$$[M]\{\ddot{x}(t)\} + [C]\{\dot{x}(t)\} + [K]\{x(t)\} = \{F(t)\} .$$

Later in this book the availability of the equations of motion is considered to be more or less trivial.

4.4 Problems

4.4.1 Problem 1

The mdof system, as shown in Fig. 4.6, consists of 5 degrees of freedom. Derive the equations of motion using

- the equations of equilibrium (Newton's law)
- Lagrange's equations

4.4.2 Problem 2

A dynamic system, as shown in Fig. 4.7, has 3 dofs; w, φ and δ. The displacement δ is with respect to line A–B. Both dofs w and φ are located in the middle of A–B. The structure in between A and B is rigid and has a mass m per unit length (kg/m). The discrete mass M (kg) is coupled at the end of the massless elastic beam with a bending stiffness EI (Nm^2). The beam is rigidly connected at point B. The complete dynamic system is supported by two springs with a spring stiffness k (N/m). The second moment of mass of the rigid beam A–B $I = \frac{1}{12}mL_1^3$ (kgm^2).

- Set up the undamped equations of motion (e.g. using Lagrange's equations).

Answer: The homogeneous equations of motion are

$$\begin{bmatrix} mL_1 + M & M\left(\frac{1}{2}l_1 + L_2\right) & M \\ M\left(\frac{1}{2}l_1 + L_2\right) & I + M\left(\frac{1}{2}l_1 + L_2\right)^2 & M\left(\frac{1}{2}l_1 + L_2\right) \\ M & M\left(\frac{1}{2}l_1 + L_2\right) & M \end{bmatrix} \begin{Bmatrix} \ddot{w} \\ \ddot{\varphi} \\ \ddot{\delta} \end{Bmatrix} + \begin{bmatrix} 2k & 0 & 0 \\ 0 & kL_1 & 0 \\ 0 & 0 & \dfrac{3EI}{L_2^3} \end{bmatrix} \begin{Bmatrix} w \\ \varphi \\ \delta \end{Bmatrix} = \begin{Bmatrix} 0 \\ 0 \\ 0 \end{Bmatrix}.$$

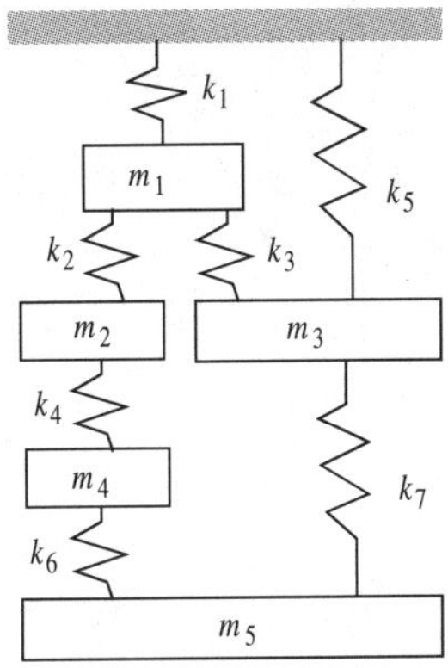

Fig. 4.6. Mdof dynamic system

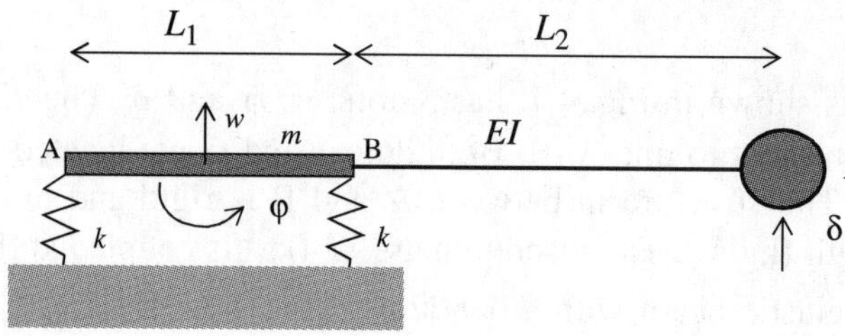

Fig. 4.7. Dynamic system with 3 dofs

5 Modal Analysis

5.1 Introduction

For structures exposed to dynamic forces the knowledge of the dynamic characteristics of these structures is of great importance. The most important intrinsic dynamic (modal) characteristics of linear dynamic systems are:

- The natural frequency
- The associated mode shape
- Damping

The natural frequencies and associated mode shapes may be analysed for both undamped and damped linear dynamic systems. In the case of damped systems we talk about damped natural frequencies and associated damped mode shapes.

In this chapter undamped and damped natural frequencies and corresponding mode shapes will be defined and related important properties will be discussed.

5.2 Undamped Linear Dynamic Systems

The modal characteristics of an undamped linear dynamic system are:

- The natural frequencies
- The associated mode shapes
- Orthogonality relations of modes (normal modes)
- Rigid body modes
- Left eigenvectors

5.2.1 Natural Frequencies and Mode Shapes

The undamped equations of motion of a linear dynamic system (structure) are

$$[M]\{\ddot{x}\} + [K]\{x\} = \{F(t)\}, \tag{5.1}$$

where

- $[M]$ the mass matrix, either diagonal or consistent with the stiffness matrix
- $[K]$ the stiffness matrix
- $\{F(t)\}$ the external (enforced) forces
- $\{x\}$ the displacement vector with n degrees of freedom
- $\{\ddot{x}\}$ the acceleration vector

For the calculation of the natural frequencies only the homogeneous equations of motion are needed, thus $\{F(t)\} = \{0\}$. We assume a harmonic solution for the displacement vector and the vector of accelerations

$$\{x\} = \{X(\omega)\}e^{j\omega t}, \tag{5.2}$$

with

- $\{X(\omega)\}$ the amplitude (dependent of ω) rotating vector in the complex plane with an angular velocity ω
- $e^{j\omega t}$ a rotating unit vector with the angular speed ω in the complex plane with the Euler's formula [Nahin 98, Spiegel 64] is $e^{j\omega t} = \cos(\omega t) + j\sin(\omega t)$ and $j = \sqrt{-1}$.

Substituting (5.2) into equation (5.1) the homogeneous equations of motion become

$$([K]-\omega^2[M])\{X(\omega)\} = ([K]-\lambda[M])\{X(\omega)\} = \{0\}. \tag{5.3}$$

Equation (5.3) has a trivial solution $\{X(\omega)\} = \{0\}$ in which we are not interested. Equation (5.3) has a non trivial solution ($\{X(\omega)\} \neq \{0\}$) only if [Strang 88]

$$\det([K]-\lambda[M]) = 0. \tag{5.4}$$

Solution of the eigenvalue problem will result in

1. The eigenvalues λ_k, $k = 1, 2, \ldots, n$ with $\lambda_k = \omega_k^2$. In general, $f_k = \frac{\omega_k}{2\pi}$ is the so-called natural frequency (Hz).

2. Associated with the eigenvalues (natural frequencies) the eigenvectors $[X]$ are calculated. $[\Phi] = [X]\{\alpha\}$ is called the modal matrix. This means that the mode shapes are scaled because the eigenvectors are defined with respect to an arbitrarily figure α. The modal matrix does obey orthogonality relations.

If the mass matrix and stiffness matrix are symmetric, $\det[K] \geq 0$ and $\det[M] > 0$, then the eigenvalues are real and $\lambda_k \geq 0$.

The undamped equations of motion are

$$[M]\{\ddot{x}\} + [K]\{x\} = \{F\}.$$

Assuming $[M]^{-1}$ does exist these equations can also be written as

$$\{\ddot{x}\} + [M]^{-1}[K]\{x\} = [M]^{-1}\{F\}.$$

Rearranging previous equations

$$\{\ddot{x}\} = -[M]^{-1}[K]\{x\} + [M]^{-1}\{F\},$$

or

$$\{\ddot{x}\} = [A]\{x\} + [M]^{-1}\{F\}. \tag{5.5}$$

The eigenvalue problem is:

$$\lambda[I]\{X\} = [A]\{X\}. \tag{5.6}$$

The solution of equations of motion can be constituted from the solution of the homogeneous differential equation complemented with a particular solution. The solution of the homogeneous part can be constructed from the eigenvalues and the associated eigenvectors. The solution becomes [Strang 88]

$$\{x\} = \sum_{i=1}^{n}(C_i e^{j\omega_i t} + D_i e^{-j\omega_i t})\{X(\lambda_i)\}, \tag{5.7}$$

or

$$\{x\} = \sum_{i=1}^{n}(A_i \cos\omega_i t + B_i \sin\omega_i t)\{X(\lambda_i)\}, \tag{5.8}$$

with

- A, B, C and D integration constants.
- n the number of dofs.

- ω_i the radian natural frequency (rad/s) with $\omega_i = \sqrt{\lambda_i}$. Here the relation of the natural frequencies of the dynamic system and the eigenvalues appears.
- $\{X(\lambda_i)\}$ the eigenvector associated with the eigenvalue λ_i.

The integration constants can be obtained using the initial conditions for the displacements and velocities at $t = 0$. The initial displacements will solve the A constants and the initial velocities the B constants. The complete solution of the displacement vector $\{x(t)\}$ becomes

$$\{x(t)\} = \sum_{i=1}^{n}(A_i\cos\omega_i t + B_i\sin\omega_i t)\{X(\lambda_i)\} + x_{\text{part}}(t), \tag{5.9}$$

where

- $x_{\text{part}}(t)$ is the particular solution of (5.5), also called the steady-state solution.

A linear dynamic system is defined by the following equations of motion

$$\begin{bmatrix}1 & 0 & 0\\0 & 2 & 0\\0 & 0 & 3\end{bmatrix}\begin{Bmatrix}\ddot{x}_1\\\ddot{x}_2\\\ddot{x}_3\end{Bmatrix} + \begin{bmatrix}20 & -20 & 0\\-20 & 50 & -30\\0 & -30 & 70\end{bmatrix}\begin{Bmatrix}x_1\\x_2\\x_3\end{Bmatrix} = \begin{Bmatrix}0\\0\\0\end{Bmatrix},\ t \geq 0,$$

or

$$[M]\{\ddot{x}\} + [K]\{x\} = \{F\}.$$

At $t = 0$ the initial displacement u_o and initial velocities v_o are

$$u_o = \begin{Bmatrix}1\\0\\0\end{Bmatrix} \text{ and } v_o = \begin{Bmatrix}1\\0\\0\end{Bmatrix}.$$

The eigenvalues and the associated eigenvectors are

$$\langle\lambda\rangle = \begin{bmatrix}4.3360 & 0 & 0\\0 & 21.9285 & 0\\0 & 0 & 42.0688\end{bmatrix},\ \langle\omega\rangle = \begin{bmatrix}2.0823 & 0 & 0\\0 & 4.6828 & 0\\0 & 0 & 6.4860\end{bmatrix}$$

and the matrix with the eigenvectors

$$[X] = \begin{bmatrix} 0.7488 & -0.8219 & 0.6245 \\ 0.5865 & 0.0793 & -0.6891 \\ 0.3087 & 0.5641 & 0.3678 \end{bmatrix}.$$

The A integration constants are calculated using the initial displacements u_o at $t = 0$ and the integration constants B are calculated with the aid of the initial velocities v_o at $t = 0$.

The A integration constants are obtained using (5.8)

$$[X]\{A\} = \{u_o\} \rightarrow \{A\} = ([X]^T[X])^{-1}[X]^T\{u_o\} = \begin{Bmatrix} 0.4880 \\ -0.5003 \\ 0.3578 \end{Bmatrix}.$$

The B integration constants are obtained using (5.8)

$$[X]\langle\omega\rangle\{B\} = \{v_o\} \rightarrow \{B\} = ([X]^T[X]\langle\omega\rangle)^{-1}[X]^T\{v_o\} = \begin{Bmatrix} 0.2343 \\ -0.1068 \\ 0.0552 \end{Bmatrix}.$$

Finally the complete solution for $\{x(t)\}$ becomes

$$\{x(t)\} = \{0.4880\cos(2.0823t) + 0.2343\sin(2.0823t)\}\begin{Bmatrix} 0.7488 \\ 0.5865 \\ 0.3087 \end{Bmatrix}$$

$$+ \{-0.5003\cos(4.6828t) - 0.1068\sin(4.6828t)\}\begin{Bmatrix} 0.8219 \\ 0.0793 \\ 0.5641 \end{Bmatrix}$$

$$+ \{0.3578\cos(6.4860t) + 0.0552\sin(6.4860t)\}\begin{Bmatrix} 0.6245 \\ -0.6891 \\ 0.3678 \end{Bmatrix}.$$

5.2.2 Orthogonality Relations of Modes

The eigenvalue problem is stated as

$$(-\lambda_i[M] + [K])\{\phi_i\} = \{0\} \tag{5.10}$$

with

- $\{\phi_i\}$ the i-th eigenvector or mode shape
- λ_i the i-th eigenvalue

Premultiplying (5.10) with the transposposition of $\{\phi_i\}$ the following equation is obtained

$$\{\phi_i\}^T(-\lambda_j[M]+[K])\{\phi_j\} = \{0\}. \tag{5.11}$$

The same equation, as (5.11), will appear if the indices of the mode shapes will interchanged

$$\{\phi_j\}^T(-\lambda_i[M]+[K])\{\phi_i\} = \{0\}. \tag{5.12}$$

The mass matrix $[M]$ and the stiffness matrix $[K]$ are symmetric thus

$$(\{\phi_i\}^T[M]\{\phi_j\})^T = \{\phi_j\}^T[M]\{\phi_i\} = \{\phi_i\}^T[M]\{\phi_j\}, \tag{5.13}$$

and also

$$(\{\phi_i\}^T[K]\{\phi_j\})^T = \{\phi_j\}^T[K]\{\phi_i\} = \{\phi_i\}^T[K]\{\phi_j\}. \tag{5.14}$$

The mass matrix is positive-definite when $\{\phi_i\}^T[M]\{\phi_j\} > 0$ and the stiffness matrix is positive-definite if $\{\phi_i\}^T[K]\{\phi_j\} > 0$, when the mode shapes are not zero vectors. Subtracting (5.12) from (5.11) the following equation is obtained

$$(\lambda_i - \lambda_j)\{\phi_i\}^T[M]\{\phi_j\} = \{0\}. \tag{5.15}$$

There are two possible solution

1. $(\lambda_i - \lambda_j) \neq 0$ then $\{\phi_i\}^T[M]\{\phi_j\} = 0$ and $\{\phi_i\}^T[K]\{\phi_j\} = 0$
2. $(\lambda_i - \lambda_j) = 0$ then $\{\phi_i\}^T[M]\{\phi_j\} \neq 0$ and $\{\phi_i\}^T[K]\{\phi_j\} \neq 0$

In general, the mode shapes are scaled such that the product

$$\{\phi_i\}^T[M]\{\phi_j\} = \delta_{ij}, \tag{5.16}$$

and from (5.11) and (5.12) one can prove

$$\{\phi_i\}^T[K]\{\phi_j\} = \lambda_i\delta_{ij}, \tag{5.17}$$

with

- δ_{ij} the Kronecker delta $i = j$ then $\delta_{ij} = 1$ and if $i \neq j$ then $\delta_{ij} = 0$.

Sometimes, the eigenvalues are repeated and then the eigenvalue problem results in multiple eigenvalues. This complicates the calculation of the modes. A discussion on multiple eigenvalues is given in [Newland 89].

The modal matrix $[\Phi] = [\phi_1, \phi_2, ...,\phi_n]$ has the following orthogonality properties with respect to the mass matrix and the stiffness matrix

$$[\Phi]^T[M][\Phi] = [I] \text{ and } [\Phi]^T[K][\Phi] = \langle\lambda\rangle, \tag{5.18}$$

- $[I]$ the identity matrix
- $\langle\lambda\rangle$ the diagonal matrix of the eigenvalues

The modes are orthogonal (orthonormal if $[\Phi]^T[M][\Phi] = [I]$) with respect to the mass and stiffness matrices. The modes with the orthogonality relations are often termed normal modes. In later chapters the orthogonality relations of modes are used to decouple the coupled equations of motion by depicting the physical displacements, velocities and accelerations on an orthogonal base of modes, the so-called modal matrix. The method to solve the equations of motions of a linear dynamical system decoupling the coupled equations is called the modal displacement method (MDM)

A linear dynamic system is defined with the following equations of motion

$$\begin{bmatrix} 1 & 0 & 0 \\ 0 & 2 & 0 \\ 0 & 0 & 3 \end{bmatrix} \begin{Bmatrix} \ddot{x}_1 \\ \ddot{x}_2 \\ \ddot{x}_3 \end{Bmatrix} + \begin{bmatrix} 20 & -20 & 0 \\ -20 & 50 & -30 \\ 0 & -30 & 70 \end{bmatrix} \begin{Bmatrix} x_1 \\ x_2 \\ x_3 \end{Bmatrix} = \begin{Bmatrix} 0 \\ 0 \\ 0 \end{Bmatrix}.$$

The eigenvalues $\{\lambda\}$ and associated mode shapes of the undamped dynamic system are

$$\{\lambda\} = \begin{Bmatrix} 4.3360 \\ 21.9285 \\ 42.0688 \end{Bmatrix} \text{ and } [\bar{\Phi}] = \begin{bmatrix} 0.7488 & -0.8219 & 0.6245 \\ 0.5865 & 0.0793 & -0.6891 \\ 0.3087 & 0.5641 & 0.3678 \end{bmatrix}.$$

The generalised mass matrix $[m] = [\bar{\Phi}]^T[M][\bar{\Phi}]$ becomes

$$[m] = \begin{bmatrix} 1.5346 & 0 & 0 \\ 0 & 1.6428 & 0 \\ 0 & 0 & 1.7453 \end{bmatrix}.$$

We normalise the modal matrix $[\Phi]$ such that $[\Phi]^T[M][\Phi] = [I]$ with $[\Phi] = \sqrt{[m]}^{-1}[\bar{\Phi}]$, thus

$$[\Phi]^T[M][\Phi] = \begin{bmatrix} 1 & 0 & 0 \\ 0 & 1 & 0 \\ 0 & 0 & 1 \end{bmatrix} \text{ and } [\Phi]^T[K][\Phi] = \begin{bmatrix} 4.3360 & 0 & 0 \\ 0 & 21.9285 & 0 \\ 0 & 0 & 42.0688 \end{bmatrix}.$$

5.2.3 Rigid-Body Modes

If the linear dynamic system is not constrained the system can move as a rigid body. This means that during the movement as a rigid body no elastic forces will occur in the dynamic system. If this is the case the stiffness matrix $[K]$ is singular (semi-positive-definite). In general, there are 6 possible motions as rigid-body; three translations and three rotations. This implies six eigenvalues $\lambda_k = 0, k = 1, 2, ..., 6$ of the eigenvalues problem $(-\lambda_i[M] + [K])\{\phi_i\} = \{0\}$ [Zurmuehl 64]. If so one can write

$$[K]\{\phi_{R,k}\} = \{0\},\ k = 1, 2, ..., 6 \tag{5.19}$$

Again it is noticed that no elastic energy will be introduced in the dynamic system. The rigid-body mode energy is defined with

$$\frac{1}{2}\{\phi_{R,k}\}^T[K]\{\phi_{R,k}\} = \{0\},\ k = 1, 2, ..., 6. \tag{5.20}$$

This illustrates that in this case the stiffness matrix is not positive-definite, but semipositive-definite. The six rigid-body modes can be calculated very easily from (5.19). The free-free dynamical system (with n degrees of freedom) is constrained at one point with 6 degrees of freedom; three translations and three rotations. The set of degrees of freedom is called the R-set. The other elastic degrees of freedom are placed in the E-set, such that $n = R + E$. The constrained R-set is determinate, so no strains will be introduced in the elastic system. The R-set consists of 6 unit displacement and rotations and those will be enforced to the dynamic system. Equation (5.19) can be written

$$\begin{bmatrix} K_{EE} & K_{ER} \\ K_{RE} & K_{RR} \end{bmatrix} \begin{Bmatrix} \Phi_{R,E} \\ I \end{Bmatrix} = \begin{Bmatrix} 0 \\ 0 \end{Bmatrix}, \tag{5.21}$$

with

- $[I]$ the identity matrix
- $[\Phi_{R,E}]$ the E-set part of the rigid-body motion

From the first equation of (5.21) the E-set part of the rigid-body mode can be solved

$$[\Phi_{R,E}] = -[K_{EE}]^{-1}[K_{ER}]. \tag{5.22}$$

The complete matrix of the 6 rigid body modes becomes

$$[\Phi_R] = \begin{bmatrix} -[K_{EE}]^{-1}[K_{ER}] \\ I \end{bmatrix}. \tag{5.23}$$

If the (5.23) is substituted into (5.21) the second equation reads

$$[\bar{K}_{RR}] = [K_{RR}] - [K_{RE}][K_{EE}]^{-1}[K_{ER}] = [0]. \tag{5.24}$$

Equation (5.24) tells us that the reduced stiffness matrix reduced to the redundant R-set degrees of freedom will vanish. The reduced matrix $[\bar{K}_{RR}]$ is very familiar to the static condensation technique, [Guyan 68].

A launch vehicle is modelled as an unconstrained bar consisting of two truss elements with in total three degrees of freedom [Przemieniecki 85]. No external forces are applied.

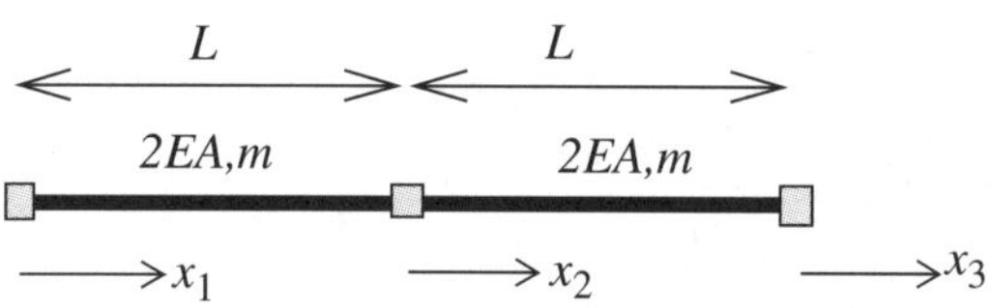

Fig. 5.1. Unconstrained bar consisting of two truss finite elements

The mass matrix $[M]$ and the stiffness matrix $[K]$ of a truss element are

$$[M] = \frac{mL}{6}\begin{bmatrix} 2 & 1 \\ 1 & 2 \end{bmatrix}, \text{ and } [K] = \frac{2AE}{L}\begin{bmatrix} 1 & -1 \\ -1 & 1 \end{bmatrix}.$$

The equations of motion of the launch vehicle are

$$\frac{mL}{6}\begin{bmatrix} 2 & 1 & 0 \\ 1 & 4 & 1 \\ 0 & 1 & 2 \end{bmatrix}\begin{Bmatrix} \ddot{x}_1 \\ \ddot{x}_2 \\ \ddot{x}_3 \end{Bmatrix} + \frac{2AE}{L}\begin{bmatrix} 1 & -1 & 0 \\ -1 & 2 & -1 \\ 0 & -1 & 1 \end{bmatrix}\begin{Bmatrix} x_1 \\ x_2 \\ x_3 \end{Bmatrix} = \begin{Bmatrix} 0 \\ 0 \\ 0 \end{Bmatrix}.$$

The eigenvalue problem becomes

$$\left(\frac{-\lambda mL}{6}\begin{bmatrix}2&1&0\\1&4&1\\0&1&2\end{bmatrix}+\frac{2AE}{L}\begin{bmatrix}1&-1&0\\-1&2&-1\\0&-1&1\end{bmatrix}\right)\begin{Bmatrix}\hat{x}_1\\\hat{x}_2\\\hat{x}_3\end{Bmatrix}=\begin{Bmatrix}0\\0\\0\end{Bmatrix}$$

$$\det\left(\frac{-\lambda mL}{6}\begin{bmatrix}2&1&0\\1&4&1\\0&1&2\end{bmatrix}+\frac{2AE}{L}\begin{bmatrix}1&-1&0\\-1&2&-1\\0&-1&1\end{bmatrix}\right)=0$$

With $\kappa^2 = \dfrac{\lambda mL^2}{12AE}$ the determinant is

$$\det\begin{bmatrix}(1-2\kappa^2)&-(1+2\kappa^2)&0\\-(1+2\kappa^2)&2(1-2\kappa^2)&-(1+2\kappa^2)\\0&-(1+2\kappa^2)&(1-2\kappa^2)\end{bmatrix}=0\,.$$

On expanding the determinant the following characteristic equation is obtained

$$6\kappa^2(1-2\kappa^2)(\kappa^2-2)=0\,.$$

The solution of the characteristics equation are three roots (eigenvalues)

1. $\kappa_1^{\,2} = 0\,,\ \lambda_1 = 0\ ,\ \omega_1 = 0$

2. $\kappa_2^{\,2} = \dfrac{1}{2}\,,\ \lambda_2 = \dfrac{6EA}{mL^2}\,,\ \omega_2 = \sqrt{\dfrac{6EA}{mL^2}}$

3. $\kappa_3^{\,2} = 2\,,\ \lambda_3 = \dfrac{24EA}{mL^2}\,,\ \omega_3 = 2\sqrt{\dfrac{6EA}{mL^2}}$

The associated modes are

$$[\Phi]=\begin{bmatrix}1&1&1\\1&0&-1\\1&-1&1\end{bmatrix}.$$

The generalised mass matrix $[m]=[\Phi]^T[M][\Phi]$, generalised stiffness matrix $[k]=[\Phi]^T[K][\Phi]$ and $\langle\lambda\rangle=[m]^{-1}[k]$ are

$$[m] = \frac{mL}{6}\begin{bmatrix} 12 & 0 & 0 \\ 0 & 4 & 0 \\ 0 & 0 & 4 \end{bmatrix} \text{ and } [k] = \frac{2AE}{L}\begin{bmatrix} 0 & 0 & 0 \\ 0 & 2 & 0 \\ 0 & 0 & 8 \end{bmatrix} \text{ and}$$

$$\langle\lambda\rangle = [m]^{-1}[k] = \frac{12AE}{mL^2}\begin{bmatrix} 0 & 0 & 0 \\ 0 & 0.5 & 0 \\ 0 & 0 & 2 \end{bmatrix}.$$

The calculation of the rigid-body mode $[\Phi_R]$ can be done using (5.23), with the displacement x_1 placed in the R-set and the displacements x_2 and x_3 placed in the E-set. After partitioning of the matrix $[k]$ we obtain

$$[K_{RR}] = \frac{2AE}{L}\begin{bmatrix} 1 \end{bmatrix},\ [K_{EE}] = \frac{2AE}{L}\begin{bmatrix} 2 & -1 \\ -1 & 1 \end{bmatrix},\ [K_{ER}] = \frac{2AE}{L}\begin{bmatrix} -1 \\ 0 \end{bmatrix}, \text{ and}$$

$$[K_{ER}] = \frac{2AE}{L}\begin{bmatrix} -1 & 0 \end{bmatrix}.$$

The rigid-body mode becomes

$$[\Phi_R] = \begin{bmatrix} -[K_{EE}]^{-1}[K_{ER}] \\ I \end{bmatrix} = \begin{bmatrix} 1 \\ 1 \\ 1 \end{bmatrix}$$

The rigid-body mode may be either extracted from the eigenvalue problem or calculated partitioning the stiffness matrix in E-set and R-set submatrices. Using the geometric information of the nodes with coordinates (x,y,z) there is a third way to calculate the rigid-body mode. The geometric matrix of a node is obtained by translations along the x-, y- and z-axis and rotations about the x-, y- and z-axis. The geometric matrix is given by (see Fig. 5.2)

$$\begin{bmatrix} u \\ v \\ w \\ \varphi_x \\ \varphi_y \\ \varphi_z \end{bmatrix} = \begin{bmatrix} 1 & 0 & 0 & 0 & z & -y \\ 0 & 1 & 0 & -z & 0 & x \\ 0 & 0 & 1 & x & -y & 0 \\ 0 & 0 & 0 & 1 & 0 & 0 \\ 0 & 0 & 0 & 0 & 1 & 0 \\ 0 & 0 & 0 & 0 & 0 & 1 \end{bmatrix}\begin{bmatrix} u_o \\ v_o \\ w_o \\ \varphi_{xo} \\ \varphi_{yo} \\ \varphi_{zo} \end{bmatrix}. \tag{5.25}$$

In fact, the geometric matrix the motion of point A with respect to origin O.

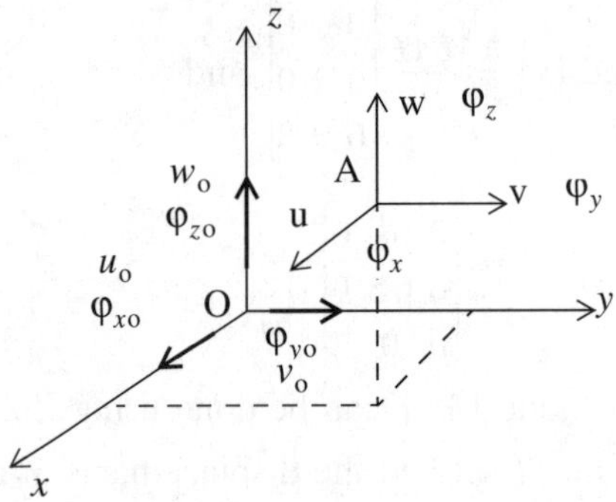

Fig. 5.2. Geometric relations between displacements and rotations

The rigid-body mode is built up from the geometric matrices of all nodes with respect to a certain origin of the coordinate system. There are six rigid-body modes.

5.2.4 Left Eigenvectors

Left eigenvectors, which can be extracted from measured frequency response functions of structures and systems convey important information on their dynamic behaviour [Bucher 97]. In this section the left eigenvectors are defined and the relation with right eigenvectors, the modes, will be explained.

The general eigenvalue problem may be written as [Zurmuehl 64]

$$[A]\{x\} = \lambda\{x\}. \tag{5.26}$$

We now define a new eigenvalue problem

$$[A]^T\{y\} = \lambda\{y\}. \tag{5.27}$$

The transposed version of (5.27) now becomes

$$\{y\}^T[A] = \lambda\{y\}^T, \tag{5.28}$$

where

- $\{y\}$ the eigenvector of the matrix $[A]^T$
- $\{y\}$ the left eigenvector of the matrix $[A]$

The eigenvalues λ of the eigenvalue problems stated in (5.26) and (5.27) are equal because [Zurmuehl 64] $\det(A - \lambda I) = \det(A^T - \lambda I)$. Zuermuhl,

[Zuermuhl 64], proved that in the case $\lambda_i \neq \lambda_j$ the eigenvectors $\{x_i\}$ and $\{y_j\}$ are orthogonal

$$\{x_i\}^T\{y_j\} = 0\,. \tag{5.29}$$

The eigenvectors $\{x_i\}$ and $\{y_i\}$ may be scaled such that

$$\{x_i\}^T\{y_i\} = \delta_{ij}\,, \tag{5.30}$$

or

$$[X]^T[Y] = [I]\,. \tag{5.31}$$

with
- $[X] = [x_1, x_2, ...]$ and
- $[Y] = [y_1, y_2, ...]$

The homogeneous equations of motion of an undamped dynamic system are written as

$$[M]\{\ddot{x}\} + [K]\{x\} = \{0\}\,,$$

and the eigenvalue problem with eigenvector $\{\phi_i\}$ is

$$([K] - \lambda[M])\{\phi_i\} = \{0\} \tag{5.32}$$

$$([M]^{-1}[K])\{\phi_i\} = \lambda\{\phi_i\}\,. \tag{5.33}$$

The left eigenvector $\{\chi_j\}$ is, from the eigenvalue problem,

$$([M]^{-1}[K])^T\{\chi_j\} = \lambda\{\chi_j\}\,, \tag{5.34}$$

or with a symmetric mass matrix $[M]$ and symmetric stiffness matrix $[K]$

$$\{\chi_j\}^T([M]^{-1}[K]) = \lambda\{\chi_j\}^T\,. \tag{5.35}$$

Equation (5.34) can be written as

$$([K][M]^{-1})\{\chi_j\} = \lambda\{\chi_j\}\,. \tag{5.36}$$

Comparing (5.32) with (5.36) it can be shown that [Bucher 97]

$$\{\chi_i\} = [M]\{\phi_i\}\,, \tag{5.37}$$

and when the left and right eigenvectors are properly scaled then

$$\{\phi_i\}^T\{\chi_j\} = \delta_{ij}, \tag{5.38}$$

with $\{\phi_i\}^T[M]\{\phi_i\} = 1$.

5.3 Damped Linear Dynamic Systems

The dynamic characteristics of a damped mechanical dynamic system are:

- The state variables
- The complex eigenvalues
- The natural frequency
- The modal damping
- The complex eigenvectors

5.3.1 The State Vector

The damped equations of motion of a damped linear dynamic system are generally written as

$$[M]\{\ddot{x}\} + [C]\{\dot{x}\} + [K]\{x\} = \{F(t)\}, \tag{5.39}$$

where

- $[M]$ the mass matrix, either diagonal or consistent with the stiffness matrix
- $[C]$ the damping matrix
- $[K]$ the stiffness matrix
- $\{F(t)\}$ the external (enforced) forces
- $\{x\}$ the displacement vector with n degrees of freedom
- $\{\dot{x}\}$ the velocity vector
- $\{\ddot{x}\}$ the acceleration vector

The damping forces are in this case proportional to the velocities.

A state vector is now introduced in which the displacements and velocities are combined

$$\{y\} = \begin{Bmatrix} x \\ \dot{x} \end{Bmatrix}. \tag{5.40}$$

The damped equations of motion can be expressed in the state matrices and vectors

$$\begin{bmatrix} I & 0 \\ 0 & M \end{bmatrix} \begin{Bmatrix} \dot{x} \\ \ddot{x} \end{Bmatrix} + \begin{bmatrix} 0 & -I \\ K & C \end{bmatrix} \begin{Bmatrix} x \\ \dot{x} \end{Bmatrix} = \begin{Bmatrix} 0 \\ F(t) \end{Bmatrix}. \tag{5.41}$$

If the mass matrix $[M]$ is not singular (5.41) may be rewritten

$$\begin{bmatrix} I & 0 \\ 0 & I \end{bmatrix} \begin{Bmatrix} \dot{x} \\ \ddot{x} \end{Bmatrix} = \begin{bmatrix} 0 & I \\ -M^{-1}K & -M^{-1}C \end{bmatrix} \begin{Bmatrix} x \\ \dot{x} \end{Bmatrix} + \begin{Bmatrix} 0 \\ M^{-1}F(t) \end{Bmatrix} \tag{5.42}$$

or as

$$\{\dot{y}\} = [A]\{y\} + \{F(t)\}. \tag{5.43}$$

The (5.41) may be arranged in another way resulting in symmetric state matrices

$$\begin{bmatrix} C & M \\ M & 0 \end{bmatrix} \begin{Bmatrix} \dot{x} \\ \ddot{x} \end{Bmatrix} + \begin{bmatrix} K & 0 \\ 0 & M \end{bmatrix} \begin{Bmatrix} x \\ \dot{x} \end{Bmatrix} = \begin{Bmatrix} F(t) \\ 0 \end{Bmatrix}. \tag{5.44}$$

The damped equations of motion of a linear dynamic system are

$$\begin{bmatrix} 0.5 & 0 & 0 \\ 0 & 1 & 0 \\ 0 & 0 & 1.5 \end{bmatrix} \begin{Bmatrix} \ddot{x}_1 \\ \ddot{x}_2 \\ \ddot{x}_3 \end{Bmatrix} + \begin{bmatrix} 0.05 & -0.05 & 0 \\ -0.05 & 0.1 & -0.05 \\ 0 & -0.05 & 0.1 \end{bmatrix} \begin{Bmatrix} \dot{x}_1 \\ \dot{x}_2 \\ \dot{x}_3 \end{Bmatrix} + \begin{bmatrix} 3 & -2 & 0 \\ -2 & 4 & -2 \\ 0 & -2 & 3 \end{bmatrix} \begin{Bmatrix} x_1 \\ x_2 \\ x_3 \end{Bmatrix} = \begin{Bmatrix} 0 \\ 1 \\ 0 \end{Bmatrix}.$$

The state vector is

$$\lfloor y \rfloor = \lfloor x_1, x_2, x_3, \dot{x}_1, \dot{x}_2, \dot{x}_3 \rfloor,$$

and the matrices $[A]$ and $\{F(t)\}$ become

$$[A] = \begin{bmatrix} 0 & 0 & 0 & 1 & 0 & 0 \\ 0 & 0 & 0 & 0 & 1 & 0 \\ 0 & 0 & 0 & 0 & 0 & 1 \\ -6 & 4 & 0 & -0.1 & 0.1 & 0 \\ 2 & -4 & 2 & 0.05 & -0.1 & 0.05 \\ 0 & 1.3333 & -2 & 0 & 0.0333 & -0.0667 \end{bmatrix}, \text{ and } \{\tilde{F}(t)\} = \begin{Bmatrix} 0 \\ 0 \\ 0 \\ 0 \\ 1 \\ 0 \end{Bmatrix}.$$

5.3.2 Eigenvalue Problem

The damped equations of motion of (5.39) are transformed with the aid of the state vector into (5.43)

$$\{\dot{y}\} = [A]\{y\} + \{\tilde{F}(t)\},$$

where

- $[A] = \begin{bmatrix} 0 & I \\ -M^{-1}K & -M^{-1}C \end{bmatrix}$ and
- $\{\tilde{F}(t)\} = \begin{Bmatrix} 0 \\ M^{-1}F(t) \end{Bmatrix}$

Assuming a solution of the state vector

$$\{y\} = \{\hat{y}\}e^{\lambda t} \tag{5.45}$$

and of the homogeneous equation

$$\{\dot{y}\} = [A]\{y\}, \tag{5.46}$$

the following eigenvalue problem is derived

$$([A] - \lambda[I])\{\hat{y}\} = \{0\}, \tag{5.47}$$

where

- λ_k the k-th complex eigenvalue and
- $\{\hat{y}\}$ the associated complex eigenvector

If the linear dynamic system is under critically damped the eigenvalues λ_k occur in pairs, the complex eigenvalue λ_k and its conjugate complex eigenvalue $\bar{\lambda}_k$,

$$\lambda_k, \bar{\lambda}_k = -\xi_k\omega_k \pm j\omega_k\sqrt{1-\xi_k^2}. \tag{5.48}$$

Equation (5.48) is the analogue of a simple damper-mass-spring system. The damping ratio ξ_k and the natural frequency ω_k can be calculated with

$$\lambda_k + \bar{\lambda}_k = -2\xi_k\omega_k, \tag{5.49}$$

in combination with

$$\lambda_k\bar{\lambda}_k = \omega_k^2. \tag{5.50}$$

The damped equations of motion of a linear dynamic system are

$$\begin{bmatrix} 0.5 & 0 & 0 \\ 0 & 1 & 0 \\ 0 & 0 & 1.5 \end{bmatrix} \begin{Bmatrix} \ddot{x}_1 \\ \ddot{x}_2 \\ \ddot{x}_3 \end{Bmatrix} + \begin{bmatrix} 0.05 & -0.05 & 0 \\ -0.05 & 0.1 & -0.05 \\ 0 & -0.05 & 0.1 \end{bmatrix} \begin{Bmatrix} \dot{x}_1 \\ \dot{x}_2 \\ \dot{x}_3 \end{Bmatrix} + \begin{bmatrix} 3 & -2 & 0 \\ -2 & 4 & -2 \\ 0 & -2 & 3 \end{bmatrix} \begin{Bmatrix} x_1 \\ x_2 \\ x_3 \end{Bmatrix} = \begin{Bmatrix} 0 \\ 1 \\ 0 \end{Bmatrix}.$$

The eigenvalues of the eigenvalue problem $([A] - \lambda[I])\{\hat{y}\} = \{0\}$ are (with $j = \sqrt{-1}$)

$$\{\lambda\} = \begin{Bmatrix} -0.0867+2.8536j \\ -0.0867-2.8536j \\ -0.0374+1.8005j \\ -0.0374-1.8005j \\ -0.0093+0.7779j \\ -0.0093-0.7779j \end{Bmatrix} \text{ and } \{\bar{\lambda}\} = \begin{Bmatrix} -0.0867-2.8536j \\ -0.0867+2.8536j \\ -0.0374-1.8005j \\ -0.0093+0.7779j \\ -0.0093-0.7779j \\ -0.0093+0.7779j \end{Bmatrix}.$$

The natural frequencies can be obtained from (5.50), $\lambda_k \bar{\lambda}_k = \omega_k^2$

$$\{\omega_k\} = \begin{Bmatrix} 2.8549 \\ 2.8549 \\ 1.8009 \\ 1.8009 \\ 0.6053 \\ 0.6053 \end{Bmatrix}.$$

The modal damping ratio is calculated from (5.51), $\lambda_k + \bar{\lambda}_k = -2\xi_k\omega_k$

$$\{\xi_k\} = \begin{Bmatrix} 0.0607 \\ 0.0607 \\ 0.0415 \\ 0.0415 \\ 0.0239 \\ 0.0239 \end{Bmatrix}.$$

5.3.3 Eigenvectors

The complex eigenvectors $\{\hat{y}_k\}$ from the eigenvalue in problem (5.47)

$$([A] - \lambda_k[I])\{\hat{y}_k\} = \{0\}$$

are gathered in the matrix of eigenvectors $[P]$.

The state vector $\{y\}$ is now depicted on the matrix of eigenvectors $[P]$

$$\{y\} = [P]\{\eta(t)\}. \tag{5.51}$$

Substituting (5.51) into (5.43) and multiplying that result by $[P]^{-1}$ the matrix equation becomes

$$\{\dot{\eta}\} = [P]^{-1}[A][P]\{\eta\} + [P]^{-1}\{\tilde{F}(t)\}. \tag{5.52}$$

From [Strang 88] the matrix of eigenvectors $[P]$ will diagonalise the matrix $[A]$ such that

$$[P]^{-1}[A][P] = \langle\lambda_k\rangle. \tag{5.53}$$

Finally, (5.52) becomes

$$\{\dot{\eta}\} = \langle\lambda_k\rangle\{\eta\} + [P]^{-1}\{\tilde{F}(t)\}. \tag{5.54}$$

The solution of (5.54) is from [D'Souza 84]

$$\{\eta(t)\} = [\Psi(t-t_0)]\{\eta(t_0)\} + \int_{t_0}^{t} [\Psi(t-\tau)][P]^{-1}\{\tilde{F}(\tau)\}d\tau, \tag{5.55}$$

where

- $[\Psi(t)] = \begin{bmatrix} e^{\lambda_1 t} & o & \dots & 0 \\ 0 & e^{\lambda_2 t} & \dots & 0 \\ . & . & \dots & 0 \\ 0 & 0 & \dots & e^{\lambda_{2n} t} \end{bmatrix}$
- n the number of degrees of freedom.

The initial conditions can be calculated using

$$\{\eta(t_0)\} = [P]^{-1}\{y(t_0)\}. \tag{5.56}$$

When the matrix of eigenvectors $[P]$ is not complete the initial conditions for the generalised coordinates are to be calculated by

$$\{\eta(t_0)\} = [P^T P]^{-1}[P]^T\{y(t_0)\}. \tag{5.57}$$

The damped equations of motion of a linear dynamic system are

$$\begin{bmatrix} 0.5 & 0 & 0 \\ 0 & 1 & 0 \\ 0 & 0 & 1.5 \end{bmatrix} \begin{Bmatrix} \ddot{x}_1 \\ \ddot{x}_2 \\ \ddot{x}_3 \end{Bmatrix} + \begin{bmatrix} 0.05 & -0.05 & 0 \\ -0.05 & 0.1 & -0.05 \\ 0 & -0.05 & 0.1 \end{bmatrix} \begin{Bmatrix} \dot{x}_1 \\ \dot{x}_2 \\ \dot{x}_3 \end{Bmatrix} + \begin{bmatrix} 3 & -2 & 0 \\ -2 & 4 & -2 \\ 0 & -2 & 3 \end{bmatrix} \begin{Bmatrix} x_1 \\ x_2 \\ x_3 \end{Bmatrix} = \begin{Bmatrix} 0 \\ 1 \\ 0 \end{Bmatrix}.$$

The eigenvectors of the eigenvalue problem $([A] - \lambda[I])\{\hat{y}\} = \{0\}$ are

$$[P] = \begin{bmatrix} -0.2707-0.1029j & -0.2707+0.1029j & -0.2684+0.2111j & -0.2684-0.2111j & .. \\ 0.1442+0.0591j & 0.1442-0.0591j & -0.1807+0.1507j & -0.1807-0.1507j & .. \\ -0.0310-0.0135j & -0.0310+0.0135j & 0.1991-0.1551j & 0.1991+0.1551j & .. \\ 0.3170-0.7634j & 0.3170+0.7634j & -0.3701-0.4911j & -0.3701+0.4911j & .. \\ -0.1813+0.4065j & -0.1813-0.4065j & -0.2646-0.3310j & -0.2646+0.3310j & .. \\ 0.0412-0.0873j & 0.0412+0.0873j & 0.2719+0.3643j & 0.2719-0.3643j & .. \end{bmatrix}$$

$$[P] = \begin{bmatrix} .... & 0.3221-0.1878j & 0.3221+0.1878j \\ .... & 0.4324-0.2567j & 0.4324+0.2567j \\ .... & 0.4115-0.2484j & 0.4115+0.2484j \\ .... & 0.1431+0.2523j & 0.1431-0.2523j \\ .... & 0.1957+0.3388j & 0.1957-0.3388j \\ .... & 0.1894+0.3224j & 0.1894-0.3224j \end{bmatrix}.$$

The matrix of eigenvectors $[P]$ will diagonalise the matrix $[A]$, (5.53),

$$[P]^{-1}[A][P] = \langle\lambda_k\rangle$$

$$\langle\lambda_k\rangle = \begin{bmatrix} -0.0867+2.8536j & 0 & 0 & 0 & .. \\ 0 & -0.0867-2.8536j & 0 & 0 & .. \\ 0 & 0 & -0.0374+1.8005j & 0 & .. \\ 0 & 0 & 0 & -0.0374-1.8005j & .. \\ 0 & 0 & 0 & 0 & .. \\ 0 & 0 & 0 & 0 & .. \end{bmatrix}$$

$$\langle\lambda_k\rangle = \begin{bmatrix} .... & 0 & 0 \\ .... & 0 & 0 \\ .... & 0 & 0 \\ .... & 0 & 0 \\ .... & -0.0093+0.7779j & 0 \\ .... & 0 & -0.0093-0.7779j \end{bmatrix}.$$

The vector $[P]^{-1}\{\tilde{F}(\tau)\}$ in (5.55) is

$$[P]^{-1}\{\tilde{F}(\tau)\} = \begin{Bmatrix} -0.1410-0.3767j \\ -0.1410+0.3767j \\ -0.1888+0.2489j \\ -0.1888-0.2489j \\ 0.2492-0.4140j \\ 0.2492+0.4140j \end{Bmatrix}.$$

5.4 Problems

5.4.1 Problem 1

The mdof system, as shown in Fig. 5.3, consists of 5 degrees of freedom.

1. Derive the equations of motion using
   - the equations of equilibrium (Newton's law)
   - Lagrange's equations

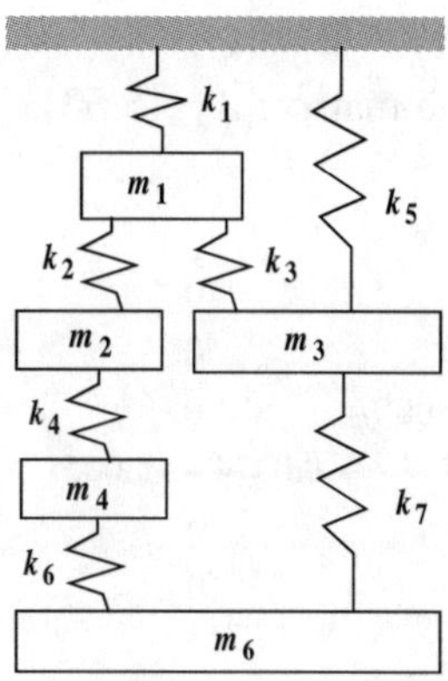

Fig. 5.3. Mdof dynamic system

The parameters have the following values;
the masses (kg), $m_1 = 3$, $m_2 = 2$, $m_3 = 4$, $m_4 = 1$ and $m_5 = 4$, the spring stiffness (N/m), $k_1 = 3x10^3$, $k_2 = 2x10^3$, $k_3 = 5x10^3$, $k_4 = 4x10^3$, $k_5 = 6x10^3$, $k_9 = 6x10^3$ and $k_7 = 1x10^3$.

2. Calculate the lowest natural frequency $\omega_{n,1}$ and associated mode shape $\{\phi_1\}$ such that $\{\phi_1\}^T[M]\{\phi_1\} = 1$.

Answer: $\omega_{n,1} = 123.4$ rad/s

5.4.2 Problem 2

A damped dynamic system obeys the following characteristics:

- mass matrix $[M] = \begin{bmatrix} 3.1486 & -2.0419 \\ -2.0419 & 1.4850 \end{bmatrix}$
- stiffness matrix $[K] = \begin{bmatrix} 119.2695 & -0.67.6742 \\ -0.67.6742 & 254.3679 \end{bmatrix}$ and
- a damping matrix $[C] = \begin{bmatrix} 0.3367 & -0.6538 \\ -0.6538 & 2.2436 \end{bmatrix}$

Questions

1. Calculate the undamped eigenvalues $\{\lambda\}$ (natural frequencies) and associated modes $[\Phi]$.
2. Calculate $[\bar{\Phi}]^T[M][\bar{\Phi}]$.
3. Scale the modal matrix $[\Phi]$ such that $[\Phi]^T[M][\Phi] = [I]$ and $[\Phi]^T[K][\Phi] = \langle\lambda\rangle$.
4. Set up the matrix $[A] = \begin{bmatrix} 0 & I \\ -M^{-1}K & -M^{-1}C \end{bmatrix}$.
5. Solve the damped eigenvalue problem $([A] - \lambda[I])\{\hat{y}\} = \{0\}$.
6. Calculate the undamped natural frequencies $\{\omega_n\}$ and the modal damping ratios $\{\xi\}$.
7. Calculate $[P]^{-1}[A][P]$.

6 Natural Frequencies, an Approximation

6.1 Introduction

Generally, when calculating the natural frequencies of complex dynamic linear systems the finite element analysis method is applied. However, it is good practice to first apply a method to approximately calculate the natural frequency of that system to get a feel for the value of the natural frequency. The system will be simplified as much as possible in order to be able to use an approximate method. In this chapter the following methods, to quickly obtain the value of the natural frequency, will be discussed:

- Static displacement method
- Rayleigh's [1] quotient [Temple 56]
- Dunkerley's equation

The theory will be illustrated with examples.

6.2 Static Displacement Method

The natural frequency of an sdof system, as shown in Fig. 6.1, is given by

$$f_n = \frac{1}{2\pi}\sqrt{\frac{k}{m}}. \tag{6.1}$$

If a 1g acceleration is acting on the mass m (kg) the inertia force mg (N) will compress the spring with a spring stiffness k (N/m)] with a static displacement x_{stat}

1. Lord Rayleigh, whose given name was John William Strutt (1942–1919)]

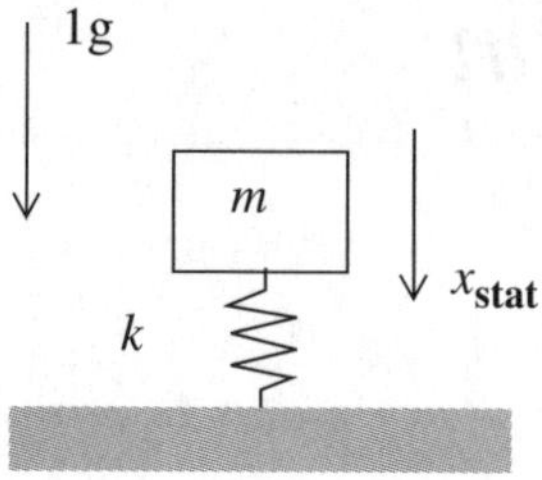

Fig. 6.1. Undamped sdof system

$$x_{stat} = \frac{mg}{k}. \tag{6.2}$$

The static displacement x_{stat} has the dimension (m). This means that we can rewrite (6.1) as follows

$$f_n = \frac{1}{2\pi}\sqrt{\frac{k}{m}} = \frac{1}{2\pi}\sqrt{\frac{g}{x_{stat}}}. \tag{6.3}$$

The approximation of the natural frequency, using the static displacement method, is only applicable if the dynamic system has a dominant lumped (discrete) mass with respect to the distributed mass.

If we calculate the static displacement Δ per 1 m/s^2 the approximation of the natural frequency is

$$f_n = \frac{1}{2\pi}\sqrt{\frac{k}{m}} = \frac{1}{2\pi}\sqrt{\frac{1}{\Delta}}. \tag{6.4}$$

A spacecraft placed on a payload adapter is such a system. The mass of the payload adapter is (much) less than the lumped mass of the spacecraft in the centre of gravity (centre of mass). The static displacement of the centre of gravity, due to the unit acceleration inertia loads, can be used to calculate the natural frequencies of the spacecraft placed on the payload adapter.

Given a spacecraft with a total mass of $M_{tot} = 2500$ kg. The centre of gravity of the spacecraft is located at $h = 1.5$ m above the interface with the conical payload adapter. The diameter at the top of the cone of the payload adapter $d = 1.2$ m. The configuration of the spacecraft is shown in Fig. 6.2. The diameter at the lower side is $D = 3$ m. The height of the cone is $H = 1.5$ m. The cone had been made of CFRP with an isotropic Young's modulus $E = 120$ GPa and a Poisson's ratio $\upsilon = 0.3$. The thickness of the cone $t = 5$ mm.

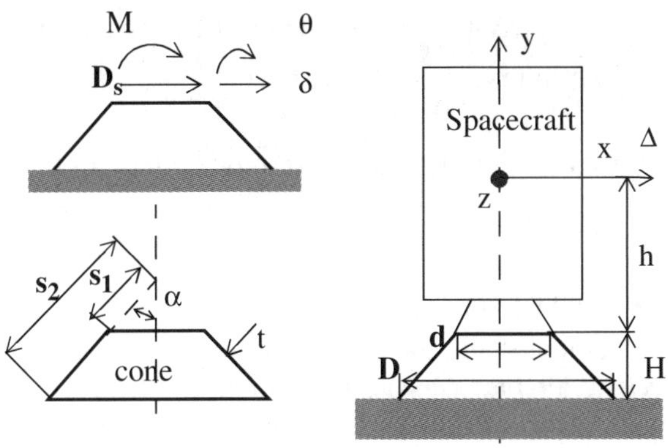

Fig. 6.2. Spacecraft mounted on a conical payload adapter

Calculate the natural frequency, associated with the first bending mode in the x–y plane, when the spacecraft has been placed on the conical payload adapter that is clamped at the lower side of the cone. The spacecraft is very well connected with the payload adapter and discontinuities do not exist.

In [Seide 72] the following influence coefficients can be found

$$\delta = \frac{1-\frac{s_1}{s_2}}{\pi E t(\sin\alpha)^3}\left\{\frac{\ln\frac{s_2}{s_1}}{1-\frac{s_1}{s_2}} - 2 + \left(1+\frac{s_1}{s_2}\right)\left[\frac{1}{2} + (1+\upsilon)(\sin\alpha)^2\right]\right\}D_s$$

$$+\frac{1-\frac{s_1}{s_2}}{\pi E t(\sin\alpha)^3}\left\{1-\left(1+\frac{s_1}{s_2}\right)\left[\frac{1}{2} + (1+\upsilon)(\sin\alpha)^2\right]\right\}\frac{M}{s_1\cos\alpha}, \tag{6.5}$$

and

$$\theta = \frac{1-\frac{s_1}{s_2}}{\pi E t s_1(\sin\alpha)^3\cos\alpha}\left\{1-\left(1+\frac{s_1}{s_2}\right)\left[\frac{1}{2} + (1+\upsilon)(\sin\alpha)^2\right]\right\}D_s$$

$$+\frac{1-\frac{s_1}{s_2}}{\pi E t s_1(\sin\alpha)^3\cos\alpha}\left\{\left(1+\frac{s_1}{s_2}\right)\left[\frac{1}{2} + (1+\upsilon)(\sin\alpha)^2\right]\right\}\frac{M}{s_1\cos\alpha}. \tag{6.6}$$

The shear force is $D_s = M_{tot}$ (N) and the bending moment $M = M_{tot}h$ (Nm). The total static displacement Δ (m), of the centre of gravity, due to 1m/s^2 acceleration in the x-direction is

$$\Delta = \delta + h\theta$$

The natural frequency f_n, corresponding to a bending mode shape in the x-direction is

$$f_n = \frac{1}{2\pi}\sqrt{\frac{1}{\Delta}} = \frac{1}{2\pi}\sqrt{\frac{1}{1.8086x10^{-5}}} = 37.42\,\text{Hz}$$

In our calculations we have neglected the mass of the payload adapter. The spacecraft on top of the payload adapter was assumed to be rigid. The influence of a flexible spacecraft on top of the flexible payload adapter can be calculated using Dunkerley's method.

6.3 Rayleigh's Quotient

We define Rayleigh's quotient as [Temple 56]

$$R(u) = \frac{\{u\}^T[K]\{u\}}{\{u\}^T[M]\{u\}}, \tag{6.7}$$

where

- $\{u\}$ an admissible vector (assumed mode shape) that fulfils the boundary conditions
- $[M]$ the positive-definite mass matrix, $\{u\}^T[M]\{u\} > 0$
- $[K]$ the stiffness matrix

The minimum value of Rayleigh's quotient $R(u)$ can be found (stationary value) when

$$\delta R(u) = 0. \tag{6.8}$$

Thus

$$\delta R(u) = \delta\{u\}^T 2\left\{\frac{[K]\{u\}}{\{u\}^T[M]\{u\}} - \frac{\{u\}^T[K]\{u\}[M]\{u\}}{(\{u\}^T[M]\{u\})^2}\right\} = 0. \tag{6.9}$$

In general, the "kinetic energy" (generalised mass) $\{u\}^T[M]\{u\} \neq 0 = m_g$, thus (6.9) can be rewritten as

$$[K]\{u\} - R(u)[M]\{u\} = 0. \tag{6.10}$$

Rayleigh's quotient is analogous to the eigenvalue problem

$$([K]-\lambda[M])\{\phi\} = 0 . \tag{6.11}$$

Rayleigh's quotient $R(u)$ is equal to the eigenvalue λ only if $\{u\} = \{\phi\}$. We normalise the mode shapes $[\Phi]$ such that

$$[\Phi]^T[M][\Phi] = [I] \text{ and } [\Phi]^T[K][\Phi] = \langle\lambda\rangle .$$

We can express the assumed vector $\{u\}$ as follows

$$\{u\} = [\Phi]\{\eta\} . \tag{6.12}$$

Equation (6.7) can then be written

$$R(u) = \frac{\{u\}^T[K]\{u\}}{\{u\}^T[M]\{u\}} = \frac{\sum_j \eta_j^2\lambda_j}{\sum_j \eta_j^2} . \tag{6.13}$$

Assume the mode shape $\{\phi_i\}$ is dominant with respect to the other mode shapes, then $\eta_j = \varepsilon_j\eta_i$ with $\varepsilon_j \ll 1$. When $\{\phi_i\} = \{\phi_1\}$ and $(\lambda_j-\lambda_1)\geq 0$ (6.13) becomes [Meirovitch 75]

$$R(u) = \frac{\sum_j \eta_j^2\lambda_j}{\sum_j \eta_j^2} = \frac{\eta_1^2\lambda_1 + \eta_1^2\sum_{j;j\neq i}\varepsilon_j^2\lambda_j}{\eta_1^2 + \eta_1^2\sum_{j;j\neq i}\varepsilon_j^2} \approx \lambda_1 + \sum_{j;j\neq i}\varepsilon_j^2(\lambda_j-\lambda_1) \geq \lambda_1 . \tag{6.14}$$

Rayleigh's quotient $R(u)$ will result in an upper bound value of the eigenvalue λ_i corresponding with the assumed mode shape $\{u\}$. Rayleigh's quotient is never below λ_1 and never above λ_n, with n the numbers of dofs, [Sprang 88].

A 1 g gravitational field is applied the masses of the system, shown in Fig. 6.3. The static displacement vector $\{x\}$ becomes

$$\{x\} = \frac{mg}{k}\left\{\begin{bmatrix} 1 & -1 & 0 \\ -1 & 3 & -2 \\ 0 & -2 & 5 \end{bmatrix}\right\}^{-1}\begin{Bmatrix} 1 \\ 1 \\ 1 \end{Bmatrix} = \frac{mg}{k}\begin{Bmatrix} 1.8533 \\ 0.8533 \\ 0.3333 \end{Bmatrix} .$$

We take as admissible vector $\{u\}$

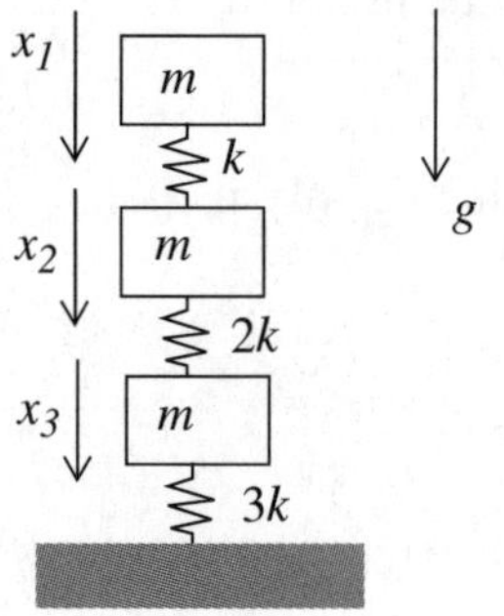

Fig. 6.3. 3 dofs sdof linear system

$$\{u\} = \begin{Bmatrix} 1.8533 \\ 0.8533 \\ 0.3333 \end{Bmatrix}.$$

Rayleigh's quotient $R(u)$ now becomes

$$R(u) = \frac{\{u\}^T[K]\{u\}}{\{u\}^T[M]\{u\}} = 0.4400\frac{k}{m}.$$

The theoretical lowest eigenvalue is $\lambda_1 = 0.41587\frac{k}{m}$.

Rayleigh's quotient of a bending beam is defined as

$$R(u) = \frac{\int_0^L EI(u')^2 dx}{\int_0^L mu^2 dx}, \tag{6.15}$$

with

- EI the bending stiffness of the beam (Nm^2)
- m the mass per unit of length (kg/m)
- $u(x)$ the assumed mode
- L the length of the beam (m)

For a beam, simply supported at both ends, we take the assumed mode $u(x)$

$$u(x) = \frac{x}{L}\left(1 - \frac{x}{L}\right).$$

Rayleigh's quotient becomes

$$R(u) = 120\frac{EI}{mL^4} \quad .$$

The theoretical value for the eigenvalue $\lambda_1 = \pi^4\frac{EI}{mL^4}$.

6.4 Dunkerley's Method

Dunkerley published his equation in 1894 [Brock 76].

The equation of Dunkerley is a method to estimate the lowest natural frequency of a dynamic system, which is composed of substructures (components) of which the lowest and lower natural frequencies are known. The damping is not involved in the equation of Dunkerley. The equation of Dunkerley will predict an accurate lowest natural frequency when this frequency is rather shifted from the next natural frequencies.

The eigenvalue problem of an undamped dynamic system can be written as

$$(-\omega^2[M] + [K])\{\phi\} = \{0\}, \tag{6.16}$$

with

- $[M]$ the mass matrix.
- $[K]$ the positive-definite stiffness matrix. The inverse of the stiffness matrix, the flexibility matrix $[G] = [K]^{-1}$ exists.
- $\{\phi\}$ the mode shape corresponding the natural frequency $\omega > 0$.

We can rewrite (6.16) as follows

$$\left(\frac{1}{\omega^2}[I]-[G][M]\right)\{\phi\} = \{0\}. \tag{6.17}$$

The solution of the determinant of (6.17), with n dofs, can be formally written as

$$\left(\frac{1}{\omega^2}-\frac{1}{\omega_1^2}\right)\left(\frac{1}{\omega^2}-\frac{1}{\omega_2^2}\right)....\left(\frac{1}{\omega^2}-\frac{1}{\omega_n^2}\right) = 0, \tag{6.18}$$

with $\frac{1}{\omega_1^2}, \frac{1}{\omega_2^2}, .., \frac{1}{\omega_2^2}$ the solution, roots, of the characteristic equation

$$\left|\left(\frac{1}{\omega^2}[I]-[G][M]\right)\right| = 0. \tag{6.19}$$

The sum of the n eigenvalues of (6.19) equals the sum of the n diagonal terms of the matrix $[G][M]$ [Strang 88]. This sum is known as the trace of $[G][M]$, thus

$$\mathrm{trace}([G][M]) = \sum_{k=1}^{n} g_{kk}m_{kk} = \sum_{k=1}^{n}\left(\frac{1}{\omega_k^2}\right) \tag{6.20}$$

To estimate the lowest natural frequency ω_1 we may neglect the contribution of the higher natural frequencies ω_k, k=2,3,.....,n. This approximation becomes more and more accurate if $\omega_1 \ll \omega_k$, k=2,3,...,n, then we obtain Dunkerley's equation

$$\frac{1}{\omega_1^2} \le \sum_{k=1}^{n} g_{kk}m_{kk}. \tag{6.21}$$

The term $g_{kk}m_{kk}$ may be interpreted as an sdof system with a discrete mass m_{kk} and a spring with spring stiffness g_{kk}^{-1}, as shown in Fig. 6.4. The natural frequency $\frac{1}{\omega_{kk}^2}$ of the equivalent sdof system is

$$\frac{1}{\omega_{kk}^2} = g_{kk}m_{kk} = \frac{m_{kk}}{g_{kk}^{-1}}. \tag{6.22}$$

Thus Dunkerley's equation, (6.21), becomes

$$\frac{1}{\omega_1^2} \le \sum_{k=1}^{n} \frac{1}{\omega_{kk}^2}. \tag{6.23}$$

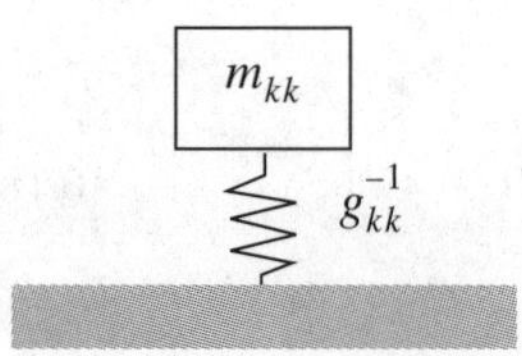

Fig. 6.4. Equivalent sdof system in Dunkerley's equation

We consider one discrete mass at a time and neglect the other masses. We calculate the flexibility term g_{kk} for that discrete mass applying a unit load.

The obtained displacement is, in fact, the flexibility g_{kk}. We shall illustrate that with an example.

We consider the dynamic system as shown in Fig. 6.3.

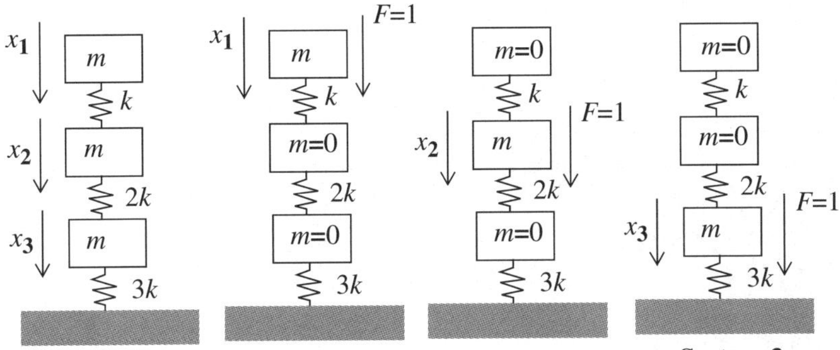

Fig. 6.5. Decomposition of the dynamic system in 3 systems

Equation (6.21) will be applied to calculate the lowest natural frequency of the complete dynamic system. The dynamic system has been decomposed in three systems; system 1, system 2 and system 3, as shown in Fig. 6.5.

Table 6.1. Example calculations Dunkerley's equation

System #	g_{kk}	m_{kk}
1	$\frac{1}{k}+\frac{1}{2k}+\frac{1}{3k}=\frac{11}{6k}$	m
2	$\frac{1}{2k}+\frac{1}{3k}=\frac{5}{6k}$	m
3	$\frac{1}{3k}=\frac{2}{6k}$	m
$\sum_{k=1}^{3} g_{kk}m_{kk}$	$\frac{3m}{k}$	$\omega_1^2=0.3333\frac{k}{m}$

The analysis procedure is illustrated in Table 6.1.

Suppose a dynamic system is built up of sdof's on top of each other, as illustrated in Fig. 6.6. then we can derive an alternative equation of Dunkerley.

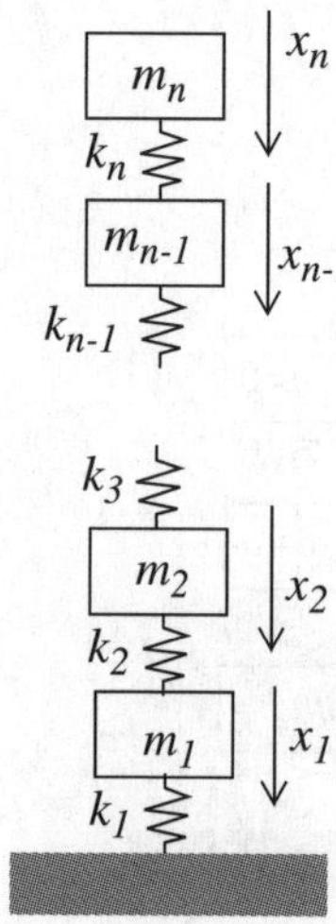

Fig. 6.6. *n* sdof dynamic systems

The diagonal terms g_{kk}, $k = 1, 2, .., n$ of the flexibility matrix $[G] = [K]^{-1}$ ($[K]$ is a positive-definite matrix), can be written as follows

$$g_{kk} = \sum_{j=1}^{k} \frac{1}{k_j} . \qquad (6.24)$$

Equation (6.21) becomes

$$\frac{1}{\omega_1^2} \le \sum_{k=1}^{n} g_{kk} m_k = \sum_{k=1}^{n} m_k \sum_{j=1}^{k} \frac{1}{k_j} = \sum_{j=1}^{n} \frac{1}{k_j} \sum_{k=j}^{n} m_k . \qquad (6.25)$$

We will apply the alternative equation of Dunkerley to the dynamic system shown in Fig. 6.5. Equation (6.25) will be applied to calculate the lowest natural frequency of a complete dynamic system. The dynamic system has been decomposed into three systems; system 1, system 2 and system 3, as shown in Fig. 6.7.

Table 6.2. Example calculations alternative Dunkerley's equation

System #	n=3, j	$\frac{1}{k_j}$	$\sum_{k=j}^{n} m_k$
1	3	$\frac{1}{k}$	m
2	2	$\frac{1}{2k}$	$2m$
3	1	$\frac{1}{3k}$	$3m$
$\sum_{j=1}^{n} \frac{1}{k_j} \sum_{k=j}^{n} m_k$		$\frac{3m}{k}$	$\omega_1^2 = 0.3333\frac{k}{m}$

The procedure to calculate the lowest natural frequency, using the alternative equation of Dunkerley, is given in Table 6.2.

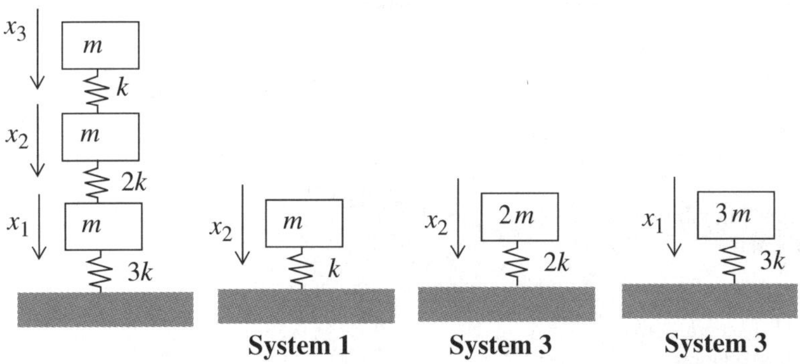

Fig. 6.7. Decomposition of dynamic system in 3 systems (alternative method)

The spacecraft mounted on the conical payload adapter is shown in Fig. 6.2. The spacecraft hardmounted at the interface between the spacecraft and the adapter has a lowest bending mode (x–y plane) of f_{sc} = 20 Hz. Calculate the lowest natural frequency of the complete system (spacecraft and adapter). Equation (6.25) will be applied to calculate the lowest natural frequency of a complete dynamic system. The dynamic system is composed of two systems; system 1 and system 2, as shown in Fig. 6.8. The procedure to calculate the lowest natural frequency, using the alternative equation of Dunkerley, is given in Table 6.3.

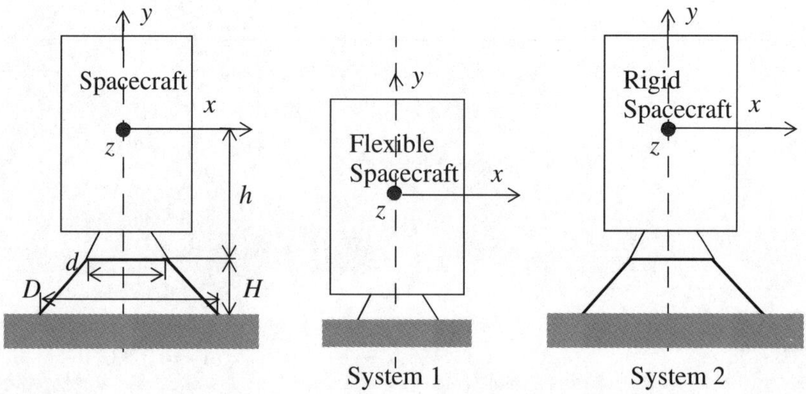

Fig. 6.8. Decomposition of dynamic system into 2 systems (alternative equation of Dunkerley)

Table 6.3. Spacecraft/payload adapter natural frequency calculations

System #	n=2, j	$\frac{(2\pi)^2}{\omega_j^2} = \frac{1}{k_j}\sum_{k=j}^{n} m_k$	$\sum_{k=j}^{n} m_k$ (kg)
1 Clamped flexible spacecraft	2	$\left(\frac{1}{f_{sc}}\right)^2 = \left(\frac{1}{20}\right)^2$	2500
2 Rigid spacecraft on payload adapter	1	$\left(\frac{1}{37.42}\right)^2$ result previous example	2500 mass of payload adapter neglected
		$(2\pi)^2 \sum_{j=1}^{n} \frac{1}{k_j} \sum_{k=j}^{n} m_k$	$f = 17.64$ Hz

6.5 Problems

6.5.1 Problem 1

An aeroplane settles 150 mm into its landing-gear springs when the aeroplane is at rest. What is the natural frequency f_n (Hz) for the vertical motion of the aeroplane with g=9.81 m/s^2 [Moretti 00]?
Answer: $f_n = 1.29$ Hz.

6.5.2 Problem 2

Demonstrate Rayleigh' method for estimating the fundamental frequency ω_n (rad/s) of a uniform cantilever of length L (m), mass-per-unit-length m (kg/m), and a bending stiffness EI. Use the assumed deflection shape $w = \frac{x^3}{L}w_{tip}$.

6.5.3 Problem 3

A dynamic system, as shown in Fig. 6.9, has 3 dofs; w, φ and δ. The displacement δ is with respect to line A–B. Both dofs w and φ are located in the middle of A–B. The structure in between A and B is rigid and has a mass m per unit of length (kg/m). The discrete mass M (kg) is coupled at the end of the massless elastic beam with a bending stiffness EI Nm^2. The beam is rigid connected at point B. The complete dynamic system is supported by two spring with a springs stiffness k (N/m). The following values shall be used: $M = 0.15$ kg, L_1=02 m, L_2=0.25 m, m=0.075 kg/m, $\frac{3EI}{L_2^3} = (2\pi 100)^2 M$ and k=10000 N/m. The second moment of mass of the rigid beam A–B $I = \frac{1}{12}mL_1^3$ (kgm^2).

1. Calculate the lowest natural frequency with the static displacement method, assuming $M \gg mL_1$.
2. Calculate the lowest natural frequency with Dunkerley's equation, using both the normal and the alternative equation.

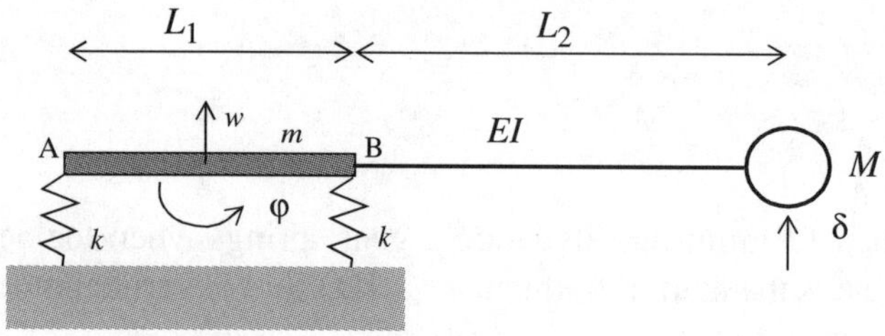

Fig. 6.9. Dynamic System with 3 dofs

3. Calculate the lowest natural frequency using Rayleigh's quotient. Hint: use the deflection mode calculated in question 1.
4. Set up the equations of motion (e.g. using Lagrange's equations) and calculate the eigenvalues and compare these results with the approximations.

The homogeneous equations of motion are

$$\begin{bmatrix} mL_1+M & M\left(\frac{1}{2}l_1+L_2\right) & M \\ M\left(\frac{1}{2}l_1+L_2\right) & I+M\left(\frac{1}{2}l_1+L_2\right)^2 & M\left(\frac{1}{2}l_1+L_2\right) \\ M & M\left(\frac{1}{2}l_1+L_2\right) & M \end{bmatrix}\begin{Bmatrix} \ddot{w} \\ \ddot{\varphi} \\ \ddot{\delta} \end{Bmatrix}+\begin{bmatrix} 2k & 0 & 0 \\ 0 & kL_1 & 0 \\ 0 & 0 & \frac{3EI}{L_2^3} \end{bmatrix}\begin{Bmatrix} w \\ \varphi \\ \delta \end{Bmatrix}=\begin{Bmatrix} 0 \\ 0 \\ 0 \end{Bmatrix}$$

and the calculated natural frequencies $f = \begin{Bmatrix} 36.0 \\ 233.1 \\ 1577.5 \end{Bmatrix}$ Hz.

6.5.4 Problem 4

A two mass system is illustrated in Fig. 6.10, [Ceasar 83]. Determine the natural frequency (Hz) of the two-mass system using.

1. the Dunkerly method
2. the Rayleigh method (use 1 g gravitation field to obtain displacement field)
3. exact method (4 dofs)

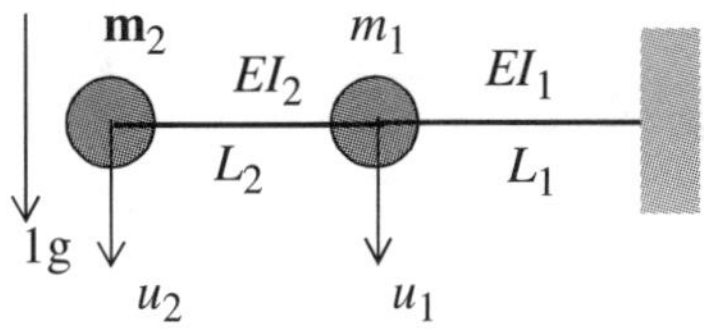

Fig. 6.10. Two-mass system

For numerical calculations use the following data:

- $E_1 = E_2 = 70$ GPa
- $I_1 = 7.5x10^{-6}$, $I_2 = 5.9x10^{-6}$ m^4
- $L_1 = L_2 = 0.5$ m

Answers: 14.02, 14.62, 14.50 Hz.

6.5.5 Problem 5

The mdof system, as shown in Fig. 6.11, consists of five degrees of freedom.

- Derive the equations of motion using
    1. the equations of equilibrium (Newton's law)
    2. Lagrange's equations

The parameters have the following values; the masses (kg), $m_1 = 3$, $m_2 = 2$, $m_3 = 4$, $m_4 = 1$ and $m_5 = 4$, the spring stiffness (N/m), $k_1 = 3x10^3$, $k_2 = 2x10^3$, $k_3 = 5x10^3$, $k_4 = 4x10^3$, $k_5 = 6x10^3$, $k_9 = 6x10^3$ and $k_7 = 1x10^3$.

- Calculate the natural frequencies and associated normal modes.
- Calculate an approximation of the lowest natural frequency $\omega_{n,1}$ using the Rayleigh method. Use an assumed mode $\{\varphi_1\}$, the static deformation vector under a 1 g gravitation field.

Answer: Exact $\omega_{n,1} = 123.4$ rad/s.

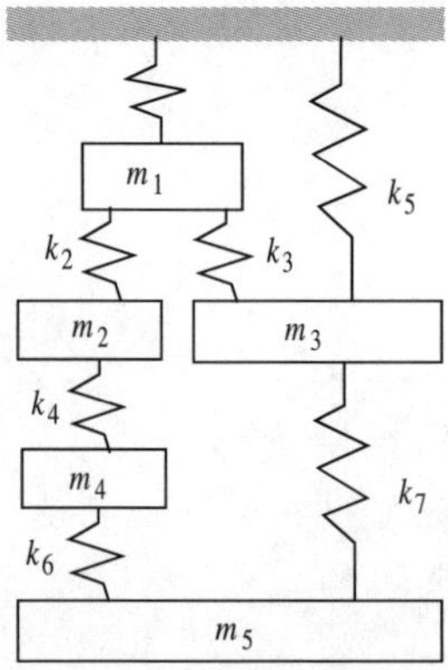

Fig. 6.11. Mdof dynamic system

7 Modal Effective Mass

7.1 Introduction

The modal effective mass is a modal dynamic property of a structure associated with the modal characteristics; natural frequencies, mode shapes, generalised masses, and participation factors. The modal effective mass is a measure to classify the importance of a mode shape when a structure is excited by base acceleration (enforced acceleration). A high effective mass will lead to a high reaction force at the base, while mode shapes with low associated modal effective mass are nearly excited by base acceleration and will give low reaction forces at the base. The effect of local modes is not well described with modal effective masses [Shunmugavel 95, Witting 96].

The modal effective mass matrix is a 6*x*6 mass matrix. Within this matrix the coupling between translations and rotations, for a certain mode shape, can be traced.

The summation over all modal effective masses will result in the mass matrix as a rigid-body.

In this chapter the theory behind the principle of the modal effective mass matrix will be discussed and the way the modal effective mass matrix can be obtained. The theory will be illustrated with an example.

7.2 Enforced Acceleration

An sdof system with a discrete mass m, a damper element c and a spring element k is placed on a moving base that is accelerated with an acceleration $\ddot{u}(t)$. The resulting displacement of the mass is $x(t)$. We introduce the natural (circular) frequency $\omega_n = \sqrt{\frac{k}{m}}$, the critical damping constant

$c_{\text{crit}} = 2\sqrt{km}$ and the damping ratio $\zeta = \frac{c}{c_{\text{crit}}}$. The amplification factor is defined as $Q = \frac{1}{2\zeta}$.

We introduce a relative motion $z(t)$, which is the displacement of the mass with respect to the base. The relative displacement is

$$z(t) = x(t) - u(t)\,. \tag{7.1}$$

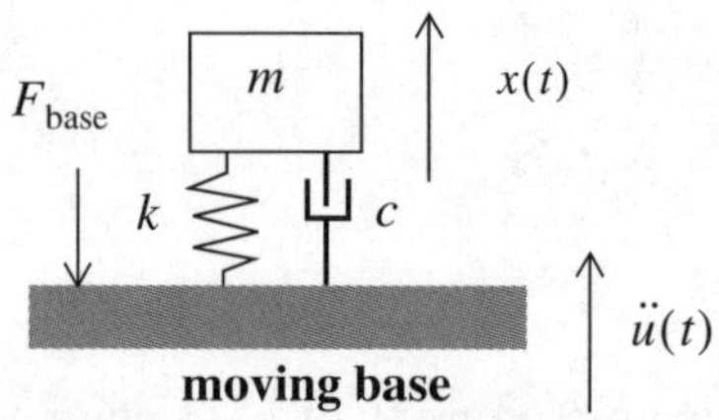

Fig. 7.1. Enforced acceleration of a damped sdof system

The equation of motion for the relative motion $z(t)$ is

$$\ddot{z}(t) + 2\zeta\omega_n\dot{z}(t) + \omega_n^2 z(t) = -\ddot{u}(t)\,. \tag{7.2}$$

The enforced acceleration of the sdof system is transformed into an external force. The absolute displacement $x(t)$ can be calculated from

$$\ddot{x}(t) = \ddot{z}(t) + \ddot{u}(t) = -2\zeta\omega_n\dot{z}(t) - \omega_n^2 z(t)\,. \tag{7.3}$$

The reaction force $F_{\text{base}}(t)$, due to the enforced acceleration $\ddot{u}(t)$, is a summation of the spring force and the damping force

$$F_{\text{base}}(t) = kz(t) + c\dot{z}(t) = -m(\ddot{z}(t) + \ddot{u}(t)) = -m\ddot{x}(t)\,. \tag{7.4}$$

Assuming harmonic vibration we can write the enforced acceleration

$$\ddot{u}(t) = \ddot{U}(\omega)e^{j\omega t}, \tag{7.5}$$

and also the relative motion $z(t)$

$$z(t) = Z(\omega)e^{j\omega t},\ \dot{z}(t) = j\omega Z(\omega)e^{j\omega t} \text{ and } \ddot{z}(t) = -\omega^2 Z(\omega)e^{j\omega t} \tag{7.6}$$

and the absolute acceleration of the sdof dynamic system is

$$\ddot{x}(t) = -\omega^2 X(\omega)e^{j\omega t}\,. \tag{7.7}$$

Equation (7.2) can be transformed in the frequency domain

$$[-\omega^2 + 2j\zeta\omega_n\omega + \omega_n^2]Z(\omega) = -\ddot{U}(\omega). \tag{7.8}$$

We are able to express the relative displacement $Z(\omega)$ in the enforced acceleration $\ddot{U}(\omega) = -\omega^2 U(\omega)$

$$Z(\omega) = \left(\frac{\omega}{\omega_n}\right)^2 H(\omega)U(\omega) \tag{7.9}$$

with

- $H(\omega) = \dfrac{1}{1-\left(\dfrac{\omega}{\omega_n}\right)^2 + 2j\zeta\left(\dfrac{\omega}{\omega_n}\right)}$ the frequency response function

Using (7.3) we can write the absolute acceleration $\ddot{X}(\omega)$ as

$$\ddot{X}(\omega) = -\omega^2[Z(\omega) + U(\omega)] = -\omega^2\left[1 + \left(\frac{\omega}{\omega_n}\right)^2 H(\omega)\right]U(\omega), \tag{7.10}$$

or

$$\ddot{X}(\omega) = \left[1 + \left(\frac{\omega}{\omega_n}\right)^2 H(\omega)\right]\ddot{U}(\omega). \tag{7.11}$$

The reaction force at the base $F_{\text{base}}(\omega)$ now becomes with the aid of (7.4)

$$F_{\text{base}}(\omega) = m\ddot{X}(\omega) = m\left[1 + \left(\frac{\omega}{\omega_n}\right)^2 H(\omega)\right]\ddot{U}(\omega). \tag{7.12}$$

In this frame the mass m is the effective mass $M_{\text{eff}} = m$. The reaction force $F_{\text{base}}(\omega)$ is proportional to the effective mass M_{eff} and the base excitation $\ddot{U}(\omega)$ multiplied by the amplification $1 + \left(\frac{\omega}{\omega_n}\right)^2 H(\omega)$. Similar relations will be derived for multi-degrees of freedom (mdof) dynamic systems.

When the excitation frequency is equal to the natural frequency of the sdof $\omega = \omega_n$, the reaction force becomes

$$|F_{base}(\omega_n)| = \left|m\left[1 + \frac{1}{2j\zeta}\right]\ddot{U}(\omega_n)\right| \approx M_{\text{eff}}Q\ddot{U}(\omega_n). \tag{7.13}$$

We write (7.12) in a dimensionless form

$$\frac{F_{\text{base}}(\omega)}{m\ddot{U}(\omega)} = \left[1 + \left(\frac{\omega}{\omega_n}\right)^2 H(\omega)\right]. \tag{7.14}$$

7.3 Modal Effective Masses of an Mdof System

The undamped (matrix) equations of motion for a free-free elastic body can be written as

$$[M]\{\ddot{x}(t)\} + [K]\{x(t)\} = \{F(t)\}. \tag{7.15}$$

We denote the external or boundary degrees of freedom with the index j and the internal degrees of freedom with the index i. The structure will be excited at the boundary dofs; 3 translations and 3 rotations.

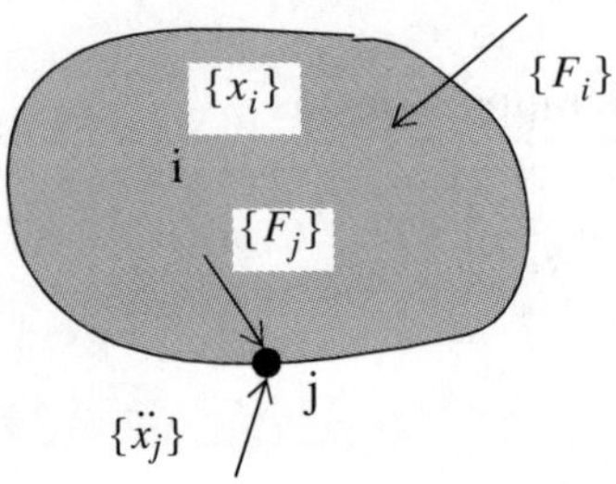

Fig. 7.2. Enforced structure

The number of boundary degrees of freedom is less than or equal to 6. The dofs and forces are illustrated in Fig. 7.2. The matrix (7.15) may be partitioned as follows

$$\begin{bmatrix} M_{ii} & M_{ij} \\ M_{ji} & M_{jj} \end{bmatrix} \begin{Bmatrix} \ddot{x}_i \\ \ddot{x}_j \end{Bmatrix} + \begin{bmatrix} K_{ii} & K_{ij} \\ K_{ji} & K_{jj} \end{bmatrix} \begin{Bmatrix} x_i \\ x_j \end{Bmatrix} = \begin{Bmatrix} F_i \\ F_j \end{Bmatrix}. \tag{7.16}$$

In [Craig 68] it is proposed to depict the displacement vector $\{x(t)\}$ on a basis of 6 rigid-body modes $[\Phi_r]$ with $\{x_j\} = [I]$ and elastic mode shapes $[\Phi_p]$ with fixed external degrees of freedom $\{x_j\} = \{0\}$ calculated from the eigenvalue problem $([K_{ii}] - \langle\lambda_p\rangle[M_{ii}])[\Phi_{ii}] = [0]$. We can express $\{x\}$ as

$$\{x\} = [\Phi_r]\{x_j\} + [\Phi_p]\{\eta_p\} = [\Phi_r, \Phi_p]\begin{Bmatrix} x_j \\ \eta_p \end{Bmatrix} = [\Psi]\{X\}. \tag{7.17}$$

The static modes can be obtained, assuming zero inertia effects, and $\{F_i\} = \{0\}$, and prescribe successively a unit displacement for the 6 boundary dofs, thus $\{x_j\} = [I]$. So we may write (7.16) as follows

$$\begin{bmatrix} K_{ii} & K_{ij} \\ K_{ji} & K_{jj} \end{bmatrix} \begin{Bmatrix} x_i \\ x_j \end{Bmatrix} = \begin{Bmatrix} 0 \\ 0 \end{Bmatrix}. \tag{7.18}$$

Enforced displacement $\{x_j\}$ will not introduce reaction forces in boundary dofs.
From the first equation of (7.18) we find for $\{x_i\}$

$$[K_{ii}]\{x_i\} + [K_{ij}][x_j] = 0, \tag{7.19}$$

hence

$$\{x_i\} = -[K_{ii}]^{-1}[K_{ij}]\{x_j\}, \tag{7.20}$$

and therefore

$$[\Phi_{ij}] = -[K_{ii}]^{-1}[K_{ij}][I] = -[K_{ii}]^{-1}[K_{ij}]. \tag{7.21}$$

The static transformation now becomes

$$\{x\} = \begin{Bmatrix} x_i \\ x_j \end{Bmatrix} = \begin{bmatrix} \Phi_{ij} \\ I \end{bmatrix}\{x_j\} = [\Phi_r]\{x_j\}. \tag{7.22}$$

Using (7.18) it follows that

$$[K][\Phi_r] = \{0\}. \tag{7.23}$$

Assuming fixed external degrees of freedom $\{x_j\} = \{0\}$ and also assuming harmonic motions $x(t) = X(\omega)e^{j\omega t}$ the eigenvalue problem can be stated as

$$([K_{ii}] - \lambda_{k,p}[M_{ii}])\{X(\lambda_{k,p})\} = \{0\}, \tag{7.24}$$

or more generally as

$$([K_{ii}] - \langle\lambda_k\rangle[M_{ii}])[\Phi_{ip}] = \{0\}, \tag{7.25}$$

with

- λ_k the eigenvalue associated with the mode shape $\{\phi_{ip,k}\}$

The internal degrees of freedom $\{x_i\}$ will be projected on the set of orthogonal mode shapes (modal matrix) $[\Phi_{ip}]$, thus

$$\{x_i\} = [\Phi_{ip}]\{\eta_p\}. \tag{7.26}$$

The modal transformation becomes

$$\{x\} = \begin{Bmatrix} x_i \\ x_j \end{Bmatrix} = \begin{bmatrix} \Phi_{ip} \\ 0 \end{bmatrix} \{\eta_p\} = [\Phi_p]\{\eta_p\}. \tag{7.27}$$

The Craig–Bampton (CB) transformation matrix $[\Psi]$ is

$$\{x\} = [\Phi_r, \Phi_p] \begin{bmatrix} x_j \\ \eta_p \end{bmatrix} = [\Psi]\{X\}, \tag{7.28}$$

with

- $[\Phi_r]$ the rigid body modes.
- $[\Phi_p]$ the modal matrix.
- $\{x_j\}$ the external or boundary degrees of freedom $(j \le 6)$.
- $\{\eta_p\}$ the generalised coordinates. In general, the number of generalised coordinates p is much less than the total number of degrees of freedom $n = i + j$, $p \ll n$.

The CB transformation (7.28) will be substituted into (7.15) assuming equal potential and kinetic energies, hence

$$[\Psi]^T[M][\Psi]\{\ddot{X}\} + [\Psi]^T[K][\Psi]\{X\} = [\Psi]^T\{F(t)\} = \{f(t)\}, \tag{7.29}$$

further elaborated we find

$$\begin{bmatrix} M_{rr} & M_{jp} \\ M_{pj} & \langle m_p \rangle \end{bmatrix} \begin{Bmatrix} \ddot{x}_j \\ \ddot{\eta}_p \end{Bmatrix} + \begin{bmatrix} \tilde{K}_{jj} & K_{jp} \\ K_{pj} & \langle k_p \rangle \end{bmatrix} \begin{Bmatrix} x_j \\ \eta_p \end{Bmatrix} = \begin{bmatrix} \Phi_{ij} & \Phi_p \\ I & 0 \end{bmatrix}^T \begin{Bmatrix} F_i \\ F_j \end{Bmatrix}, \tag{7.30}$$

with

- $[M_{rr}]$ the 6x6 rigid body mass matrix with respect to the boundary dofs
- $[\tilde{K}_{jj}]$ the Guyan reduced stiffness matrix (j-set)
- $\langle m_p \rangle$ the diagonal matrix of generalised masses, $\langle m_p \rangle = [\Phi_p]^T[M][\Phi_p]$
- $\langle k_p \rangle$ the diagonal matrix of generalised stiffnesses,

  $$\langle k_p \rangle = [\Phi_p]^T K[\Phi_p] = \langle \lambda_p \rangle \langle m_p \rangle = \langle \omega_p^2 \rangle \langle m_p \rangle$$
- $[K_{ip}] = [\Phi_{ij}]^T[K_{ii}][\Phi_p] + [K_{ji}][\Phi_p] = (-[K_{ij}]^T[K_{ii}]^{-1}[K_{ii}] + [K_{ji}])[\Phi_p] = [0]$
- $[K_{pi}] = [K_{ip}]^T = [0]$
- $[\tilde{K}_{jj}] = [\Phi_r]^T[K][\Phi_r] = [0]$

Thus (7.30) becomes

$$\begin{bmatrix} M_{rr} & L^T \\ L & \langle m_p \rangle \end{bmatrix} \begin{Bmatrix} \ddot{x}_j \\ \ddot{\eta}_p \end{Bmatrix} + \begin{bmatrix} 0 & 0 \\ 0 & \langle m_p \lambda_p \rangle \end{bmatrix} \begin{Bmatrix} x_j \\ \eta_p \end{Bmatrix} = \begin{bmatrix} \Phi_{ij} & \Phi_p \\ I & 0 \end{bmatrix}^T \begin{Bmatrix} 0 \\ F_j \end{Bmatrix} = \begin{Bmatrix} F_j \\ 0 \end{Bmatrix}, \quad (7.31)$$

with

- $[M_{jp}] = [\Phi_r]^T [M][\Phi_p] = [L]^T$, $[L]^T$ is the matrix with the modal participation factors, $L_{kl} = \{\phi_{r,k}\}^T [M]\{\phi_{p,l}\}$, $k = 1,2,...,6$, $l = 1,2,...,p$. The matrix of modal participation factors couples the rigid-body modes $[\Phi_r]$ with the elastic modes $[\Phi_p]$.
- $\{F_i\} = \{0\}$ No internal loads are applied.

Introducing the modal damping ratios ζ_p we can rewrite (7.31) as follows

$$\begin{bmatrix} M_{rr} & L^T \\ L & \langle m_p \rangle \end{bmatrix} \begin{Bmatrix} \ddot{x}_j \\ \ddot{\eta}_p \end{Bmatrix} + \begin{bmatrix} 0 & 0 \\ 0 & \langle 2 m_p \zeta_p \omega_p \rangle \end{bmatrix} \begin{Bmatrix} \dot{x}_j \\ \dot{\eta}_p \end{Bmatrix} + \begin{bmatrix} 0 & 0 \\ 0 & \langle m_p \lambda_p \rangle \end{bmatrix} \begin{Bmatrix} x_j \\ \eta_p \end{Bmatrix} = \begin{Bmatrix} F_j \\ 0 \end{Bmatrix} \quad (7.32)$$

Equation (7.32) can be divided into two equations

$$[M_{rr}]\{\ddot{x}_j\} + [L]^T \{\ddot{\eta}_p\} = \{0\}, \quad (7.33)$$

and

$$[L]\{\ddot{x}_j\} + \langle m_p \rangle \{\ddot{\eta}_p\} + \langle 2 m_p \zeta_p \omega_p \rangle \{\dot{\eta}_p\} + \langle m_p \lambda_p \rangle \{\eta_p\} = \{0\}. \quad (7.34)$$

Equations(7.33) and (7.34), when transformed in the frequency domain, give

$$[M_{rr}]\{\ddot{X}_j\} + [L]^T \{\ddot{\Pi}_p\} = \{F_j\}, \quad (7.35)$$

and

$$[L]\{\ddot{X}_j\} + \langle m_p \rangle \{\ddot{\Pi}_p\} + \langle 2 m_p \zeta_p \omega_p \rangle \{\dot{\Pi}_p\} + \langle m_p \lambda_p \rangle \{\Pi_p\} = \{0\}, \quad (7.36)$$

with

- $x(t) = X e^{j\omega t}$, $\ddot{X} = -\omega^2 X$
- $\eta(t) = \Pi e^{j\omega t}$, $\dot{\Pi} = j\omega\Pi$ and $\ddot{\Pi} = -\omega^2 \Pi$

- $F(t) = \hat{F}e^{j\omega t}$

With (7.36) we express $\{\Pi_p\}$ in $\{X_j\}$

$$m_p[-\omega^2 + 2j\zeta_p\omega_p\omega + \omega_p^2]\Pi_p = -[L_p]\{\ddot{X}_j\}, \tag{7.37}$$

with

- $[L_k] = \{\phi_{p,k}\}^T[M][\Phi_r]$ 1x6 vector with modal participation factors
- $L_{kj} = \{\phi_{p,k}\}^T[M]\{\Phi_{r,j}\}$ participation factor with $k = 1,2,...,p$ and $j = 1,2,...,6$

Thus (7.37) becomes

$$\Pi_k = -\frac{[L_k]\{\ddot{X}_j\}}{m_k\omega_k^2}\left[\frac{1}{1-\left(\frac{\omega}{\omega_k}\right)^2 + 2j\zeta_k\frac{\omega}{\omega_k}}\right] = -\frac{[L_k]\{\ddot{X}_j\}}{m_k\omega_k^2}H_k\left(\frac{\omega}{\omega_k}\right). \tag{7.38}$$

Equation (7.38) will be substituted into (7.35) giving

$$[M_{rr}]\{\ddot{X}_j\} + [L_1^T, ..., L_p^T]\left\{\left(\frac{\omega}{\omega_k}\right)^2\frac{[L_k]}{m_k}H_k\left(\frac{\omega}{\omega_k}\right)\right\}\{\ddot{X}_j\} = \{\hat{F}_j\}, k=1,2,...,p \tag{7.39}$$

$$\left[[M_{rr}] + \sum_{k=1}^{p}\frac{[L_k]^T[L_k]}{m_k}\left\{\left(\frac{\omega}{\omega_k}\right)^2 H_k\left(\frac{\omega}{\omega_k}\right)\right\}\right]\{\ddot{X}_j\} = \{\hat{F}_j\}. \tag{7.40}$$

We can prove that

$$[M_{rr}] = \sum_{k=1}^{p}\frac{[L_k]^T[L_k]}{m_k}, \tag{7.41}$$

because

$$[M_{rr}] = [\Phi_r]^T[M][\Phi_p]([\Phi_{ip}]^T[M_{ii}][\Phi_{ip}])^{-1}[\Phi_p]^T[M][\Phi_r] = [\Phi_r]^T[M][\Phi_r], \tag{7.42}$$

or

$$[M_{rr}] = [\Phi_{ij}^T, I][M]\begin{bmatrix}\Phi_{ip}\\0\end{bmatrix}([\Phi_{ip}]^T[M_{ii}][\Phi_{ip}])^{-1}[\Phi_{ip}^T, 0][M]\begin{bmatrix}\Phi_{ij}\\0\end{bmatrix} = [\Phi_r]^T[M][\Phi_r].$$

Assuming the inverse of $[\Phi_{ip}]$ exists. We define the modal effective mass $[M_{\text{eff},k}]$ as follows

$$[M_{\text{eff},k}] = \frac{[L_k]^T[L_k]}{m_k}, \tag{7.43}$$

with

- $[L_k] = \{\phi_{p,k}\}^T[M][\Phi_r]$
- $m_k = \{\phi_{p,k}\}^T[M]\{\phi_{p,k}\}$

The summation over all modal effective masses $[M_{\text{eff},k}]$ will result in the rigid- body mass matrix $[M_{rr}]$ with respect to $\{x_j\}$. (7.41) becomes

$$[M_{rr}]= \sum_{k=1}^{p=i}[M_{\text{eff},k}], \tag{7.44}$$

Therefore (7.40) can be written

$$\left[\sum_{k=1}^{p}[M_{\text{eff},k}]\left\{1+\left(\frac{\omega}{\omega_k}\right)^2 H_k\left(\frac{\omega}{\omega_k}\right)\right\}\right]\{\ddot{X}_j\} = \{\hat{F}_j\} \tag{7.45}$$

Equation (7.45) can be decomposed into modal reaction forces $\{F_{\text{base},k}\}$

$$\sum_{k=1}^{p}\{F_{\text{base},k}\} = \{\hat{F}_j\}, \tag{7.46}$$

with

$$\{F_{\text{base},k}\} = [M_{\text{eff},k}]\left\{1+\left(\frac{\omega}{\omega_k}\right)^2 H_k\left(\frac{\omega}{\omega_k}\right)\right\}\{\ddot{X}_j\}. \tag{7.47}$$

Equation (7.47) is very similar to (7.12).

For the dynamic system, as illustrated in Fig. 7.3, the effective masses $[M_{\text{eff},k}]$ will be calculated. The parameters m (kg) and k (N/m) are, respectively, $m = 1$ and $k = 100000$. The set of internal dofs is $\{x_i\} = \{x_1, x_2, x_3, x_4, x_5, x_6, x_7\}^T$ and the boundary dof is $\{x_j\} = \{x_8\}$.

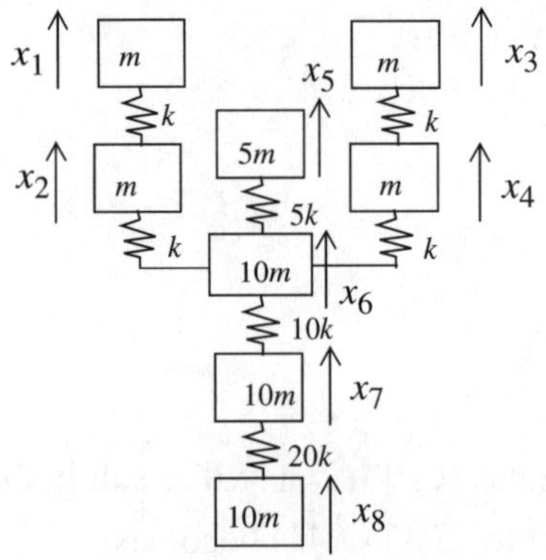

Fig. 7.3. 8 dofs dynamic system

The following procedure will be followed

1. $\{x_j\}$ will be assigned, $x_j = x_8$

2. Calculate the rigid-body modes $[\Phi_r] = \begin{bmatrix} -[K_{ii}]^{-1}[K_{ij}] \\ I \end{bmatrix}$, $x_j = x_8 = 1$

3. Fix the dofs $\{x_j\}$, $x_j = x_8 = 0$

4. Calculate the natural frequencies and associated mode shapes $[\Phi_p]$, $x_j = x_8 = 0$

5. Assemble $[\Psi] = [\Phi_r \Phi_p]$

6. Calculate $[\Psi]^T[M][\Psi]$ and $[\Psi]^T[K][\Psi]$

7. Calculate the modal effective masses per mode $[M_{\text{em},k}] = \dfrac{[L_k]^T[L_k]}{m_k}$

8. Calculate the summation of modal effective masses $[M_{rr}] = \sum_{k=1}^{p} [M_{\text{em},k}]$

The rigid-body mode $\{\phi_r\}$, is with respect to $x_8 = x_j = 1$, and the natural frequencies and associated mode shapes are with respect to $x_8 = x_j = 0$ are

$$\{f_n\} = \begin{Bmatrix} 24.4522 \\ 31.1052 \\ 36.6716 \\ 64.4657 \\ 81.4344 \\ 82.0637 \\ 95.9164 \end{Bmatrix}, \; [\phi_r] = \begin{Bmatrix} 1 \\ 1 \\ 1 \\ 1 \\ 1 \\ 1 \\ 1 \\ 1 \end{Bmatrix}$$

$$[\Phi_p] = \begin{bmatrix} 0.5347 & -0.6015 & -0.5781 & -0.3363 & -0.3717 & 0.3630 & -0.1343 \\ 0.4075 & -0.3717 & -0.2712 & 0.2155 & 0.6015 & -0.6022 & 0.3534 \\ 0.5347 & 0.6015 & -0.5781 & -0.3363 & 0.3717 & 0.3630 & -0.1343 \\ 0.4075 & 0.3717 & -0.2712 & 0.2155 & -0.6015 & -0.6022 & 0.3534 \\ 0.2407 & 0 & 0.3831 & -0.6458 & 0 & -0.0202 & 0.1681 \\ 0.1835 & 0 & 0.1797 & 0.4137 & 0 & 0.0336 & -0.4425 \\ 0.0664 & 0 & 0.0728 & 0.3044 & 0 & 0.0984 & 0.7001 \\ 0 & 0 & 0 & 0 & 0 & 0 & 0 \end{bmatrix}.$$

The mass matrix $[\Psi]^T[M][\Psi]$ and the stiffness matrix $[\Psi]^T[K][\Psi]$ become

$$[\Psi]^T[M][\Psi] = \begin{bmatrix} 39 & 5.5874 & 0 & 2.7421 & 3.7104 & 0 & 0.7400 & 3.8552 \\ 5.5874 & 1.5746 & 0 & 0 & 0 & 0 & 0 & 0 \\ 0 & 0 & 1.0000 & 0 & 0 & 0 & 0 & 0 \\ 2.7421 & 0 & 0 & 1.9255 & 0 & 0 & 0 & 0 \\ 3.7104 & 0 & 0 & 0 & 5.0429 & 0 & 0 & 0 \\ 0 & 0 & 0 & 0 & 0 & 1.0000 & 0 & 0 \\ 0.7400 & 0 & 0 & 0 & 0 & 0 & 1.0989 & 0 \\ 3.8552 & 0 & 0 & 0 & 0 & 0 & 0 & 7.2863 \end{bmatrix}$$

$$[\Psi]^T[K][\Psi] = 10^6 \begin{bmatrix} 0 & 0 & 0 & 0 & 0 & 0 & 0 & 0 \\ 0 & 0.0374 & 0 & 0 & 0 & 0 & 0 & 0 \\ 0 & 0 & 0.0382 & 0 & 0 & 0 & 0 & 0 \\ 0 & 0 & 0 & 0.1022 & 0 & 0 & 0 & 0 \\ 0 & 0 & 0 & 0 & 0.8274 & 0 & 0 & 0 \\ 0 & 0 & 0 & 0 & 0 & 0.2618 & 0 & 0 \\ 0 & 0 & 0 & 0 & 0 & 0 & 0.2922 & 0 \\ 0 & 0 & 0 & 0 & 0 & 0 & 0 & 2.6464 \end{bmatrix}.$$

The results of the calculations are summarised in Table 7.1.

Table 7.1. Calculation of the modal effective masses

Mode shape #	Natural frequency (Hz)	Modal participation factor $[L_k]^T$	Generalised masses $[m_k]$	Modal effective mass $[M_{\text{em},k}]$ (kg)
1	24.5422	5.5874	1.5746	19.8271
2	31.1052	0.0000	1.0000	0.0000
3	36.6716	2.7421	1.9255	3.9048
4	64.4657	3.7104	5.0429	2.7300
5	81.4344	0.0000	1.0000	0.0000
6	82.0637	0.7400	1.0989	0.4983
7	95.9164	3.8552	7.2863	2.0398
	Total mass (without $m_8 = 10m$)			29.0000

The mass $m_8 = 10m$ (connected to dof x_8) is eliminated because the elastic modes are with respect to x_8.

It appears that the modal effective mass of the first mode shape is already 68.37% of the total mass of 29 kg. The second and the fifth mode shapes have zero modal effective mass. Modes with zero modal effective mass cannot be excited in the case of enforced acceleration.

The absolute value of the normalised base force $\left|\frac{F_{\text{base}}(\omega)}{\ddot{X}(\omega)}\right|$ can be written as

$$\left|\frac{F_{\text{base}}(\omega)}{\ddot{X}(\omega)}\right| = \left|\sum_{k=1}^{7}[M_{em,k}]\left\{1+\left(\frac{\omega}{\omega_k}\right)^2 H_k\left(\frac{\omega}{\omega_k}\right)\right\}\right|,$$

and the calculations are illustrated in Fig. 7.4.

7.4 Problems

7.4.1 Problem 1

The problem is defined in the example Fig. 7.3, however apply a large mass $M_{\text{lm}} = 10^6$ kg at x_8. This method is discussed in [Appel 92].

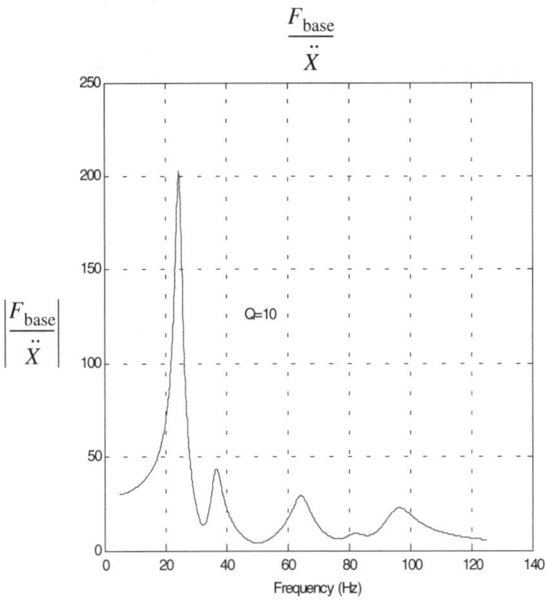

Fig. 7.4. $\left|\frac{F_{base}(f)}{\ddot{X}(f)}\right| = \left|\sum_{k=1}^{7}[M_{em,k}]\left\{1+\left(\frac{f}{f_k}\right)^2 H_k\left(\frac{f}{f_k}\right)\right\}\right|$

Calculate the following parameters:

- The free-free mode shapes $[\phi]$.
- Calculate $\frac{(M_{lm}\phi_{i,8})^2}{\phi_i^T[M]\phi_i}$ per mode "i" and compare the results with the modal effective masses as calculated in the example. What is your conclusion?

7.4.2 Problem 2

A cantilevered beam, as illustrated in Fig. •, has two discrete masses with mass m. The distance between the masses and the clamped interface is L.

We assume a mode shape $\phi(x) = \left(\frac{x}{L}\right)^3$.

Using $\phi(x)$ calculate:

- the natural frequency using Rayleigh's Quotient.
- the generalised mass

- the modal participation factor with respect to clamped position A
- the effective mass

Answers: $\omega = 3.44\sqrt{\dfrac{EI}{ml^3}}$, $\dfrac{65}{64}m$, $\dfrac{9}{8}m$ and $\dfrac{81}{65}m$.

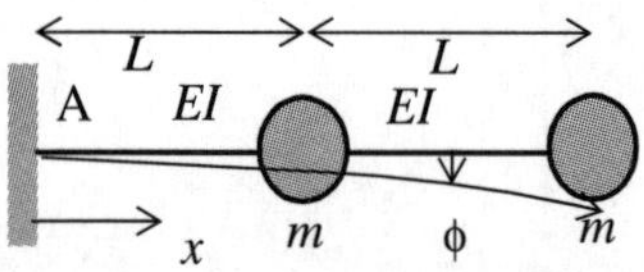

Fig. 7.5. Cantilevered beam

8 Response Analysis

8.1 Introduction

The analysis of responses of a mdof linear dynamic system due to either dynamic forces or to enforced motions; displacements, velocities and acceleration will be discussed in this chapter. The general equations of motions are set up and a partitioning between internal and boundary dofs had been made in order to solve the internal dofs because boundary motions were applied in combination with forces. When solving the equations of motion a distinction has been made between relative motions, motions with respect to the base, and absolute motions. Also a distinction had been made between redundant and non redundant boundaries. The equations are applicable to solve the responses both in the time and the frequency domain.

In general, the equation of motions, in particular the internal dofs, are solved using the mode displacement method (MDM), because the full damping characteristics, a full damping matrix, are not readily available. The theory will be illustrated with an example. The chapter will end with a set of problems to be solved by the reader.

8.2 Forces and Enforced Acceleration

Forces and enforced acceleration will be discussed. An example of enforced motion is the acceleration at the interface between the satellite (spacecraft) and the launch vehicle. The enforced acceleration of a structure can be treated in three different methods:

- Relative motions: The absolute motion $\{x(t)\}$ of the dynamic is separated into a relative motion $\{z(t)\}$ with respect to the base and the motion at the base $\{x_b(t)\}$. The absolute motion $\{x(t)\}$ is the summa-

tion of the motion at the base and the relative motion $\{x(t)\} = \{z(t)\} + \{x_b(t)\}$. The dynamic system has fixed free boundary conditions, either determinate or nondeterminate.
- Absolute motions: The absolute motions $\{x(t)\}$ of the dynamic system are calculated in a direct manner as a result of forces $\{F(t)\}$ and enforced motion $\{x_b(t)\}$. The dynamic system has fixed free boundary conditions.
- Large-mass approach: The dynamic system is a free-free structure, however, attached to the interface is a very large mass to introduce as a force the enforced motion $\{x_b(t)\}$ to calculate the absolute motions $\{x(t)\}$.

The damped equations of motion of a mdof dynamic system are

$$[M]\{\ddot{x}(t)\} + [C]\{\dot{x}(t)\} + [K]\{x(t)\} = \{F(t)\}. \tag{8.1}$$

8.2.1 Relative Motions

With the relative-motion approach the absolute motion $\{x(t)\}$ is build up from the relative motion $\{z(t)\}$ and the enforced motion $\{x_b(t)\}$. The relative motion $\{z(t)\}$ is with respect to the fixed base. The absolute displacements $\{x(t)\}$ may be partitioned in a set of internal dofs $\{x_i\}$ and a set of boundary dofs $\{x_b\}$. The boundary dofs are equal to the enforced displacements $\{x_b(t)\}$. The absolute displacements $\{x(t)\}$ become

$$\{x\} = \begin{Bmatrix} x_i \\ x_b \end{Bmatrix}. \tag{8.2}$$

The same expression exists for the velocities $\{\dot{x}\}$

$$\{\dot{x}\} = \begin{Bmatrix} \dot{x}_i \\ \dot{x}_b \end{Bmatrix}, \tag{8.3}$$

and for the acceleration $\{\ddot{x}\}$

$$\{\ddot{x}\} = \begin{Bmatrix} \ddot{x}_i \\ \ddot{x}_b \end{Bmatrix}. \tag{8.4}$$

The static relation between the internal dofs $\{x_i\}$ and the boundary dofs $\{x_b\}$ can be calculated from

$$\begin{bmatrix} K_{ii} & K_{ib} \\ K_{bi} & K_{bb} \end{bmatrix} \begin{Bmatrix} \bar{x}_i \\ x_b \end{Bmatrix} = \begin{Bmatrix} 0 \\ \bar{F}_b \end{Bmatrix}. \tag{8.5}$$

The stiffness matrix $[K]$ is also partitioned with respect to the internal dofs and the boundary dofs. The reaction force $\{\bar{F}_b\}$ is due to the enforced displacements $\{x_b\}$. From the first equation of (8.5) we obtain the static internal displacements $\{x_i\}$ due to the boundary static displacements $\{x_b\}$

$$\{\bar{x}_i\} = -[K_{ii}]^{-1}[K_{ib}]\{x_b\} = [G_{ib}]\{x_b\}. \tag{8.6}$$

The absolute displacement $\{x\}$ can be written as a summation of the relative displacements $\{z\}$ and the static displacements caused by the boundary displacements.

$$\{x\} = \begin{Bmatrix} x_i \\ x_b \end{Bmatrix} = \begin{Bmatrix} z \\ 0 \end{Bmatrix} + \begin{Bmatrix} \bar{x}_i \\ x_b \end{Bmatrix} = \begin{Bmatrix} z \\ 0 \end{Bmatrix} + \begin{bmatrix} G_{ib} \\ I \end{bmatrix} \{x_b\}. \tag{8.7}$$

The matrix $[\Phi_c] = \begin{bmatrix} G_{ib} \\ I \end{bmatrix}$ consists of the so-called "contraint modes", [Craig 68].

We will substitute (8.2), (8.3) and (8.4) into (8.1) and also partitioning the force vector $\{F\}$ in internal and boundary loads.

$$\begin{bmatrix} M_{ii} & M_{ib} \\ M_{bi} & M_{bb} \end{bmatrix} \begin{Bmatrix} \ddot{x}_i \\ \ddot{x}_b \end{Bmatrix} + \begin{bmatrix} C_{ii} & C_{ib} \\ C_{bi} & C_{bb} \end{bmatrix} \begin{Bmatrix} \dot{x}_i \\ \dot{x}_b \end{Bmatrix} + \begin{bmatrix} K_{ii} & K_{ib} \\ K_{bi} & K_{bb} \end{bmatrix} \begin{Bmatrix} x_i \\ x_b \end{Bmatrix} = \begin{Bmatrix} F_i \\ F_b \end{Bmatrix}. \tag{8.8}$$

We will depict the relative motion $\{z\}$ on the modal base $[\Phi_i]$ as follows

$$\{z\} = [\Phi_i]\{\eta_i\} \tag{8.9}$$

with

- $[\Phi_i]$ the modal base obtained from the eigenvalue problem $([K_{ii}] - \omega_k^2[M_{ii}])\{\phi_{ik}\} = \{0\}$, with $[\Phi_i] = [\phi_{i1}, \phi_{i2}, ...]$ and ω_k^2 the k-th eigenvalue.
- $\{\eta_i\}$ are the generalised coordinates

The mode shapes $[\Phi_i]$ obey the orthogonality relations with respect to the mass matrix $[M_{ii}]$ and the stiffness matrix $[K_{ii}]$

$$\left.\begin{matrix} [\Phi_i]^T[M_{ii}][\Phi_i] \\ [\Phi_i]^T[K_{ii}][\Phi_i] \end{matrix}\right\} = \left\{\begin{matrix} \langle m_{ik}\rangle \\ \langle \omega_k^2 m_{ik}\rangle \end{matrix}\right. . \tag{8.10}$$

After the base transformation we assume a diagonal matrix of modal damping ratios, hence

$$[\Phi_i]^T[C_{ii}][\Phi_i] = \langle 2\zeta_k\omega_k m_{ik}\rangle , \tag{8.11}$$

with ζ_k the modal damping ratio associated with mode $\{\phi_{i,k}\}$.
Equation (8.7) can be written

$$\{x\} = \begin{Bmatrix} x_i \\ x_b \end{Bmatrix} = \begin{Bmatrix} z \\ 0 \end{Bmatrix} + \begin{Bmatrix} \bar{x}_i \\ x_b \end{Bmatrix} = \begin{bmatrix} \Phi_i & G_{ib} \\ 0 & I \end{bmatrix}\begin{Bmatrix} \eta_i \\ x_b \end{Bmatrix} = [\Psi]\{X\} . \tag{8.12}$$

We will rewrite (8.1) as follows

$$[\Psi]^T[M][\Psi]\{\ddot{X}\} + [\Psi]^T[C][\Psi]\{\dot{X}\} + [\Psi]^T[M][\Psi]\{X\} = [\Psi]^T\{F(t)\} , \tag{8.13}$$

with

- $[\Psi]^T[M][\Psi] = \begin{bmatrix} \Phi_i^T M_{ii}\Phi_i & \Phi_i^T M_{ii}G_{ib} + \Phi_i^T M_{ib} \\ G_{ib}^T M_{ii}\Phi_i + M_{bi}\Phi_i & G_{ib}^T M_{ii}G_{ib} + G_{ib}^T M_{ib} + M_{bi}G_{ib} + M_{bb} \end{bmatrix}$
- $[\Psi]^T[C][\Psi] = \begin{bmatrix} \Phi_i^T C_{ii}\Phi_i & \Phi_i^T C_{ii}G_{ib} + \Phi_i^T C_{ib} \\ G_{ib}^T C_{ii}\Phi_i + C_{bi}\Phi_i & G_{ib}^T C_{ii}G_{ib} + G_{ib}^T C_{ib} + C_{bi}G_{ib} + C_{bb} \end{bmatrix}$
- $[\Psi]^T[K][\Psi] = \begin{bmatrix} \Phi_i^T K_{ii}\Phi_i & \Phi_i^T K_{ii}G_{ib} + \Phi_i^T K_{ib} \\ G_{ib}^T K_{ii}\Phi_i + K_{bi}\Phi_i & G_{ib}^T K_{ii}G_{ib} + G_{ib}^T K_{ib} + K_{bi}G_{ib} + K_{bb} \end{bmatrix}$
- $[\Phi_i^T K_{ii}G_{ib} + \Phi_i^T K_{ib}] = [0]$

and, furthermore we may write

- $[\overline{M}_{bb}] = [G_{ib}^T M_{ii} G_{ib} + G_{ib}^T M_{ib} + M_{bi} G_{ib} + M_{bb}]$
- $[\overline{M}_{ib}] = [\Phi_i^T M_{ii} G_{ib} + \Phi_i^T M_{ib}]$
- $[\overline{M}_{bi}] = [G_{ib}^T M_{ii} \Phi_i + M_{bi} \Phi_i] = [\Phi_i^T M_{ii} G_{ib} + \Phi_i^T M_{ib}]^T$
- $[\overline{K}_{bb}] = [G_{ib}^T K_{ii} G_{ib} + G_{ib}^T K_{ib} + K_{bi} G_{ib} + K_{bb}]$
- $[\overline{C}_{ib}] = [\Phi_i^T C_{ii} G_{ib} + \Phi_i^T C_{ib}]$
- $[\overline{C}_{bi}] = [G_{ib}^T C_{ii} \Phi_i + C_{bi} \Phi_i] = [\Phi_i^T C_{ii} G_{ib} + \Phi_i^T C_{ib}]^T$
- $[\overline{C}_{bb}] = [G_{ib}^T C_{ii} G_{ib} + G_{ib}^T C_{ib} + C_{bi} G_{ib} + C_{bb}]$
- $[\overline{K}_{ib}] = [\Phi_i^T K_{ii} G_{ib} + \Phi_i^T K_{ib}] = [0]$
- $[K_{bi}] = [G_{ib}^T K_{ii} \Phi_i + K_{bi} \Phi_i] = [\Phi_i^T K_{ii} G_{ib} + \Phi_i^T K_{ib}]^T = [0]$
- $[\overline{K}_{bb}] = [G_{ib}^T K_{ii} G_{ib} + G_{ib}^T K_{ib} + K_{bi} G_{ib} + K_{bb}]$

Equation (8.13) becomes, applying the orthogonality relations of modes with respect to the mass and stiffness matrix

$$\begin{bmatrix} \langle m_{ik} \rangle & \overline{M}_{ib} \\ \overline{M}_{bi} & \overline{M}_{bb} \end{bmatrix} \begin{Bmatrix} \ddot{\eta}_i \\ \ddot{x}_b \end{Bmatrix} + \begin{bmatrix} \langle 2\zeta_k \omega_k m_{ik} \rangle & \overline{C}_{ib} \\ \overline{C}_{bi} & \overline{C}_{bb} \end{bmatrix} \begin{Bmatrix} \dot{\eta}_i \\ \dot{x}_b \end{Bmatrix} + \begin{bmatrix} \langle \omega_k^2 m_{ik} \rangle & 0 \\ 0 & \overline{K}_{bb} \end{bmatrix} \begin{Bmatrix} \eta_i \\ x_b \end{Bmatrix} = [\Psi]^T \begin{Bmatrix} F_i \\ F_b \end{Bmatrix} \tag{8.14}$$

and the generalised forces become

$$[\Psi]^T \begin{Bmatrix} F_i \\ F_b \end{Bmatrix} = \begin{bmatrix} \Phi_i & G_{ib} \\ 0 & I \end{bmatrix}^T \begin{Bmatrix} F_i \\ F_b \end{Bmatrix} = \begin{Bmatrix} \Phi_i^T F_i \\ G_{ib}^T F_i + F_b \end{Bmatrix}. \tag{8.15}$$

Thus, finally, the first part of (8.14) becomes

$$\langle m_{ik} \rangle \{\ddot{\eta}_i(t)\} + \langle 2\zeta_k \omega_k m_{ik} \rangle \{\dot{\eta}_i(t)\} + \langle \omega_k^2 m_{ik} \rangle \{\eta_i(t)\}$$

$$= \{\Phi_i\}^T \{F_i(t)\} - [\overline{M}_{ib}]\{\ddot{x}_b(t)\} - [\overline{C}_{ib}]\{\dot{x}_b(t)\} \; . \tag{8.16}$$

Equation (8.16) can be either solved analytically or using numeric schemes like the Wilson-θ, the Newmark-beta methods.

If the enforced acceleration $\{\ddot{x}_b\}$ is specified the velocity $\{\dot{x}_b\}$ must be obtained by integration of the acceleration $\{\ddot{x}_b\}$

$$\{\dot{x}_b(t)\} = \{\dot{x}_b(t_0)\} + \int_{t_0}^{t} \{\ddot{x}_b(t)\}dt\,. \tag{8.17}$$

With a time increment Δt we can use the trapezoidal approximation of the integral [Schwarz 89] in (8.17)

$$\{\dot{x}_b(t_n)\} = \{\dot{x}_b(t_0)\} + \Delta t\left(\frac{1}{2}\{\ddot{x}_b(t_0)\} + \sum_{j=1}^{n-1}\{\ddot{x}_b(t_j)\} + \frac{1}{2}\{\ddot{x}_b(t_n)\}\right), \tag{8.18}$$

with

- $t_j = t_0 + j\Delta t$
- $t_n = t_0 + n\Delta t = t$
- $\{\dot{x}_b(t_{n+1})\} = \{\dot{x}_b(t_n)\} + \frac{\Delta t}{2}(\{\ddot{x}_b(t_n)\} + \{\ddot{x}_b(t_{n+1})\})$

In the frequency domain we have

- $\{\eta_i\} = \{\Pi_i(\omega)\}e^{j\omega t}$, $\{\dot{\Pi}_i\} = j\omega\{\Pi(\omega)\}$ and $\{\ddot{\Pi}_i\} = -\omega^2\{\Pi(\omega)\}$
- $\{x_b\} = \{X_b(\omega)\}e^{j\omega t}$, $\{\ddot{X}_b\} = -\omega^2\{X_b(\omega)\}$ or $\{\ddot{x}_b\} = \{\ddot{X}_b(\omega)\}e^{j\omega t}$,
  $\{\dot{X}_b\} = \dfrac{\{\ddot{X}_b(\omega)\}}{j\omega} = -j\dfrac{\{\ddot{X}_b(\omega)\}}{\omega}$ and $\{X_b\} = \dfrac{-\{\ddot{X}_b(\omega)\}}{\omega^2}$
- $\{F_i\} = \{F_i(\omega)\}e^{j\omega t}$

If (8.16) is transformed in the frequency domain we obtain

$$[\langle(\omega_k^2-\omega^2)m_{ik}\rangle + \langle 2j\zeta_k\omega_k\omega m_{ik}\rangle]\{\Pi_i(\omega)\}$$

$$= \Phi_i^T F_i(\omega) - \left([\bar{M}_{ib}] - \frac{j}{\omega}[\bar{C}_{ib}]\right)\{\ddot{X}_b(\omega)\}\ . \tag{8.19}$$

A dynamic system is shown in Fig. 8.1. The number of dofs is 10; x_1, φ_1, x_2, φ_2, x_3, φ_3, x_4, φ_4, x_5 and x_6. The springs with a spring stiffness k_5 and k_6 are coupled between the displacements $x_3 - x_5$ and $x_5 - x_6$ respectively. The set of boundary dofs $\{x_b\} = \left\{ x_1\ x_2\ x_4 \right\}^T$ and the internal dofs are $\{x_i\} = \left\{ \varphi_1\ \varphi_2\ x_3\ \varphi_3\ \varphi_4\ x_5\ x_6 \right\}^T$.

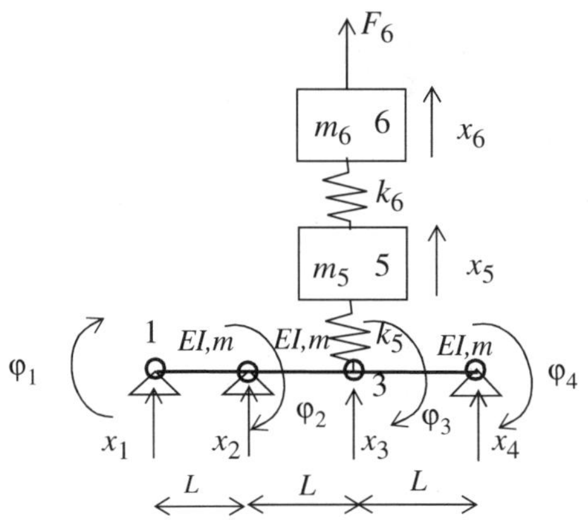

Fig. 8.1. Dynamic system enforced with loads and acceleration

The enforced accelerations are $\{\ddot{x}_b\} = \left\{ \ddot{x}_1 \ \ddot{x}_2 \ \ddot{x}_4 \right\}^T$.

A dynamic load F_6 is applied to dof x_6.

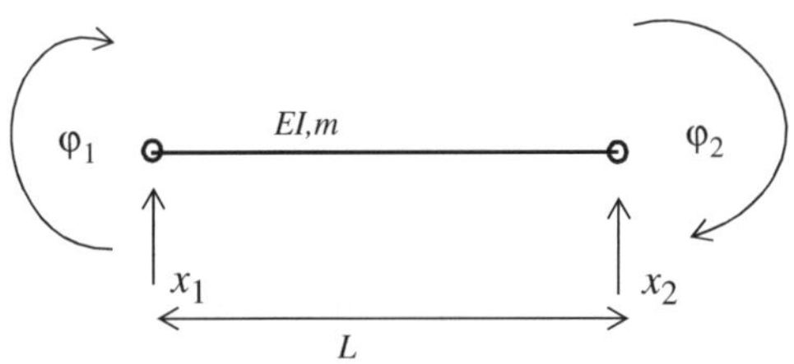

Fig. 8.2. Bar element

The stiffness matrix and mass matrix of a bar shown in Fig. 8.2 are from [Craig 81, Cook 89].

$$[K_{\text{bar}}] = \frac{EI}{L^3}\begin{bmatrix} 12 & 6L & -12 & 6L \\ 6L & 4L^2 & -6L & 2L^2 \\ -12 & -6L & 12 & -6L \\ 6L & 2L^2 & -6L & 4L^2 \end{bmatrix}, \ [M_{\text{bar}}] = \frac{mL}{420}\begin{bmatrix} 156 & 22L & 54 & -13L \\ 22L & 4L^2 & 13L & -3L^2 \\ 54 & 13L & 156 & -22L \\ -13L & -3L^2 & -22L & 4L^2 \end{bmatrix}.$$

The total stiffness matrix $[K]$ and the total mass matrix $[M]$ of the mdof linear dynamic system, as illustrated in Fig. 8.1, can now be set up. The total stiffness matrix $[K]$ is

$$[K] = \frac{EI}{L^3}\begin{bmatrix} 12 & 6L & -12 & 6L & 0 & 0 & 0 & 0 & 0 & 0 \\ 6L & 4L^2 & -6L & 2L^2 & 0 & 0 & 0 & 0 & 0 & 0 \\ -12 & -6L & 24 & 0 & -12 & 6L & 0 & 0 & 0 & 0 \\ 6L & 2L^2 & 0 & 8L^2 & -6L & 2L^2 & 0 & 0 & 0 & 0 \\ 0 & 0 & -12 & -6L & 24+\dfrac{k_5L^3}{EI} & 0 & -12 & 6L & -\dfrac{k_5L^3}{EI} & 0 \\ 0 & 0 & 6L & 2L^2 & 0 & 8L^2 & -6L & 2L^2 & 0 & 0 \\ 0 & 0 & 0 & 0 & -12 & -6L & 12 & -6L & 0 & 0 \\ 0 & 0 & 0 & 0 & 6L & 2L^2 & -6L & 4L^2 & 0 & 0 \\ 0 & 0 & 0 & 0 & -\dfrac{k_5L^3}{EI} & 0 & 0 & 0 & \dfrac{(k_5+k_6)L^3}{EI} & -\dfrac{k_6L^3}{EI} \\ 0 & 0 & 0 & 0 & 0 & 0 & 0 & 0 & -\dfrac{k_6L^3}{EI} & \dfrac{k_6L^3}{EI} \end{bmatrix},$$

and the total mass matrix $[M]$ becomes

$$[M] = \frac{ml}{420}\begin{bmatrix} 156 & 22L & 54 & -13L & 0 & 0 & 0 & 0 & 0 & 0 \\ 22L & 4L^2 & 13L & -3L^2 & 0 & 0 & 0 & 0 & 0 & 0 \\ 54 & -6L & 312 & 0 & 54 & -13L & 0 & 0 & 0 & 0 \\ -13L & -3L^2 & 0 & 8L^2 & 13L & -3L^2 & 0 & 0 & 0 & 0 \\ 0 & 0 & 54 & -6L & 312 & 0 & 54 & -13L & 0 & 0 \\ 0 & 0 & -13L & -3L^2 & 0 & 8L^2 & 13L & -3L^2 & 0 & 0 \\ 0 & 0 & 0 & 0 & 54 & 13L & 156 & -22L & 0 & 0 \\ 0 & 0 & 0 & 0 & -13L & -3L^2 & -22L & 4L^2 & 0 & 0 \\ 0 & 0 & 0 & 0 & 0 & 0 & 0 & 0 & \dfrac{420m_5}{ml} & 0 \\ 0 & 0 & 0 & 0 & 0 & 0 & 0 & 0 & 0 & \dfrac{420m_6}{ml} \end{bmatrix}$$

The submatrices $[K_{ii}]$, $[K_{bb}]$, $[K_{ib}] = [K_{bi}]^T$, $[M_{ii}]$, $[M_{bb}]$ and $[M_{ib}] = [M_{bi}]^T$ are given as.

$$[K_{ii}] = \frac{EI}{L^3}\begin{bmatrix} 4L^2 & 2L^2 & 0 & 0 & 0 & 0 & 0 \\ 2L^2 & 8L^2 & -6L & 2L^2 & 0 & 0 & 0 \\ 0 & -6L & 24+\dfrac{k_5L^3}{EI} & 0 & 6L & -\dfrac{k_5L^3}{EI} & 0 \\ 0 & 2L^2 & 0 & 8L^2 & 2L^2 & 0 & 0 \\ 0 & 0 & 6L & 2L^2 & 4L^2 & 0 & 0 \\ 0 & 0 & -\dfrac{k_5L^3}{EI} & 0 & 0 & \dfrac{(k_5+k_6)L^3}{EI} & -\dfrac{k_6L^3}{EI} \\ 0 & 0 & 0 & 0 & 0 & -\dfrac{k_6L^3}{EI} & \dfrac{k_6L^3}{EI} \end{bmatrix},$$

$$[K_{bb}] = \frac{EI}{L^3}\begin{bmatrix} 12 & -12 & 0 \\ -12 & 24 & 0 \\ 0 & 0 & 12 \end{bmatrix} \text{ and } [K_{bi}] = [K_{ib}]^T = \frac{EI}{L^3}\begin{bmatrix} 6L & 6L & 0 & 0 & 0 & 0 & 0 \\ -6L & 0 & -12 & 6L & 0 & 0 & 0 \\ 0 & 0 & -12 & -6L & -6L & 0 & 0 \end{bmatrix}.$$

Furthermore for the mass submatrices we get

$$[M_{ii}] = \frac{ml}{420}\begin{bmatrix} 4L^2 & -3L^2 & 0 & 0 & 0 & 0 & 0 \\ -3L^2 & 8L^2 & 13L & -3L^2 & 0 & 0 & 0 \\ 0 & -6L & 312 & 0 & -13L & 0 & 0 \\ 0 & -3L^2 & 0 & 8L^2 & -3L^2 & 0 & 0 \\ 0 & 0 & -13L & -3L^2 & 4L^2 & 0 & 0 \\ 0 & 0 & 0 & 0 & 0 & \dfrac{420m_5}{ml} & 0 \\ 0 & 0 & 0 & 0 & 0 & 0 & \dfrac{420m_6}{ml} \end{bmatrix}$$

and

$$[M_{bb}] = \frac{ml}{420}\begin{bmatrix} 156 & 54 & 0 \\ 54 & 312 & 0 \\ 0 & 0 & 156 \end{bmatrix},$$

with

$$[M_{bi}] = [M_{ib}]^T = \frac{ml}{420}\begin{bmatrix} 22L & -13L & 0 & 0 & 0 & 0 & 0 \\ -6L & 0 & 54 & -13L & 0 & 0 & 0 \\ 0 & 0 & 54 & 13L & -22L & 0 & 0 \end{bmatrix}.$$

Indeterminate (redundant) Interface

If the interface $\{x_b(t)\}$ forms a redundant set of displacements we have to use (8.16) to obtain the responses.

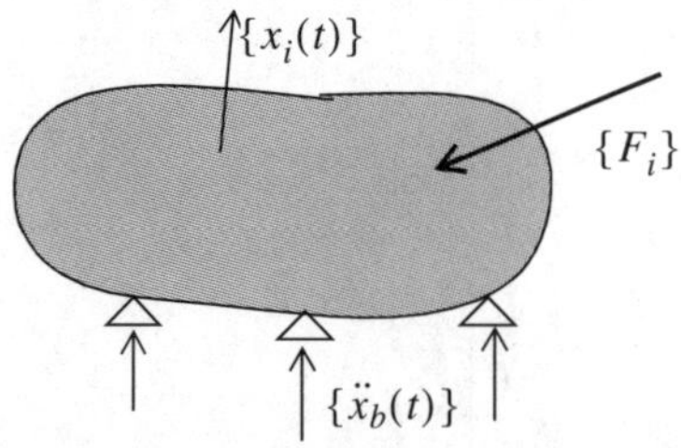

Fig. 8.3. Forces and enforced acceleration (redundant set)

$$\langle m_{ik}\rangle\{\ddot{\eta}_i(t)\} + \langle 2\zeta_k\omega_k m_{ik}\rangle\{\dot{\eta}_i(t)\} + \langle\omega_k^2 m_{ik}\rangle\{\eta_i(t)\} = [\Phi_i]^T\{F_i(t)\} - [\overline{M}_{ib}]\{\ddot{x}_b(t)\} - [\overline{C}_{ib}]\{\dot{x}_b(t)\}\ .$$

In the case of modal damping $[\Phi_i]^T[C_{ii}][\Phi_i] = \langle 2\zeta_k\omega_k m_{ik}\rangle$ (8.11) there is no damping matrix $[C]$ available and we assume no damping with respect to the constraint modes $[\Phi_c]$, hence $[\overline{C}_{ib}] = [0]$, $[\overline{C}_{bi}] = [0]$, $[\overline{C}_{bb}] = [0]$ and (8.16) becomes

$$\langle m_{ik}\rangle\{\ddot{\eta}_i(t)\} + \langle 2\zeta_k\omega_k m_{ik}\rangle\{\dot{\eta}_i(t)\} + \langle\omega_k^2 m_{ik}\rangle\{\eta_i(t)\} = [\Phi_i]^T\{F_i(t)\} - [\overline{M}_{ib}]\{\ddot{x}_b(t)\},$$

or

$$\langle m_{ik}\rangle\{\ddot{\eta}_i(t)\} + \langle 2\zeta_k\omega_k m_{ik}\rangle\{\dot{\eta}_i(t)\} + \langle\omega_k^2 m_{ik}\rangle\{\eta_i(t)\} = [\Phi_i]^T\{F_i(t)\} - \begin{bmatrix}\Phi_i\\0\end{bmatrix}^T[M][\Phi_c]\{\ddot{x}_b(t)\}. \tag{8.20}$$

Equation (8.20) can be solved either analytically or by means of numeric schemes like the Houbolt and the Newmark-beta methods.

In the frequency domain we can now solve $\Pi_{ik}(\omega)$, with $\eta_{ik}(t) = \Pi_{ik}(\omega)e^{j\omega t}$,

$$\Pi_{ik}(\omega) = \frac{\{\phi_{ik}\}^T\{F_i(\omega)\} - \begin{Bmatrix}\phi_{ik}\\0\end{Bmatrix}^T[M][\Phi_c]\{\ddot{X}_b(\omega)\}}{m_{ik}(\omega_k^2 - \omega^2 + 2j\zeta_k\omega_k\omega)},\ k = 1, 2, \ldots . \tag{8.21}$$

The relative motions $\{z\}$ can now be calculated, with $\{z(t)\} = Z(\omega)e^{j\omega t}$, using (8.29) $Z(\omega) = [\Phi_i]\{\Pi_i(\omega)\}$

$$Z(\omega) = [\Phi_i]\left\langle\frac{1}{m_{ik}(\omega_k^2-\omega^2+2j\zeta_k\omega_k\omega)}\right\rangle\left([\Phi_i]^T\{F_i(\omega)\}-\begin{bmatrix}\Phi_i\\0\end{bmatrix}^T[M][\Phi_c]\{\ddot{X}_b(\omega)\}\right). \tag{8.22}$$

The absolute motion $\{x(t)\} = \{X(\omega)\}e^{j\omega t}$ can now be calculated with (8.7) and is

$$\{X(\omega)\} = \begin{Bmatrix}Z(\omega)\\0\end{Bmatrix}+\begin{bmatrix}G_{ib}\\I\end{bmatrix}\{X_b(\omega)\} = \begin{Bmatrix}Z(\omega)\\0\end{Bmatrix}+[\Phi_c]\{X_b(\omega)\}. \tag{8.23}$$

The physical parameters, involved in the dynamic problem as shown in Fig. 8.1, are set to:
$E = 70x10^9$ N/m^2, $I = 6x10^{-10}$ m^4, $m = 0.5$ kg/m, $m_5 = 0.04$ kg, $m_6 = 0.02$ kg, $k_5 = 54000$ N/m], $k_6 = 27000$ N/m and $L = 0.1$ m.

The elastic modes $\begin{bmatrix}\Phi_i\\0\end{bmatrix}$ and constraint modes $[\Phi_c]$ are presented, with the sequence of the 10 dofs; x_1, φ_1, x_2, φ_2, x_3, φ_3, x_4, φ_4, x_5 and x_6.

The first three lowest natural frequencies, assuming $\{x_b\} = \{0\}$, calculated are: $\lfloor f_n \rfloor = \lfloor 124.8, 254.7, 498.9 \rfloor$ Hz. The associated vibration modes, such that the generalised mass matrix is a unit matrix, and the contraint modes are:

$$\begin{bmatrix}\Phi_i\\0\end{bmatrix} = \begin{bmatrix}0.0000 & 0.0000 & 0.0000\\-1.5489 & 2.6060 & -21.2465\\0.0000 & 0.0000 & 0.0000\\3.0844 & -5.1185 & 39.6527\\0.3832 & -0.6238 & 4.4993\\0.7745 & -1.3030 & 10.6233\\0.0000 & 0.0000 & 0.0000\\-6.1688 & 10.2370 & -79.3054\\3.0398 & -3.8905 & -0.7894\\5.5834 & 4.3370 & 0.1257\end{bmatrix}\quad [\Phi_c] = \begin{bmatrix}1.0000 & 0.0000 & 0.0000\\-11.6667 & 12.5000 & -0.8333\\0.0000 & 1.0000 & 0.0000\\-6.6667 & 5.0000 & 1.6667\\-0.2500 & 0.8750 & 0.3750\\0.8333 & -6.2500 & 5.4167\\0.0000 & 0.0000 & 1.0000\\3.3333 & -10.0000 & 6.6667\\-0.2500 & 0.8750 & 0.3750\\-0.2500 & 0.8750 & 0.3750\end{bmatrix}.$$

Determinate Interface

When the enforced acceleration takes place at one node “A“, with three translations and three rotations the contraint modes are, in fact, the six rigid modes $[\Phi_c] = [\Phi_r]$ with respect to the node “A“. This is called a nonredundant base excitation. This is shown in Fig. 8.4.

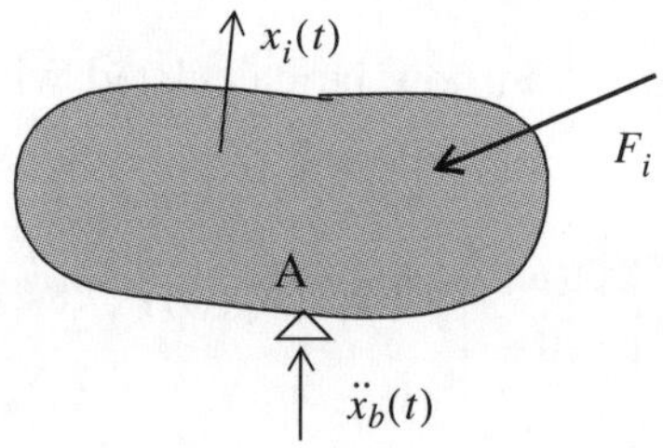

Fig. 8.4. Enforced acceleration at one node “A“(nonredundant)

This means that $[K][\Phi_r] = [0]$ or no strain energy is stored in the elastic body with the stiffness matrix $[K]$ or $[\bar{K}_{bb}] = [\Phi_r][K][\Phi_r] = [0]$. The same applies to the damping matrix $[C]$. No energy will be dissipated if the dynamic body moves as a rigid body, with $[C][\Phi_r] = [0]$ and $[\bar{C}_{bb}] = [\Phi_r][C][\Phi_r] = [0]$, hence $[\bar{C}_{ib}] = [0]$ and $[\bar{C}_{bi}] = [0]$. becomes now

$$\begin{bmatrix} \langle m_{ik} \rangle & \bar{M}_{ib} \\ \bar{M}_{bi} & \bar{M}_{bb} \end{bmatrix} \begin{Bmatrix} \ddot{\eta}_i \\ \ddot{x}_b \end{Bmatrix} + \begin{bmatrix} \langle 2\zeta_k \omega_k m_{ik} \rangle & 0 \\ 0 & 0 \end{bmatrix} \begin{Bmatrix} \dot{\eta}_i \\ \dot{x}_b \end{Bmatrix} + \begin{bmatrix} \langle \omega_k^2 m_{ik} \rangle & 0 \\ 0 & 0 \end{bmatrix} \begin{Bmatrix} \eta_i \\ x_b \end{Bmatrix} = [\Psi]^T \begin{Bmatrix} F_i \\ F_b \end{Bmatrix}. \tag{8.24}$$

The mass matrix $\bar{M}_{ib}$ is also called the matrix of the modal participation factors $[L]^T$ and can be expressed as

$$[\bar{M}_{ib}] = [L]^T = \begin{bmatrix} \Phi_i \\ 0 \end{bmatrix}^T [M][\Phi_r]. \tag{8.25}$$

The matrix of the modal participation factors is with respect to node “A“.

The first equation of can be written as

$$\langle m_{ik} \rangle \{\ddot{\eta}_i(t)\} + \langle 2\zeta_k \omega_k m_{ik} \rangle \{\dot{\eta}_i(t)\} + \langle \omega_k^2 m_{ik} \rangle \{\eta_i(t)\}$$

$$= [\Phi_i]^T\{F_i(t)\} - \begin{bmatrix}\Phi_i \\ 0\end{bmatrix}^T [M][\Phi_r]\{\ddot{x}_b(t)\}. \quad (8.26)$$

Equation (8.26) can be solved either analytically or by means of numeric schemes like the Houbolt and the Newmark-beta methods.

Transforming into the frequency domain and using (8.25) we obtain the following matrix equations of motion.

$$[\langle(\omega_k^2-\omega^2)m_{ik}\rangle + \langle 2j\zeta_k\omega_k\omega m_{ik}\rangle]\{\Pi_i(\omega)\} = [\Phi_i]^T\{F_i(\omega)\}-[L]^T\{\ddot{X}_b(\omega)\}, \quad (8.27)$$

We can now solve $\Pi_{ik}(\omega)$

$$\Pi_{ik}(\omega) = \frac{\{\phi_{ik}\}^T\{F_i(\omega)\}-[L_k]^T\{\ddot{X}_b(\omega)\}}{m_{ik}(\omega_k^2-\omega^2 + 2j\zeta_k\omega_k\omega)}, \quad k = 1, 2, \ldots . \quad (8.28)$$

The relative motions $\{z\}$ can now be calculated, with $\{z(t)\} = Z(\omega)e^{j\omega t}$,

$$Z(\omega) = [\Phi_i]\{\Pi_i(\omega)\} \quad (8.29)$$

$$Z(\omega) = [\Phi_i]\langle\frac{1}{m_{ik}(\omega_k^2-\omega^2 + 2j\zeta_k\omega_k\omega)}\rangle([\Phi_i]^T\{F_i(\omega)\}-[L]^T\{\ddot{X}_b(\omega)\}). \quad (8.30)$$

The absolute motion $\{x(t)\} = \{X(\omega)\}e^{j\omega t}$ can now be calculated with (8.7)

$$\{X(\omega)\} = \begin{Bmatrix} Z(\omega) \\ 0 \end{Bmatrix} + \begin{bmatrix} G_{ib} \\ I \end{bmatrix}\{X_b(\omega)\} = \begin{Bmatrix} Z(\omega) \\ 0 \end{Bmatrix} + [\Phi_r]\{X_b(\omega)\}. \quad (8.31)$$

We will calculate the rigid body vector $\{\Phi_r\}$ with respect to $\{x_1, x_2, x_4\}$ assuming

$$\{\ddot{x}_b\} = \begin{Bmatrix} \ddot{x}_1 \\ \ddot{x}_2 \\ \ddot{x}_4 \end{Bmatrix} = \{T\}\ddot{u} = \begin{Bmatrix} 1 \\ 1 \\ 1 \end{Bmatrix}\ddot{u},$$

then the rigid-body mode $[\Phi_r]$ are presented, with the sequence of the 10 dofs; x_1, φ_1, x_2, φ_2, x_3, φ_3, x_4, φ_4, x_5 and x_6.

$$\{\Phi_r\} = [\Phi_c]\{T\} = \begin{Bmatrix} 1.0000 \\ 0.0000 \\ 1.0000 \\ 0.0000 \\ 1.0000 \\ 0.0000 \\ 1.0000 \\ 0.0000 \\ 1.0000 \\ 1.0000 \end{Bmatrix}.$$

The natural frequencies and associated modes are the same as for the undetermined case.

8.2.2 Absolute Motions

We start with (8.8)

$$\begin{bmatrix} M_{ii} & M_{ib} \\ M_{bi} & M_{bb} \end{bmatrix}\begin{Bmatrix} \ddot{x}_i \\ \ddot{x}_b \end{Bmatrix} + \begin{bmatrix} C_{ii} & C_{ib} \\ C_{bi} & C_{bb} \end{bmatrix}\begin{Bmatrix} \dot{x}_i \\ \dot{x}_b \end{Bmatrix} + \begin{bmatrix} K_{ii} & K_{ib} \\ K_{bi} & K_{bb} \end{bmatrix}\begin{Bmatrix} x_i \\ x_b \end{Bmatrix} = \begin{Bmatrix} F_i \\ F_b \end{Bmatrix},$$

thus

$$[M_{ii}]\{\ddot{x}_i\} + [C_{ii}]\{\dot{x}_i\} + [K_{ii}]\{x_i\}$$

$$= (\{F_i\} - ([M_{ib}]\{\ddot{x}_b\} + [C_{ib}]\{\dot{x}_b\} + [K_{ii}]\{x_b\})) \tag{8.32}$$

We will depict the internal motion $\{x_i\}$ on the modal base $[\Phi_i]$ as follows

$$\{x_i\} = [\Phi_i]\{\eta_i\}, \tag{8.33}$$

with

- $[\Phi_i]$ the modal base obtained from the eigenvalue problem $([K_{ii}] - \omega_k^2[M_{ii}])\{\phi_{ik}\} = \{0\}$, with $[\Phi_i] = [\phi_{i1}, \phi_{i2}, ...]$ and ω_k^2 the k-th eigenvalue.
- $\{\eta_i\}$ are the generalised coordinates.

The mode shapes $[\Phi_i]$ obey the orthogonality relations with respect to the mass matrix $[M_{ii}]$ and the stiffness matrix $[K_{ii}]$

$$\left.\begin{matrix}[\Phi_i]^T[M_{ii}][\Phi_i] \\ [\Phi_i]^T[K_{ii}][\Phi_i]\end{matrix}\right\} = \left\{\begin{matrix}\langle m_{ik}\rangle \\ \langle \omega_k^2 m_{ik}\rangle\end{matrix}\right. . \tag{8.34}$$

After the base transformation we assume a diagonal matrix of modal damping ratios, hence

$$[\Phi_i]^T[C_{ii}][\Phi_i] = \langle 2\zeta_k\omega_k m_{ik}\rangle . \tag{8.35}$$

We will substitute (8.33) into (8.32), premultiply by $[\Phi_i]^T$ and obtain the following equation, assuming $[C_{ib}]\{\dot{x}_b\} = (0)$,

$$[\Phi_i]^T[M_{ii}][\Phi_i]\{\ddot{\eta}_i\} + [\Phi_i]^T[C_{ii}][\Phi_i]\{\dot{\eta}_i\} + [\Phi_i]^T[K_{ii}][\Phi_i]\{\eta_i\}$$

$$= ([\Phi_i]^T\{F_i\} - [\Phi_i]^T([M_{ib}]\{\ddot{x}_b\} + [K_{ii}]\{x_b\})) . \tag{8.36}$$

Using the orthogonality relation from (8.34) we get

$$\langle m_{ik}\rangle\{\ddot{\eta}_i(t)\} + \langle 2\zeta_k\omega_k m_{ik}\rangle\{\dot{\eta}_i(t)\} + \langle\omega_k^2 m_{ik}\rangle\{\eta_i(t)\}$$

$$= [\phi_i]^T\{F_i\} - [\phi_i]^T([M_{ib}]\{\ddot{x}_b\} + [K_{ii}]\{x_b\}) . \tag{8.37}$$

Equation (8.37) can be either solved analytically or using numeric schemes like the Wilson-θ, or the Newmark-beta methods.

If the enforced acceleration $\{\ddot{x}_b\}$ is specified the displacement $\{x_b\}$ must be obtained by integration of the acceleration $\{\ddot{x}_b\}$ (8.17). The displacements $\{x_b(t)\}$ can be obtained with

$$\{x_b(t)\} = \{x_b(t_0)\} + \int_{t_0}^{t}\{\dot{x}_b(t)\}dt . \tag{8.38}$$

With a time increment Δt we can use the trapezoidal approximation of the integral [Schwarz 89] in (8.38)

$$\{x_b(t_n)\} = \{x_b(t_0)\} + \Delta t\left(\frac{1}{2}\{\dot{x}_b(t_0)\} + \sum_{j=1}^{n-1}\{\dot{x}_b(t_j)\} + \frac{1}{2}\{\dot{x}_b(t_n)\}\right), \tag{8.39}$$

with

- $t_j = t_0 + j\Delta t$
- $t_n = t_0 + n\Delta t = t$

- $\{x_b(t_{n+1})\} = \{x_b(t_n)\} + \frac{\Delta t}{2}(\{\dot{x}_b(t_n)\} + \{\dot{x}_b(t_{n+1})\})$

  In the frequency domain we have

- $\{\eta_i\} = \{\Pi_i(\omega)\}e^{j\omega t}$, $\{\dot{\Pi}_i\} = j\omega\{\Pi(\omega)\}$ and $\{\ddot{\Pi}_i\} = -\omega^2\{\Pi(\omega)\}$
- $\{x_b\} = \{X_b(\omega)\}e^{j\omega t}$, $\{\ddot{X}_b\} = -\omega^2\{X_b(\omega)\}$ or $\{\ddot{x}_b\} = \{\ddot{X}_b(\omega)\}e^{j\omega t}$,

  $\{\dot{X}_b\} = \frac{\{\ddot{X}_b(\omega)\}}{j\omega} = -j\frac{\{\ddot{X}_b(\omega)\}}{\omega}$ and $\{X_b\} = \frac{-\{\ddot{X}_b(\omega)\}}{\omega^2}$

We can now solve $\Pi_{ik}(\omega)$

$$\Pi_{ik}(\omega) = \frac{\{\phi_{ik}\}^T F_i(\omega) - \{\phi_{ik}\}^T\left([M_{ib}]\{\ddot{X}_b(\omega)\} - \frac{1}{\omega^2}[K_{ii}]\{\ddot{X}_b(\omega)\}\right)}{m_{ik}(\omega_k^2 - \omega^2 + 2j\zeta_k\omega_k\omega)}, \quad k = 1, 2, \ldots \tag{8.40}$$

The absolute motion $\{x(t)\} = \{X(\omega)\}e^{j\omega t}$ can now be calculated

$$\{X(\omega)\} = \begin{Bmatrix} X_i(\omega) \\ X_b(\omega) \end{Bmatrix} = \begin{bmatrix} [\Phi_i] \\ I \end{bmatrix} \begin{Bmatrix} \Pi_i(\omega) \\ X_b(\omega) \end{Bmatrix}. \tag{8.41}$$

8.2.3 Large-Mass Approach

The large-mass approach is illustrated in the following example. To introduce the enforced acceleration a very large mass $\widehat{M}$ has been attached to the sdof system. This large-mass is loaded with a dynamic load $F = \widehat{M}\ddot{u}$ with $\ddot{u}$ the enforced acceleration. The dynamic system is shown in Fig. 8.5. The sdof system has a discrete mass m, a spring with a spring constant k and a modal damping ζ. The undamped equations of motion are

$$\begin{bmatrix} m & 0 \\ 0 & \widehat{M} \end{bmatrix} \begin{Bmatrix} \ddot{x}_1(t) \\ \ddot{x}_2(t) \end{Bmatrix} + \begin{bmatrix} k & -k \\ -k & k \end{bmatrix} \begin{Bmatrix} x_1(t) \\ x_2(t) \end{Bmatrix} = \begin{Bmatrix} 0 \\ \widehat{M}\ddot{u} \end{Bmatrix}, \tag{8.42}$$

or

$$[M]\{\ddot{x}(t)\} + [K]\{x(t)\} = \{F(t)\}. \tag{8.43}$$

The modal damping ζ will be introduced after the decoupling of the equations motion with the modal displacement method.

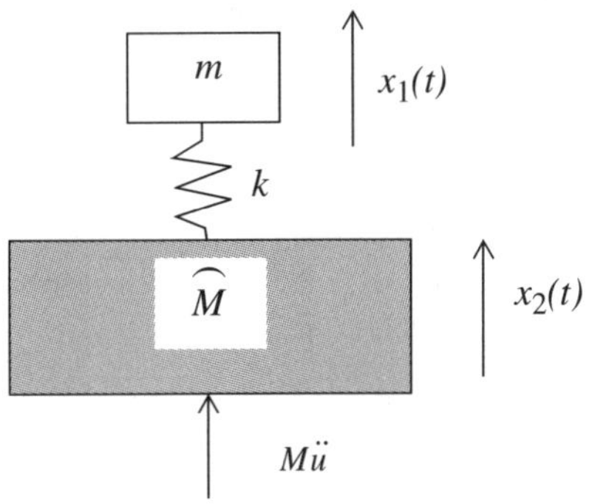

Fig. 8.5. Frequency-response analysis, large-mass approach

The undamped eigenvalue problem of the dynamic system can be written as

$$\left(-\lambda\begin{bmatrix} m & 0 \\ 0 & \widehat{M} \end{bmatrix} + \begin{bmatrix} k & -k \\ -k & k \end{bmatrix}\right)\left\{\begin{matrix} \varphi_1 \\ \varphi_2 \end{matrix}\right\} = \left\{\begin{matrix} 0 \\ 0 \end{matrix}\right\}. \tag{8.44}$$

The nontrivial solution can be found if

$$\det\left(-\lambda\begin{bmatrix} m & 0 \\ 0 & \widehat{M} \end{bmatrix} + \begin{bmatrix} k & -k \\ -k & k \end{bmatrix}\right) = 0, \tag{8.45}$$

thus the following equation must be solved

$$(k-\lambda\widehat{M})(k-\widehat{\lambda}M)-k^2 = 0, \tag{8.46}$$

with two roots

$$\lambda_1 = 0, \quad \lambda_2 = \omega_2^2 = \frac{k(m+\widehat{M})}{m\widehat{M}}. \tag{8.47}$$

The eigenvectors (mode shapes) associated with the eigenvalues λ_1 and λ_2 are

$$[\Phi] = \begin{bmatrix} 1 & 1 \\ 1 & -\frac{m}{\widehat{M}} \end{bmatrix}. \tag{8.48}$$

The associated generalised masses become

$$\langle mg\rangle = [\Phi]^T[M][\Phi] = \begin{bmatrix} m+\widehat{M} & 0 \\ 0 & m\left(1+\frac{m}{\widehat{M}}\right) \end{bmatrix}, \tag{8.49}$$

and the corresponding generalised stiffnesses are

$$\langle kg\rangle = \{\lambda mg\} = [\Phi]^T[K][\Phi] = \begin{bmatrix} 0 & 0 \\ 0 & k\left(1+\frac{m}{\widehat{M}}\right)^2 \end{bmatrix} = \begin{bmatrix} 0 & 0 \\ 0 & k\left(\frac{m+\widehat{M}}{m\widehat{M}}\right)m\left(1+\frac{m}{\widehat{M}}\right) \end{bmatrix}. \tag{8.50}$$

Applying the mode displacement method (MDM) with

$$\{x(t)\} = [\Phi]\{\eta(t)\}, \tag{8.51}$$

then

$$[\Phi]^T[M][\Phi]\{\ddot{\eta}(t)\} + [\Phi]^T[M][\Phi]\{\eta(t)\} = [\Phi]^T\{F(t)\}. \tag{8.52}$$

With the introduction of the “ad hoc” modal damping ζ the two uncoupled damped equations of motion for the generalised coordinates $\{\eta(t)\}$ become

$$\langle mg\rangle\{\ddot{\eta}(t)\} + \langle 2\zeta mg\sqrt{\lambda}\rangle\{\dot{\eta}(t)\} + \langle kg\rangle\{\eta(t)\} = [\Phi]^T\{F(t)\} = \{f(t)\}, \tag{8.53}$$

or

$$\begin{bmatrix} m+\widehat{M} & 0 \\ 0 & m\left(1+\frac{m}{\widehat{M}}\right) \end{bmatrix}\begin{Bmatrix} \ddot{\eta}_1(t) \\ \ddot{\eta}_2(t) \end{Bmatrix} + \begin{bmatrix} 0 & 0 \\ 0 & 2\zeta\sqrt{k\left(\frac{m+\widehat{M}}{m\widehat{M}}\right)}m\left(1+\frac{m}{\widehat{M}}\right) \end{bmatrix}\begin{Bmatrix} \dot{\eta}_1(t) \\ \dot{\eta}_2(t) \end{Bmatrix}$$

$$+\begin{bmatrix} 0 & 0 \\ 0 & k\left(\frac{m+\widehat{M}}{m\widehat{M}}\right)m\left(1+\frac{m}{\widehat{M}}\right) \end{bmatrix}\begin{Bmatrix} \eta_1(t) \\ \eta_2(t) \end{Bmatrix} = \begin{Bmatrix} \widehat{M}\ddot{u} \\ -m\ddot{u} \end{Bmatrix}. \tag{8.54}$$

Premultilpying (8.54) with the inverse matrix of the generalised masses

$$\begin{bmatrix}1 & 0\\0 & 1\end{bmatrix}\begin{Bmatrix}\ddot{\eta}_1(t)\\\ddot{\eta}_2(t)\end{Bmatrix}+\begin{bmatrix}0 & 0\\0 & 2\zeta\omega_2\end{bmatrix}\begin{Bmatrix}\dot{\eta}_1(t)\\\dot{\eta}_2(t)\end{Bmatrix}+\begin{bmatrix}0 & 0\\0 & \omega_2^2\end{bmatrix}\begin{Bmatrix}\eta_1(t)\\\eta_2(t)\end{Bmatrix}$$

$$=\begin{bmatrix}m+\widehat{M} & 0\\0 & m\left(1+\dfrac{m}{\widehat{M}}\right)\end{bmatrix}^{-1}\begin{Bmatrix}\widehat{M}\ddot{u}\\-m\ddot{u}\end{Bmatrix}=\left(\frac{1}{\dfrac{m}{\widehat{M}}+1}\right)\begin{Bmatrix}1\\-1\end{Bmatrix}\ddot{u}. \tag{8.55}$$

Solutions in the frequency domain are

$$\{\eta(t)\}=\{\Pi(\omega)\}e^{j\omega t},\ \{\ddot{u}(t)\}=\{\ddot{U}(\omega)\}e^{j\omega t} \tag{8.56}$$

$$\left(-\omega^2\begin{bmatrix}1 & 0\\0 & 1\end{bmatrix}+\begin{bmatrix}0 & 0\\0 & 2j\zeta\omega\omega_2\end{bmatrix}+\begin{bmatrix}0 & 0\\0 & \omega_2^2\end{bmatrix}\right)\begin{Bmatrix}\Pi_1(\omega)\\\Pi_2(\omega)\end{Bmatrix}=\frac{1}{\dfrac{m}{\widehat{M}}+1}\begin{Bmatrix}1\\-1\end{Bmatrix}\ddot{U}(\omega) \tag{8.57}$$

The solutions for the generalised coordinates become

$$\Pi_1(\omega)=-\frac{1}{\omega^2}\left(\frac{1}{\dfrac{m}{\widehat{M}}+1}\right)\ddot{U}(\omega), \tag{8.58}$$

and

$$\Pi_2(\omega)=\frac{-\left(\dfrac{m}{\widehat{M}}+1\right)\ddot{U}(\omega)}{\omega_2^2-\omega^2+2j\zeta\omega\omega_2}=-\frac{1}{\dfrac{m}{\widehat{M}}+1}H(\omega)\ddot{U}(\omega). \tag{8.59}$$

The physical displacement vector $\{X(\omega)\}$

$$\{X(\omega)\}=[\Phi]\{\Pi(\omega)\}=\begin{bmatrix}1 & 1\\1 & -\dfrac{m}{\widehat{M}}\end{bmatrix}\begin{Bmatrix}\Pi_1(\omega)\\\Pi_2(\omega)\end{Bmatrix}=\begin{bmatrix}\Pi_1(\omega)+\Pi_2(\omega)\\\Pi_1(\omega)-\dfrac{m}{\widehat{M}}\Pi_2(\omega)\end{bmatrix} \tag{8.60}$$

or

$$\{X(\omega)\} = \begin{Bmatrix} \Pi_1(\omega) + \Pi_2(\omega) \\ \Pi_1(\omega) - \dfrac{m}{\widehat{M}}\Pi_2(\omega) \end{Bmatrix} = \frac{1}{\dfrac{m}{\widehat{M}} + 1}\begin{Bmatrix} -\dfrac{1}{\omega^2} - H(\omega) \\ \left(-\dfrac{1}{\omega^2} + \dfrac{m}{\widehat{M}}H(\omega)\right) \end{Bmatrix}\ddot{U}(\omega), \quad (8.61)$$

and for the accelerations we may write

$$\{\ddot{X}(\omega)\} = -\omega^2\{X(\omega)\} = \left(\frac{1}{\dfrac{m}{\widehat{M}} + 1}\right)\begin{Bmatrix} 1 + \omega^2 H(\omega) \\ \left(1 - \omega^2\dfrac{m}{\widehat{M}}H(\omega)\right) \end{Bmatrix}\ddot{U}(\omega). \quad (8.62)$$

Assume $\widehat{M} \gg m$, thus $\dfrac{m}{\widehat{M}} \ll 1$ (8.62) becomes

$$\{\ddot{X}(\omega)\} = -\omega^2\{X(\omega)\} = \begin{Bmatrix} 1 + \omega^2 H(\omega) \\ 1 \end{Bmatrix}\ddot{U}(\omega), \quad (8.63)$$

thus

$$\ddot{X}_1(\omega) = [1 + \omega^2 H(\omega)]\ddot{U}(\omega) \quad (8.64)$$

with

$$\lim_{\omega \to \infty} \ddot{X}_1(\omega) = 0, \quad (8.65)$$

and

$$\ddot{X}_2(\omega) = \ddot{U}(\omega). \quad (8.66)$$

The attached mass $\widehat{M}$ will be of order $O(10^6 m \ldots 10^8 m)$ in order to obtain reliable responses.

The results of the large-mass approach are checked with the relative motion approach, as illustrated hereafter (Fig. 8.6).

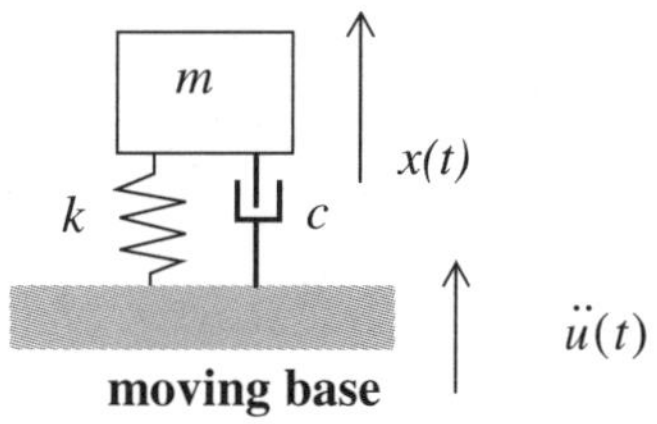

Fig. 8.6. Sdof system with enforced acceleration

With the introduction of the relative motion $z(t)$,

$$z(t) = x(t) - u(t), \tag{8.67}$$

we can write the equation of motion of the sdof system

$$\ddot{z}(t) + 2\zeta\omega_n\dot{z}(t) + \omega_n^2 z(t) = -\ddot{u}(t) \tag{8.68}$$

The enforced acceleration of the sdof system is transformed into an external force. The absolute acceleration $\dot{x}(t)$ is

$$\ddot{x}(t) = \ddot{z}(t) + \ddot{u}(t) = -2\zeta\omega_n\dot{z}(t) - \omega_n^2 z(t) \tag{8.69}$$

The response and base excitation transformed from the time domain into the frequency domain is

$$z(t) = Z(\omega)e^{j\omega t}, \ \ddot{u}(t) = \ddot{U}(\omega)e^{j\omega t}. \tag{8.70}$$

The complex relative motion $Z(\omega)$ can be expressed in $\ddot{U}(\omega)$

$$Z(\omega) = -\frac{\ddot{U}(\omega)}{\omega_n^2 - \omega^2 + 2j\zeta\omega\omega_n} = -H(\omega)\ddot{U}(\omega), \tag{8.71}$$

and

$$\dot{Z}(\omega) = -j\omega Z(\omega) = -j\omega H(\omega)\ddot{U}(\omega), \tag{8.72}$$

and

$$\ddot{Z}(\omega) = -\omega^2 Z(\omega) = \omega^2 H(\omega)\ddot{U}(\omega). \tag{8.73}$$

The absolute acceleration $\ddot{X}(\omega)$ becomes

$$\ddot{X}(\omega) = \{2j\zeta\omega_n\omega H(\omega) + \lambda H(\omega)\}\ddot{U}(\omega) = \{2j\zeta\omega_n\omega + \omega_n^2\}H(\omega)\ddot{U}(\omega), \tag{8.74}$$

and the absolute displacement $X(\omega)$

$$X(\omega)= Z(\omega)+U(\omega) = \left[-\frac{1}{\omega^2}-H(\omega)\right]\ddot{U}(\omega), \tag{8.75}$$

and

$$\ddot{X}(\omega)= \ddot{Z}(\omega)+\ddot{U}(\omega) = [1+\omega^2 H(\omega)]\ddot{U}(\omega) , \tag{8.76}$$

with

$$\lim_{\omega \to \infty} \ddot{X}(\omega) = 0 . \tag{8.77}$$

8.3 Problems

8.3.1 Problem 1

A linear dynamic system, shown in Fig. 8.7, consist of 4 dofs; x_1, x_2, x_3 and x_4. The enforced acceleration are $\lfloor \ddot{x}_b \rfloor = \lfloor \ddot{x}_1, \ddot{x}_4 \rfloor = \lfloor e^{j\omega t}, e^{j\omega t - \theta} \rfloor$ m/s^2 and the internal dofs are $\lfloor x_i \rfloor = \lfloor x_2, x_3 \rfloor$. All masses are equal $m_1 = m_2 = m_3 = m_4 = 1$ kg and all stiffness are equal $k_1 = k_2 = k_3 = 1000$ N/m. The modal damping ratio for all modes is $\zeta = 0.02$.

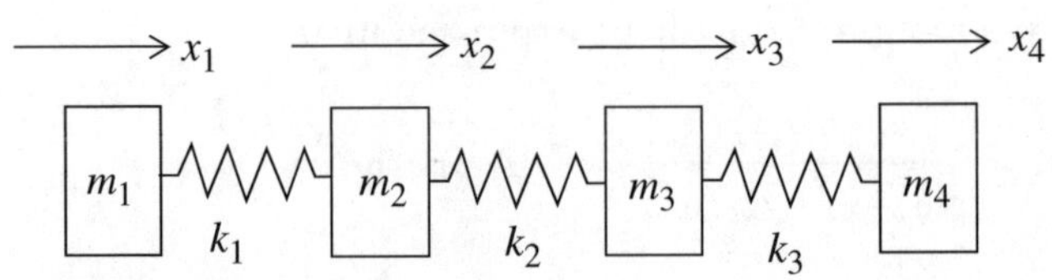

Fig. 8.7. 4 dofs linear dynamic system

Perform the following activities:

- Set up the total mass matrix $[M]$
- Set up the total stiffness matrix $[K]$
- Set up the submatrices $[M_{ii}]$, $[M_{bb}]$, $[M_{ib}] = [M_{bi}]^T$
- Set up the submatrices $[K_{ii}]$, $[K_{bb}]$, $[K_{ib}] = [K_{bi}]^T$
- Calculate eigenvalues and associated eigenvectors from eigenvalue problem $([K_{ii}] - \omega_{ik}^2[M_{ii}])\{\phi_{ik}\}$

- Set up the modal matrix $[\Phi_i]$
- Calculate the constraint modes $[\Phi_c]$
- Calculate vector $\{\ddot{z}\} = \{\ddot{Z}(\omega)\}e^{j\omega t}$, $\theta = 0, \frac{\pi}{2}, \pi$, (8.23)
- Calculate the absolute acceleration
  $\{\ddot{x}\} = \{\ddot{X}(\omega)\}e^{j\omega t} = (\{\ddot{Z}(\omega)\} + \{\ddot{X}_b\})e^{j\omega t}$, $\theta = 0, \frac{\pi}{2}, \pi$, (8.23)
- Calculate directly the absolute acceleration $\{\ddot{x}\} = \{\ddot{X}(\omega)\}e^{j\omega t}$,
  $\theta = 0, \frac{\pi}{2}, \pi$ (8.41)

8.3.2 Problem 2

A linear dynamic system (shown in Fig. 8.8) consists of 3 dofs; x_1, x_2 and x_3.

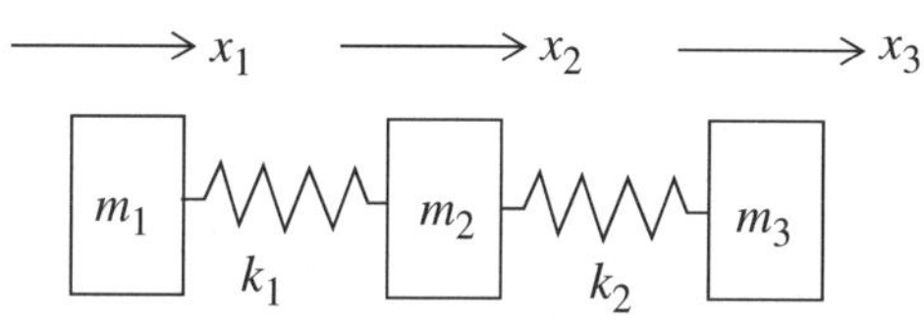

Fig. 8.8. 3 dofs linear dynamic system

The enforced acceleration are $\ddot{x}_b = \ddot{x}_1 = e^{j\omega t}$ m/s^2 and the internal dofs are $\lfloor x_i \rfloor = \lfloor x_2, x_3 \rfloor$. The masses are $m_1 = 2m_2 = 2m_3 = 2$ kg and the stiffnesses are $k_1 = 2k_2 = 2000$ N/m. The modal damping ratios for all modes are $\zeta = 0.05$.

- Set up the total mass matrix $[M]$
- Set up the total stiffness matrix $[K]$
- Set up the submatrices $[M_{ii}]$, $[M_{bb}]$, $[M_{ib}] = [M_{bi}]^T$
- Set up the submatrices $[K_{ii}]$, $[K_{bb}]$, $[K_{ib}] = [K_{bi}]^T$
- Calculate eigenvalues and associated eigenvectors from eigenvalue problem $([K_{ii}] - \omega_{ik}^2[M_{ii}])\{\phi_{ik}\}$
- Set up the modal matrix $[\Phi_i]$
- Calculate the constraint modes $[\Phi_r]$

- Calculate vector $\{\ddot{z}\} = \{\ddot{Z}(\omega)\}e^{j\omega t}$, (8.30)
- Calculate the absolute acceleration

  $\{\ddot{x}\} = \{\ddot{X}(\omega)\}e^{j\omega t} = (\{\ddot{Z}(\omega)\} + \{\ddot{X}_b\})e^{j\omega t}$, (8.31)
- Calculate directly the absolute acceleration $\{\ddot{x}\} = \{\ddot{X}(\omega)\}e^{j\omega t}$, (8.41)

8.3.3 Problem 3

Solve the previous problem by replacing the mass m_1 by a large mass $M_1 = 10^6$ kg and apply a force F_1 at x_1, which is $F_1 = M_1 e^{j\omega t}$, as illustrated in Fig. 8.9.

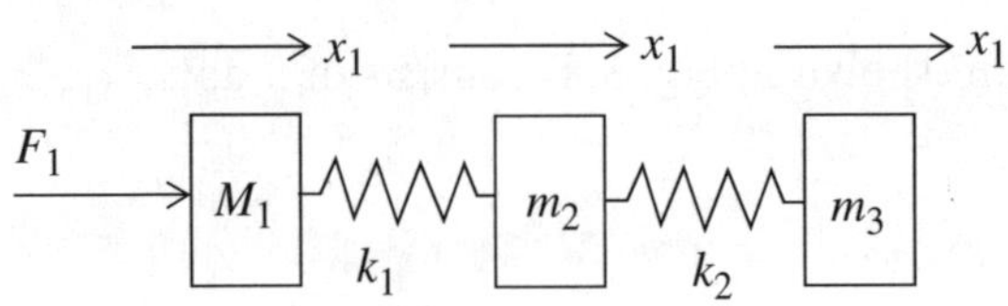

Fig. 8.9. 3 dofs linear dynamic system

The masses are $m_2 = m_3 = 1$ kg and the stiffnesses are $k_1 = 2k_2 = 2000$ N/m]. The modal damping ratios for all elastic modes are $\zeta = 0.05$.
Use the MDM method!

9 Transient-Response Analysis

9.1 Introduction

Transient-response analysis is the solution of a linear sdof or linear mdof system in the time domain. For linear mdof dynamic systems with the aid of the modal superposition the mdof system can be broken down into a series of uncoupled sdof dynamic systems. For a very few cases the analytical solution of the second-order differential equation, in the time domain, may be obtained and numerical methods are needed to solve the sdof and the mdof dynamic systems. Often, the numerical solution schemes are time-integration methods. The time-integration methods may have fixed or non-fixed (sliding) time increments per time integration step, and solve the equation of motion numerically for every time step, taking into account the initial values either for sdof or mdof dynamic systems. The sdof dynamic system may be written as

$$\ddot{x}(t) + 2\zeta\omega_n\dot{x}(t) + \omega_n^2 x(t) = \frac{F(t)}{m} = f(t). \tag{9.1}$$

The linear mdof dynamic systems is represented by the following matrix equations of motion

$$[M]\{\ddot{x}(t)\} + [C]\{\dot{x}(t)\} + [K]\{x(t)\} = \{F(t)\}, \tag{9.2}$$

the coupled linear equations can be decoupled using the mode displacement method (MDM) or mode superposition method. The physical displacement vector $x(t)$ is expressed as follows

$$x(t) = [\Phi]\{\eta(t)\},$$

with

- $\{\eta(t)\}$ the vector of generalised coordinates

The modal matrix $[\Phi] = [\phi_1, \phi_2, \ldots, \phi_n]$ has the following orthogonality properties with respect to the mass matrix and the stiffness matrix

$$[\Phi]^T[M][\Phi] = \langle m \rangle \text{ and } [\Phi]^T[K][\Phi] = \langle \omega_n^2 m \rangle$$

- $\langle m \rangle$ the diagonal matrix of generalised or modal masses
- $\langle \omega_n^2 m \rangle$ the diagonal matrix of the eigenvalues multiplied by the generalised mass

Making use of the orthogonality properties of the modal matrix, the equations of motion are expressed in the generalised coordinates, generalised masses, eigenvalues and generalised forces

$$\langle m_i \rangle \{\ddot{\eta}(t)\} + \langle c_i \rangle \{\dot{\eta}(t)\} + \langle \omega_i^2 m_i \rangle \{\eta(t)\} = f(t),$$

with

- $f(t)$ the vector of generalised forces.
- $\langle c_i \rangle$ the diagonal matrix of the generalised damping. This means the damping matrix $[C]$ consists of proportional damping. In general we will add modal viscous damping to the uncoupled equations of motion of the generalised coordinates, $\frac{c_i}{m_i} = 2\zeta_i\omega_i$, $c_i = \{\phi_i\}^T[C]\{\phi_i\}$ on an ad hoc basis.
- $m_i = \{\phi_i\}^T[M]\{\phi_i\}$ the generalised mass associated with mode $\{\phi_i\}$.
- $\zeta_i = \frac{c_i}{2\sqrt{k_i m_i}}$ the modal damping ratio.
- $k_i = \{\phi_i\}^T[K]\{\phi_i\} = \omega_i^2 m_i$ the generalised stiffness.
- ω_i^2 the eigenvalue of the eigenvalue problem

  $([K] - \omega_i^2[M])\{\phi_i\} = \{0\}$

Finally, the uncoupled equations of motion for the generalised coordinates enforced with the generalised forces become

$$\ddot{\eta}_i(t) + 2\zeta_i\omega_i\dot{\eta}(t) + \omega_i^2\eta_i(t) = \frac{\{\phi_i\}^T\{F(t)\}}{m_i} = f_i(t),\ i=1,2,\ldots. \tag{9.3}$$

If the damped natural (circular) frequency ω_{di} (rad/s) is defined as

$$\omega_{di} = \omega_i\sqrt{1-\zeta_i^2},$$

then the theoretical solutions of (9.3) are

$$\eta_i(t) = \eta_{io}e^{-\zeta\omega_i t}\left(\cos\omega_{di}t + \frac{\zeta_i}{\sqrt{1-\zeta_i^2}}\sin\omega_{di}t\right)$$

$$+\dot{\eta}_{io}e^{-\zeta\omega_i t}\frac{\sin\omega_{di}t}{\omega_{di}} + \int_0^t e^{-\zeta\omega_i\tau}\frac{\sin\omega_{di}\tau}{\omega_{di}}f_i(t-\tau)d\tau\,, \tag{9.4}$$

with initial conditions

- $\{\eta_o\} = ([\Phi]^T[\Phi])^{-1}[\Phi]\{x_o\}$ and
- $\{\dot{\eta}_o\} = ([\Phi]^T[\Phi])^{-1}[\Phi]\{\dot{x}_o\}$

In most cases no closed form solution of (9.4) exists and therefore we will solve (9.1), or (9.2) or (9.3) numerically. Many numerical time integration methods are described in the literature, e.g. [Abramowitz 70, Babuska 66, Chopra 95, Chung 93, Dokainish 89, Ebeling 97, Hughes 83, Kreyszig 93, Petyt 90, Pilkey 94, Schwarz 89, Subbraraj 89, Wood 90].

9.2 Numerical Time Integration

We define a series of time steps:

$$t_0 = 0,\ t_1 = \Delta t,\ t_2 = 2\Delta t,\ldots,t_{n-1} = (n-1)\Delta t,\ t_n = n\Delta t\,. \tag{9.5}$$

The displacement vector $\{d_n\}$, the velocity vector $\{v_n\}$ and the acceleration vector $\{a_n\}$ are approximations of $\{x(t_n)\}$, $\{\dot{x}(t_n)\}$ and $\{\ddot{x}(t_n)\}$, respectively, conform to (9.1).

9.2.1 Discrete Solution Convolution Integral

The closed form solution of (9.1), with the initial conditions $x(0) = x_o$ and $\dot{x}(0) = v_o$, is more or less illustrated in (9.4)

$$x(t) = x_o e^{-\zeta\omega_n t}\left(\cos\omega_d t + \frac{\zeta}{\sqrt{1-\zeta^2}}\sin\omega_d t\right)$$

$$+ v_o e^{-\zeta\omega_n t}\frac{\sin\omega_d t}{\omega_d} + \int_0^t e^{-\zeta\omega_n\tau}\frac{\sin\omega_d\tau}{m\omega_d}F(t-\tau)d\tau\,. \tag{9.6}$$

The damped natural (circular) frequency ω_d (rad/s) is defined as

$$\omega_d = \omega_n\sqrt{1-\zeta^2}.$$

The last part in the right-hand side of (9.6) is called the convolution integral with the impulse response function $h(\tau) = e^{-\zeta\omega_n\tau}\dfrac{\sin\omega_d\tau}{m\omega_d}$ and the external force $F(t-\tau)$, viz. $\int_0^t h(\tau)F(t-\tau)d\tau$. Assuming zero initial condition the solution of (9.1) can be written as

$$x(t) = \int_0^t h(\tau)F(t-\tau)d\tau. \tag{9.7}$$

The final time t and the running time τ are expressed as follows $t = t_j = \tau_j = j\Delta t$ (see also (9.5)). The convolution integral in (9.7) will approximated by a summation [Meirovitch 97]

$$x(t) = \int_0^t h(\tau)F(t-\tau)d\tau = \int_0^t h(t-\tau)F(\tau)d\tau = \sum_{k=0}^{j} \Delta t h\{(j-k)\Delta t\}F(k\Delta t) \tag{9.8}$$

$$d_n = \sum_{k=0}^{j} \Delta t h\{(j-k)\Delta t\}F(k\Delta t) = \sum_{k=0}^{j} \Delta t h_{j-k}F_k. \tag{9.9}$$

For example,

- $j = 0$; $\dfrac{d_0}{\Delta t} = h_0F_0$
- $j = 1$; $\dfrac{d_1}{\Delta t} = h_1F_0 + h_0F_1$
- ....
- $j = n$; $\dfrac{d_n}{\Delta t} = h_nF_0 + h_{n-1}F_1 + ... + h_1F_{n-1} + h_0F_n$

It is clear that the number of operations increases with each sampling t_n. The numerical solution of the convolution integral must be placed in the area of the academic solutions and is not very practical for real-life problems. It is better to apply recurrence-matrix methods, for which the last solution $\{x_{n+1}\}$ is only dependent upon the penultimate solution $\{x_n\}$, hence

$$\{x_{n+1}\} = \{x(t+[n+1]\Delta t)\} = [A_n]\{x_n\} = [A_n]\{x(t+n\Delta t)\}, \tag{9.10}$$

where the matrix $[A_n]$ is called the amplification matrix. For linear systems the recurrent relation in (9.10) can be written as

$$\{x_{n+1}\} = [A_n]\{x_n\} = [A_n][A_{n-1}]......[A_0]\{x_0\} = [A]^n\{x_0\}, \tag{9.11}$$

with
- $[A_n] = [A_{n-1}] = ... = [A_1] = [A_0] = [A]$

9.2.2 Explicit Time-Integration Method

The explicit time-integration method is the solution of equation(s) of motion at time $t+\Delta t$ and the solution is obtained by considering the equilibrium conditions at time t, and such integration schemes do not require the inversion of the stiffness matrix in the step-by-step solution. Hence, the method requires no storage of matrices if the diagonal (lumped or generalised masses) mass matrix is used. The explicit methods are conditionally stable and require small time steps to be employed to insure stability.

In the next section the following explicit time-integration methods will be discussed:
- the central difference method
- the Runge–Kutta methods (for first- and second-order differential equations)

9.2.3 Implicit Time-Integration Methods

In the implicit time-integration methods the equations for the displacements at the current time step involve the velocities and accelerations at the current step itself, $t+\Delta t$. Hence the determination of the displacements at $t+\Delta t$ involves the solution of the structural stiffness matrix at that time step. However, many implicit methods are unconditionally stable for linear analysis and maximum time step length. In the next section the following implicit time-integration methods will be discussed:
- the Houbolt method
- the Wilson-θ method
- the Newmark method (explicit or implicit depending on the choice of the parameters α, γ and β)

9.2.4 Stability

The difference equation

$$\{x_{n+1}\} = [A]\{x_n\} \tag{9.12}$$

is [Strang 88]

- stable if all eigenvalues satisfy $|\lambda_i| < 1$
- neutrally stable if some $|\lambda_i| = 1$ and the other $|\lambda_i| < 1$
- unstable if at least one eigenvalue has $|\lambda_i| > 1$

The matrix $[A]$ is called the amplification matrix.

In the case of unconditionally stable integration methods the solution remains bounded for any time step Δt. For conditionally stable integration methods the solution remains bounded only if Δt is smaller than or equal to a certain critical value Δt_{crit}.

9.3 Explicit Time-Integration

In this section the following explicit time integration methods will be reviewed:

- the central difference method
- the Runge–Kutta methods (for first- and second-order differential equations)

9.3.1 Central Difference Method

This method is based on the finite difference approximation, [Chopra 95, Pilkey 94], of the time derivatives displacement. The central difference method is one of the most widely used among explicit techniques in large-scale structural dynamics programs.

The time step is Δt. The central difference expressions for the velocity $\dot{x}(t)$ and the acceleration $\ddot{x}(t)$ at time t are

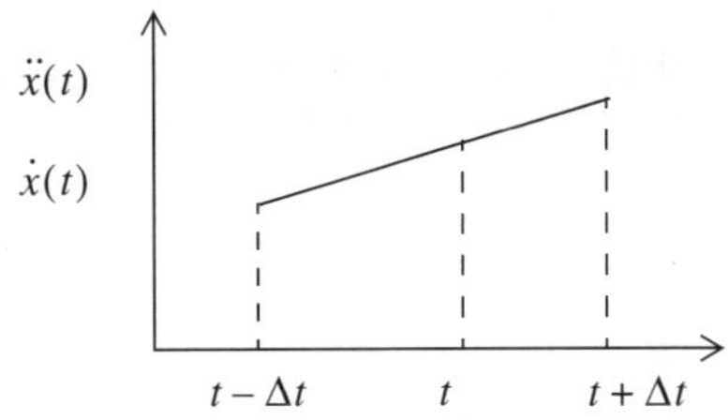

Fig. 9.1. Illustration central difference method

$$\dot{x}(t) = \frac{x(t+\Delta t)-x(t-\Delta t)}{2\Delta t} = \frac{d_{n+1}-d_{n-1}}{2\Delta t} = v_n, \tag{9.13}$$

and

$$\ddot{x}(t) = \frac{\dot{x}\left(t+\frac{1}{2}\Delta t\right)-\dot{x}\left(t-\frac{1}{2}\Delta t\right)}{\Delta t}, \tag{9.14}$$

or

$$\ddot{x}(t) = \frac{x(t+\Delta t)-2x(t)+x(t-\Delta t)}{\Delta t^2} = \frac{d_{n+1}-2d_n+d_{n-1}}{\Delta t^2} = a_n. \tag{9.15}$$

The approximations of the velocity is denoted with v_n, the acceleration with a_n and the displacement d_n. Equations (9.13) and (9.14) are more or less illustrated in Fig. 9.1. In vector notation we may write

$$\{v_n\} = \frac{\{d_{n+1}\}-\{d_{n-1}\}}{2\Delta t}. \tag{9.16}$$

and

$$\{a_n\} = \frac{\{d_{n+1}\}-2\{d_n\}+\{d_{n-1}\}}{\Delta t^2}. \tag{9.17}$$

The equation of motion for the time $t = n\Delta t$ is

$$[M]\{a_n\}+[C]\{v_n\}+[K]\{d_n\} = \{F_n\} \tag{9.18}$$

with the initial conditions

$$\{d_0\} = \{x(0)\} \tag{9.19}$$

$$\{v_0\} = \{\dot{x}(0)\} \tag{9.20}$$

$$\{a_0\} = [M]^{-1}[\{F(0)\} - [C]\{v_0\} - [K]]\{d_0\}. \tag{9.21}$$

From (9.16) and (9.17) substituted into (9.18) we can derive the following equation

$$\left(\frac{1}{\Delta t^2}[M] + \frac{1}{2\Delta t}[C]\right)\{d_{n+1}\} = \{F_n\}$$

$$-\left([K] - \frac{2}{\Delta t^2}[M]\right)\{d_n\} - \left(\frac{1}{\Delta t^2}[M] - \frac{1}{2\Delta t}[C]\right)\{d_{n-1}\}. \tag{9.22}$$

From (9.16) and (9.17) we can derive the following equation

$$\{d_{n-1}\} = \{d_n\} - \Delta t\{v_n\} + \frac{\Delta t^2}{2}\{a_n\}. \tag{9.23}$$

Thus for $n = 0$

$$\{d_{-1}\} = \{d_0\} - \Delta t\{v_0\} + \frac{\Delta t^2}{2}\{a_0\}. \tag{9.24}$$

With use of (9.24) we are able to start the recurrence procedure of the central difference method.

The central difference method is very efficient when the mass matrix $[M]$ is a lumped (diagonal) mass matrix and the damping matrix $[C]$ is diagonal too. This is the case when the modal displacement method will be applied (see (9.3))

The numerical process of the central difference method will be stable for time steps smaller than $\Delta t \le \frac{2}{\omega_{max}} = \frac{1}{\pi f_{max}}$ [Dokainish 89, Chopra 95]. The truncation error is of $O(\Delta t^4)$.

9.3.2 Runge–Kutta Formulae for First-Order Differential Equations

The classical fourth-order Runge–Kutta method [Abramowitz 70, Babuska 66, Dokainish 89, Kreyszig 93, Schwarz 89] has long been popular and is often recommended for accurate numerical computations of solutions for ordinary differential equations. The Runge–Kutta formulas are written as

$$y' = f(x, y) \ , \ y(x_o) = y_o \tag{9.25}$$

$$x_{n+1} = x_n + h \tag{9.26}$$

$$y_{n+1} = y_n + h\phi_f(x, y, h) \tag{9.27}$$

$$\phi_f(x, y, h) = w_1k_1 + w_2k_2 + w_3k_3 + w_4k_4, \tag{9.28}$$

with

- $k_1 = f(x_n, y_n)$
- $k_2 = f(x_n + \alpha_2 h, y_n + \beta_{2,1}hk_1)$
- $k_3 = f(x_n + \alpha_3 h, y_n + \beta_{3,1}hk_1 + \beta_{3,2}hk_2)$
- $k_4 = f(x_n + \alpha_4 h, y_n + \beta_{4,1}hk_1 + \beta_{4,2}hk_2 + \beta_{4,3}hk_3)$

In Table 9.1 the constants for three different Runge–Kutta methods are given. The truncation error is of $O(\Delta t^5)$.

The Runge–Kutta time-integration method is a one-step and a four-stage method.

Table 9.1. Family of Runge-Kutta integration methods

Coefficients	Standard Runge–Kutta Formula	Three over eight Runge–Kutta formula	Gill's formula
w_1	$\frac{1}{6}$	$\frac{1}{8}$	$\frac{1}{6}$
w_2	$\frac{1}{3}$	$\frac{3}{8}$	$\frac{1}{3}\left(1 - \frac{1}{\sqrt{2}}\right)$
w_3	$\frac{1}{3}$	$\frac{3}{8}$	$\frac{1}{3}\left(1 + \frac{1}{\sqrt{2}}\right)$
w_4	$\frac{1}{6}$	$\frac{1}{8}$	$\frac{1}{6}$
α_2	$\frac{1}{2}$	$\frac{1}{3}$	$\frac{1}{2}$
α_3	$\frac{1}{2}$	$\frac{2}{3}$	$\frac{1}{2}$
α_4	1	1	1
$\beta_{2,1}$	$\frac{1}{2}$	$\frac{1}{3}$	$\frac{1}{2}$
$\beta_{3,1}$	0	$-\frac{1}{3}$	$\frac{1}{2}(\sqrt{2} - 1)$

Table 9.1. Family of Runge-Kutta integration methods (Continued)

Coefficients	Standard Runge–Kutta Formula	Three over eight Runge–Kutta formula	Gill's formula
$\beta_{3,2}$	$\frac{1}{2}$	1	$1-\frac{1}{\sqrt{2}}$
$\beta_{4,1}$	0	1	0
$\beta_{4,2}$	0	−1	$-\frac{1}{\sqrt{2}}$
$\beta_{4,3}$	1	1	$1+\frac{1}{\sqrt{2}}$

The equation of motion of an sdof dynamic system has been derived in a previous chapter. We assume underdamped damping ratio $(\zeta < 1)$ characteristics.

$$\ddot{x}(t) + 2\zeta\omega_n\dot{x}(t) + \omega_n^2 x(t) = \frac{F(t)}{m} = f(t). \tag{9.29}$$

We will now introduce the space state variables $y(t)$

$$y(t) = \begin{Bmatrix} y_1(t) \\ y_2(t) \end{Bmatrix} = \begin{Bmatrix} x(t) \\ \dot{x}(t) \end{Bmatrix}. \tag{9.30}$$

Rearranging (9.29), we get the space state equations of motion

$$\begin{Bmatrix} \dot{y}_1(t) \\ \dot{y}_2(t) \end{Bmatrix} = \begin{bmatrix} 0 & 1 \\ -\omega_n^2 & -2\zeta\omega_n \end{bmatrix} \begin{Bmatrix} y_1(t) \\ y_2(t) \end{Bmatrix} + \begin{Bmatrix} 0 \\ f(t) \end{Bmatrix}. \tag{9.31}$$

In the case of (9.2) (9.31) can be written as

$$\begin{Bmatrix} \dot{x} \\ \ddot{x} \end{Bmatrix} = \begin{bmatrix} [0] & [I] \\ -[M]^{-1}[C] & -[M]^{-1}[K] \end{bmatrix} \begin{Bmatrix} \dot{x} \\ x \end{Bmatrix} + [M]^{-1} \begin{Bmatrix} 0 \\ F \end{Bmatrix}, \tag{9.32}$$

and can be solved by the Runge-Kutta time integration method.

$$\begin{Bmatrix} \dot{y}_1 \\ \dot{y}_2 \end{Bmatrix} = \begin{bmatrix} [0] & [I] \\ -[M]^{-1}[C] & -[M]^{-1}[K] \end{bmatrix} \begin{Bmatrix} y_1 \\ \dot{y}_2 \end{Bmatrix} + [M]^{-1} \begin{Bmatrix} 0 \\ F \end{Bmatrix}. \tag{9.33}$$

9.3.3 Runge–Kutta–Nyström Method for S-O Differential Equations

The Runge–Kutta–Nyström method is a fourth-order method and is an extension to the general Runge–Kutta methods for first-order differential equations as discussed in the previous section [Abramowitz 70, Kreyszig 93]. The procedure is given here.

$$y'' = f(x, y, y'),\ y(x_o) = y_o \text{ and } y'(x_o) = y'_o \tag{9.34}$$

$$x_{n+1} = x_n + h \tag{9.35}$$

$$y'_{n+1} = y'_n + \frac{1}{3}(k_1 + 2k_2 + 2k_3 + k_4) \tag{9.36}$$

$$y_{n+1} = y_n + h\left(y'_n + \frac{1}{3}(k_1 + k_2 + k_3)\right), \tag{9.37}$$

with

- $k_1 = \frac{1}{2}hf(x_n, y_n, y'_n)$
- $K = \frac{1}{2}h\left(y'_n + \frac{1}{2}k_1\right)$
- $k_2 = \frac{1}{2}hf\left(x_n + \frac{1}{2}h, y_n + K, y'_n + k_1\right)$
- $k_3 = \frac{1}{2}hf\left(x_n + \frac{1}{2}h, y_n + K, y'_n + k_2\right)$
- $L = h(y'_n + k_3)$
- $k_4 = \frac{1}{2}hf(x_n + h, y_n + L, y'_n + 2k_3)$

9.4 Implicit Time Integration

In this section the following implicit time-integration methods will be described:

- the Houbolt method

- the Wilson-θ method
- Newmark-beta method (explicit or implicit depending on the choice of the parameters α, γ and β)

9.4.1 Houbolt Method

The Houbolt recurrence-matrix solution, to calculate dynamic responses of dynamic systems is proposed in [Houbolt 50, Subbaraj 89, Pilkey 94]. For an sdof system the equation of motion is given by (9.29)

$$\ddot{x}(t) + 2\zeta\omega_n\dot{x}(t) + \omega_n^2 x(t) = \frac{F(t)}{m} = f(t).$$

To solve this equation Houbolt proposed the following difference equations for the velocity $\dot{x}_n$ and the acceleration $\ddot{x}_n$. The approximations of the velocity is denoted with v_n, the acceleration with a_n and the displacement d_n. The numerical solution for the velocity is

$$v_n = \frac{1}{6\Delta t}[11d_n - 18d_{n-1} + 9d_{n-2} - 2d_{n-3}], \tag{9.38}$$

and the numerical solution of the acceleration can be obtained from

$$a_n = \frac{1}{\Delta t^2}[2d_n - 5d_{n-1} + 4d_{n-2} - d_{n-3}]. \tag{9.39}$$

The substitution of (9.38) and (9.39) into (9.29) results in the following relation to solve the displacement d_{n+1}

$$a_{n+1} + 2\zeta\omega_n v_{n+1} + \omega_n^2 d_{n+1} = \frac{F_{n+1}}{m} = f_{n+1}, \tag{9.40}$$

or

$$\left[\frac{2}{\Delta t^2} + \frac{22\zeta\omega_n}{6\Delta t} + \omega_n^2\right]d_{n+1} = f_{n+1} + \left[\frac{5}{\Delta t^2} + \frac{6\zeta\omega_n}{\Delta t}\right]d_n$$

$$-\left[\frac{4}{\Delta t^2} + \frac{3\zeta\omega_n}{\Delta t}\right]d_{n-1} + \left[\frac{1}{\Delta t^2} + \frac{2\zeta\omega_n}{3\Delta t}\right]d_{n-2}, \tag{9.41}$$

or (9.41) written in a shorter way

$$k_{n+1}d_{n+1} = f_{n+1} + k_n d_n - k_{n-1}d_{n-1} + k_{n-2}d_{n-2}. \tag{9.42}$$

In matrix notation

$$\left(\frac{2}{\Delta t^2}[M] + \frac{11}{6\Delta t}[C] + [K]\right)\{d_{n+1}\} = \{F_{n+1}\}$$

$$+\left(\frac{5}{\Delta t^2}[M]+\frac{3}{\Delta t}[C]\right)\{d_n\}-\left(\frac{4}{\Delta t^2}[M]+\frac{3}{2\Delta t}[C]\right)\{d_{n-1}\}$$

$$+\left(\frac{1}{\Delta t^2}[M]+\frac{1}{3\Delta t}[C]\right)\{d_{n-2}\}\ , \tag{9.43}$$

or (9.43) written in a more compact way

$$[K_{n+1}]\{d_{n+1}\} = \{F_{n+1}\}+[K_n]\{d_n\}-[K_{n-1}]\{d_{n-1}\}+[K_{n-2}]\{d_{n-2}\}\,. \tag{9.44}$$

To solve the initial-value problem Houbolt also proposed the following relations for v_n and a_n

$$v_n = \frac{1}{6\Delta t}[2d_{n+1}+3d_n-6d_{n-1}+2d_{n-2}] \tag{9.45}$$

$$a_n = \frac{1}{\Delta t^2}[d_{n+1}-2d_n+4d_{n-1}]\,. \tag{9.46}$$

The initial values are defined as; $d_0 = x_o$, $v_0 = \dot{x}_0 = v_o$ and the associated acceleration can be obtained from (9.29) and is

$$a_0 = f_0-2\zeta\omega_n v_o-\omega_n^2 x_o\,. \tag{9.47}$$

For $n = 0$ the displacements d_{-1} and d_{-2} can be expressed in d_0, d_1, v_o and a_0, using (9.45) and (9.46), and become

$$d_{-1} = 2d_0-d_1+\Delta t^2 a_0 \tag{9.48}$$

$$d_{-2} = 9d_0-8d_1+6\Delta t v_0+6\Delta t^2 a_0\,. \tag{9.49}$$

The starting point of the recurrence solution is $n = 0$. The values of d_{-1} and d_{-2} are known and expressed in the initial conditions, d_0 and d_1, thus (9.41) becomes

$$(k_{n+1}-k_{n-1}+8k_{n-2})d_1 = f_1+(k_n-2k_{n-1}+9k_{n-2})d_0$$

$$-(k_{n-1}-6k_{n-2})\Delta t^2 a_0+6\Delta t k_{n-2}v_o \tag{9.50}$$

or in matrix notation

$$([K_{n+1}]-[K_{n-1}]+8[K_{n-2}])\{d_1\} = \{f_1\}+([K_n]-2[K_{n-1}]+9[K_{n-2}])\{d_0\}$$

$$-([K_{n-1}]-6[K_{n-2}])\Delta t^2\{a_0\}+6\Delta t K_{n-2}\{v_0\}\,. \tag{9.51}$$

Now we may proceed with the recurrence solution for d_2 or $\{d_2\}$ and so on with (9.42) or (9.44).

The Houbolt method, which is an implicit method, is unconditionally stable, so there is no critical time step limit Δt. The most significant drawback associated with the Houbolt integrator is the algoritmic damping, which is inherent in the numerical procedure, and is introduced into the response when large time steps are used. In linear problems this damping can cause the transient response to decay so severely that the static response is obtained, [Subbraraj 89]. The truncation error is of $O(\Delta t^4)$.

9.4.2 Wilson–theta Method

The Wilson-θ method is also discussed in [Pilkey 94, Subbraraj 89].

For any τ so that $t \le \tau \le t + \theta\Delta t$ we have assuming a linear acceleration

$$\ddot{x}_{t+\tau} = \ddot{x}_t + \frac{\tau}{\theta\Delta t}(\ddot{x}_{t+\theta\Delta t} - \ddot{x}_t)\,. \tag{9.52}$$

Equation (9.52) is illustrated in Fig. 9.2.

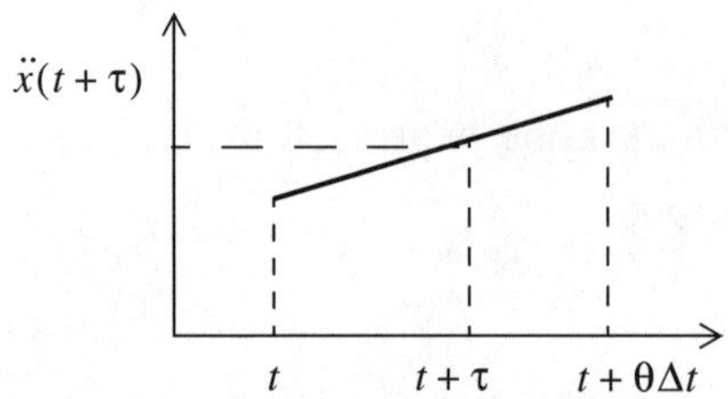

Fig. 9.2. Illustration calculation $\ddot{x}(t+\tau)$

The velocity $\dot{x}_{t+\tau}$ can be obtained by integrating (9.52) with respect to time

$$\dot{x}_{t+\tau} = \int_0^\tau \ddot{x}_{t+\alpha}d\alpha = \dot{x}_t + \ddot{x}_t\tau + \frac{\tau^2}{2\theta\Delta t}(\ddot{x}_{t+\theta\Delta t} - \ddot{x}_t)\,, \tag{9.53}$$

and the displacement $x_{t+\tau}$ is obtained by integrating (9.53) again with respect to time

$$x_{t+\tau} = \int_0^\tau \dot{x}_{t+\alpha}d\alpha = x_t + \dot{x}_t\tau + \frac{\ddot{x}_t\tau^2}{2} + \frac{\tau^3}{6\theta\Delta t}(\ddot{x}_{t+\theta\Delta t} - \ddot{x}_t)\,. \tag{9.54}$$

If we substitute for $\tau = \theta\Delta t$ into (9.53) and (9.54) the result will be

$$v_{n+\theta} = \dot{x}_{t+\theta\Delta t} = \dot{x}_t + \frac{\theta\Delta t}{2}(\ddot{x}_{t+\theta\Delta t} + \ddot{x}_t), \tag{9.55}$$

and

$$d_{n+\theta} = x_{t+\theta\Delta t} = x_t + \dot{x}_t\theta\Delta t + \frac{(\theta\Delta t)^2}{6}(\ddot{x}_{t+\theta\Delta t} + 2\ddot{x}_t). \tag{9.56}$$

Equation (9.2) can be written as

$$[M]\{a_{n+\theta}\} + [C]\{v_{n+\theta}\} + [K]\{d_{n+\theta}\} = \{F(t_{n+\theta})\}, \tag{9.57}$$

with initial conditions

$$\{d_0\} = \{x(0)\} \tag{9.58}$$

$$\{v_0\} = \{\dot{x}(0)\} \tag{9.59}$$

$$\{a_0\} = [M]^{-1}[\{F(0)\} - [C]\{v_0\} - [K]]\{d_0\} \tag{9.60}$$

$$\{F(t_{n+\theta})\} = \{F(t_n)\} + \theta[\{F(t_{n+1})\} - \{F(t_n)\}] \tag{9.61}$$

$$\{v_{n+\theta}\} = \{v_n\} + \frac{\theta\Delta t}{2}(\{a_{n+\theta}\} + \{a_n\}) \tag{9.62}$$

$$\{d_{n+\theta}\} = \{d_n\} + \theta\Delta t\{v_n\} + \frac{(\theta\Delta t)^2}{6}(\{a_{n+\theta}\} + 2\{a_n\}). \tag{9.63}$$

For linear structures the Wilson-θ method is unconditionally stable for $\theta \geq 1.39$, [Subbraraj 89]. If we substitute (9.61), (9.62) and (9.63) into (9.60) the accelerations $\{a_{n+\theta}\}$ will be solved.

$$\left([M] + \frac{\theta\Delta t}{2}[C] + \frac{(\theta\Delta t)^2}{6}[K]\right)\{a_{n+\theta}\} = \{F_n\} + \theta(\{F_{n+1}\} - \{F_n\})$$

$$-([K])\{d_n\} - ([C] + \theta\Delta t[K])\{v_n\} - \left(\frac{\theta\Delta t}{2}[C] + \frac{(\theta\Delta t)^2}{3}[K]\right)\{a_n\} \quad . \tag{9.64}$$

The accelerations, velocities and displacements can now be solved using (9.52), (9.53) and (9.54), respectively, with $\tau = \Delta t$

$$\{a_{n+1}\} = \{a_n\} + \frac{1}{\theta}[\{a_{n+\theta}\} - \{a_n\}] \tag{9.65}$$

$$\{v_{n+1}\} = \{v_n\} + \{a_n\}\Delta t + \frac{\Delta t}{2\theta}[\{a_{n+\theta}\} - \{a_n\}] \tag{9.66}$$

$$\{d_{n+1}\} = \{d_n\} + \{v_n\}\Delta t + \{a_n\}\frac{\Delta t^2}{2} + \frac{\Delta t^2}{6\theta}[\{a_{n+\theta}\} - \{a_n\}]. \tag{9.67}$$

We may rewrite (9.63) such that

$$\{a_{n+\theta}\} = \frac{6}{(\theta\Delta t)^2}(\{d_{n+\theta}\} - \{d_n\}) - \frac{6}{\theta\Delta t}\{v_n\} - 2\{a_n\}. \tag{9.68}$$

Substituting (9.68) in (9.62) we obtain

$$\{v_{n+\theta}\} = \frac{3}{\theta\Delta t}(\{d_{n+\theta}\} - \{d_n\}) - 2\{v_n\} - \frac{\theta\Delta t}{2}\{a_n\}. \tag{9.69}$$

If we substitute (9.68), (9.69) and (9.63) into (9.60) the displacements $\{d_{n+\theta}\}$ are solved.

$$\begin{aligned}\left(\frac{6}{(\theta\Delta t)^2}[M] + \frac{3}{\theta\Delta t}[C] + [K]\right)\{d_{n+\theta}\} &= \{F_n\} + \theta(\{F_{n+1}\} - \{F_n\}) \\ &+ \left(\frac{6}{(\theta\Delta t)^2}[M] + \frac{3}{\theta\Delta t}[C]\right)\{d_n\} + \left(\frac{6}{\theta\Delta t}[M] + 2[C]\right)\{v_n\} + \left(2[M] + \frac{\theta\Delta t}{2}[C]\right)\{a_n\},\end{aligned} \tag{9.70}$$

With (9.68) substituted into (9.65) we obtain

$$\{a_{n+1}\} = \frac{6}{\theta^3\Delta t^2}(\{d_{n+\theta}\} - \{d_n\}) - \frac{6}{\theta^2\Delta t}\{v_n\} + \left(1 - \frac{3}{\theta}\right)\{a_n\}. \tag{9.71}$$

Using (9.63) with $\theta = 1$ the velocities at time t_{n+1} can be solved

$$\{v_{n+1}\} = \{v_n\} + \frac{\Delta t}{2}[\{a_{n+1}\} + \{a_n\}]. \tag{9.72}$$

Using (9.62) with $\theta = 1$ the displacements at time t_{n+1} can be solved

$$\{d_{n+1}\} = \{d_n\} + \{v_n\}\Delta t + \frac{\Delta t^2}{6}[\{a_{n+1}\} + 2\{a_n\}]. \tag{9.73}$$

9.4.3 Newmark–beta Method

The Newmark-beta method, is described in [Belytschko 00, Hughes 83, Pilkey 94, Subbraraj 89]. The equations of motion of a damped mdof system are given by (9.1) and/or (9.3). The basic form of the Newmark-beta method is given by

$$\{d_{n+1}\} = \{d_n\} + \{v_n\}\Delta t + \left[\left(\frac{1}{2} - \beta\right)\{a_n\} + \beta\{a_{n+1}\}\right]\Delta t^2 \tag{9.74}$$

$$\{v_{n+1}\} = \{v_n\} + [(1-\gamma)\{a_n\} + \gamma\{a_{n+1}\}]\Delta t. \tag{9.75}$$

We will rearrange (9.74) as

$$\{a_{n+1}\} = \frac{1}{\beta\Delta t^2}[\{d_{n+1}\} - \{d_n\}] - \frac{1}{\beta\Delta t}\{v_n\} - \left(\frac{1}{2\beta} - 1\right)\{a_n\}. \tag{9.76}$$

Substituting (9.76) into (9.75) we obtain

$$\{v_{n+1}\} = \frac{\gamma}{\beta\Delta t}[\{d_{n+1}\} - \{d_n\}] + \left(1 - \frac{\gamma}{\beta}\right)\{v_n\} - \left(\frac{\gamma}{2\beta} - 1\right)\Delta t\{a_n\}. \tag{9.77}$$

The equation of motion for time $t = (n+1)\Delta t$ is

$$[M]\{a_{n+1}\} + [C]\{v_{n+1}\} + [K]\{d_{n+1}\} = \{F_{n+1}\} \tag{9.78}$$

with initial conditions

$$\{d_0\} = \{x(0)\} \tag{9.79}$$

$$\{v_0\} = \{\dot{x}(0)\} \tag{9.80}$$

$$\{a_0\} = [M]^{-1}[\{F(0)\} - [C]\{v_0\} - [K]]\{d_0\}, \tag{9.81}$$

in which $n = 0, 1, ..., N-1$ is the number of time steps and Δt is the time step.
The acceleration vector $\{a_{n+1}\}$ can be solved if (9.74) and (9.75) are substituted into (9.78) using the appropriate initial conditions or start values.

$$([M] + \gamma\Delta t[C] + \beta\Delta t^2[K])\{a_{n+1}\} = \{F_{n+1}\}$$

$$- [K]\{d_n\} - ([C] + \Delta t[K])\{v_n\} - \left((1-\gamma)\Delta t[C] + \left(\frac{1}{2} - \beta\right)\Delta t^2[K]\right)\{a_n\}. \tag{9.82}$$

If we substitute (9.76) and (9.77) into (9.78) we can solve $\{d_{n+1}\}$

$$\left(\frac{1}{\beta\Delta t^2}[M] + \frac{\gamma}{\beta\Delta t}[C] + [K]\right)\{d_{n+1}\} = \{F_{n+1}\}$$

$$\left(\frac{1}{\beta\Delta t^2}[M] + \frac{\gamma}{\beta\Delta t}[C]\right)\{d_n\} + \left(\frac{1}{\beta\Delta t}[M] + \left(\frac{\gamma}{\beta} - 1\right)[C]\right)\{v_n\}$$

$$+\left(\left(\frac{1}{2\beta} - 1\right)[M] + \Delta t\left(\frac{\gamma}{2\beta} - 1\right)[C]\right)\{a_n\}. \tag{9.83}$$

The Newmark-beta method is unconditionally stable for a linear system when

- $\gamma \geq \frac{1}{2}$

- $\beta \geq \frac{(2\gamma+1)^2}{16}$

Positive algoritmic damping is introduced if $\gamma > 0.5$ and negative algoritmic damping leading to an unbounded response if $\gamma < 0.5$. Thus, in most applications $\gamma = 0.5$ is used. If $\gamma > 0.5$ and $\beta < \frac{\gamma}{2}$ the following condition must be met

$$\omega_n t \leq \Omega_{crit} = \frac{\zeta(\gamma - 0.5) + \left[\frac{\gamma}{2} - \beta + \zeta^2(\gamma - 0.5)^2\right]^{\frac{1}{2}}}{\left(\frac{\gamma}{2} - \beta\right)},$$

where ζ is the damping ratio. The damping ratio may be obtained assuming $\langle 2\zeta_i\omega_i\rangle = [\Phi]^T[C][\Phi]$ with $[\Phi]$ as the modal base.

Variations of the parameters γ and β will lead to other well-known time integrators very much related to the Newmark-beta method [Subbraraj 89].

Table 9.2. Properties of Newmark-beta method

Method	Type	γ	β	Stability	Order of accuracy
Average acceleration (trapezoidal rule)	Implicit	0.5	0.25	unconditional	$O(\Delta t^2)$
Linear acceleration	Implicit	0.5	1/6	$\Omega_{crit} = 2\sqrt{3}$	$O(\Delta t^2)$
Central difference[a]	Explicit	0.5	0	$\Omega_{crit} = 2$	$O(\Delta t^2)$

a. The acceleration vector $\{a_{n+1}\}$ will be solved instead of the displacement vector $\{d_{n+1}\}$.

9.4.4 The Hughes, Hilber and Taylor (HHT) alpha–Method

The HHT method, the α-method, is described in [Belytschko 00]. The HHT method is based on the Newmark-beta method. The equations of motion of a damped mdof system are given by (9.1) and/or (9.3). The HHT method is a one-step, three-stage, numerically dissipative integration method. The HHT method improves numerical dissipation for high-frequencies without degrading the accuracy as much and combines minimum low-frequency dissipation with high- frequency dissipation. The basic form of the HHT-α method is given by

$$\{d_{n+1}\} = \{d_n\} + \{v_n\}\Delta t + \left[\left(\frac{1}{2} - \beta\right)\{a_n\} + \beta\{a_{n+1}\}\right]\Delta t^2 \tag{9.84}$$

$$\{v_{n+1}\} = \{v_n\} + [(1-\gamma)\{a_n\} + \gamma\{a_{n+1}\}]\Delta t \tag{9.85}$$

$$[M]\{a_{n+1}\} + [C]\{v_{n+\alpha}\} + [K]\{d_{n+\alpha}\} = \{F(t_{n+\alpha})\} \tag{9.86}$$

$$\{d_0\} = \{x(0)\} \tag{9.87}$$

$$\{v_0\} = \{\dot{x}(0)\} \tag{9.88}$$

$$\{a_0\} = [M]^{-1}[\{F(0)\} - [C]\{v_0\} - [K]]\{d_0\}, \tag{9.89}$$

where

$$\{d_{n+\alpha}\} = (1+\alpha)\{d_{n+1}\} - \alpha\{d_n\} \tag{9.90}$$

$$\{v_{n+\alpha}\} = (1+\alpha)\{v_{n+1}\} - \alpha\{v_n\} \tag{9.91}$$

$$\{t_{n+\alpha}\} = (1+\alpha)\{t_{n+1}\} - \alpha\{t_n\} \tag{9.92}$$

$$\{F(t_{n+\alpha})\} = (1+\alpha)\{F(t_{n+1})\} - \alpha\{F(t_n)\}, \tag{9.93}$$

in which $n = 0, 1, ..., N-1$ is the number of time steps and Δt is the time step.
The HHT method is unconditionally stable for a linear system when

- $\alpha \in \left[-\frac{1}{3}, 0\right]$
- $\gamma = \frac{1-2\alpha}{2}$
- $\beta = \frac{(1-\alpha)^2}{4}$

9.4.5 The Wood, Bossak and Zienkiewicz (WBZ) alpha–Method

The WBZ method, the α-method, is described in [Wood 90]. This method is based on the Newmark-beta method. The equations of motion of a damped mdof system are given by (9.1) and/or (9.3). The WBZ method is a one-step, three-stage, numerically dissipative integration method. The WBZ method improves numerical dissipation for high-frequencies without degrading the accuracy as much. It combines minimum low-frequency dissipation with high-frequency dissipation. The basic form of the WBZ-α method is given by

$$\{d_{n+1}\} = \{d_n\} + \{v_n\}\Delta t + \left[\left(\frac{1}{2} - \beta\right)\{a_n\} + \beta\{a_{n+1}\}\right]\Delta t^2 \tag{9.94}$$

$$\{v_{n+1}\} = \{v_n\} + [(1-\gamma)\{a_n\} + \gamma\{a_{n+1}\}]\Delta t \tag{9.95}$$

$$[M]\{a_{n+\alpha_B}\} + [C]\{v_{n+1}\} + [K]\{d_{n+1}\} = \{F(t_{n+1})\} \tag{9.96}$$

$$\{d_0\} = \{x(0)\} \tag{9.97}$$

$$\{v_0\} = \{\dot{x}(0)\} \tag{9.98}$$

$$\{a_0\} = [M]^{-1}[\{F(0)\} - [C]\{v_0\} - [K]]\{d_0\}, \tag{9.99}$$

where

$$\{a_{n+\alpha_B}\} = (1-\alpha_B)\{a_{n+1}\} + \alpha_B\{a_n\}, \tag{9.100}$$

in which $n = 0, 1, ..., N-1$ is the number of time steps and Δt is the time step.

The WBZ method is unconditionally stable for a linear system when

- $\alpha_\beta \in \left[-\frac{1}{3}, 0\right]$
- $\gamma = \dfrac{1-2\alpha_\beta}{2}$
- $\beta = \dfrac{(1-\alpha_\beta)^2}{4}$

9.4.6 The Generalised–alpha Algorithm

The generalised-α algorithm is described in [Chung 93]. The equations of motion of a damped mdof system are given by (9.1) and or (9.3). The generalised-α algorithm is a one-step, three-stage, numerically dissipative integration method. The algorithm combines minimum low-frequency dissipation with high-frequency dissipation. The time integration is unconditionally stable and possesses second order accuracy. The basic form of the generalised-α algorithm is given by

$$\{d_{n+1}\} = \{d_n\} + \{v_n\}\Delta t + \left[\left(\frac{1}{2} - \beta\right)\{a_n\} + \beta\{a_{n+1}\}\right]\Delta t^2 \tag{9.101}$$

$$\{v_{n+1}\} = \{v_n\} + [(1-\gamma)\{a_n\} + \gamma\{a_{n+1}\}]\Delta t \tag{9.102}$$

$$[M]\{a_{n+1-\alpha_m}\} + [C]\{v_{n+1-\alpha_f}\} + [K]\{d_{n+1-\alpha_f}\} = \{F(t_{n+1-\alpha_f})\} \tag{9.103}$$

$$\{d_0\} = \{x(0)\} \tag{9.104}$$

$$\{v_0\} = \{\dot{x}(0)\} \tag{9.105}$$

$$\{a_0\} = [M]^{-1}[\{F(0)\} - [C]\{v_0\} - [K]]\{d_0\}, \tag{9.106}$$

where

$$\{d_{n+1-\alpha_f}\} = (1-\alpha_f)\{d_{n+1}\} + \alpha_f\{d_n\} \tag{9.107}$$

$$\{v_{n+1-\alpha_f}\} = (1-\alpha_f)\{v_{n+1}\} + \alpha_f\{v_n\} \tag{9.108}$$

$$\{a_{n+1-\alpha_m}\} = (1-\alpha_m)\{a_{n+1}\} + \alpha_m\{a_n\} \tag{9.109}$$

$$\{t_{n+1-\alpha_f}\} = (1-\alpha_f)\{t_{n+1}\} + \alpha_f\{t_n\} \tag{9.110}$$

$$\{F(t_{n+1-\alpha_f})\} = (1-\alpha_f)\{F(t_{n+1})\} + \alpha_f\{F(t_n)\}, \tag{9.111}$$

in which $n = 0, 1, ..., N-1$ are the number of time steps and Δt is the time step.
The generalised-α algorithm is unconditionally stable for a linear system when

- $\alpha_m \le \alpha_f \le \frac{1}{2}$
- $\beta \ge \frac{1}{4} + \frac{1}{2}(\alpha_f - \alpha_m)$

The generalised-α algorithm is second-order accurate, providing

- $\gamma = \frac{1}{2} - \alpha_m + \alpha_f$

9.5 Piecewise Linear Method

In the Chap. "shock-response spectrum" the piecewise linear method is described in great detail to calculate the shock response spectrum of time responses. This is a time integrator only for an sdof dynamic system. The piecewise linear method is based on theoretical solution of the equation of motion, however, the applied loads are linearised in the time frame of interest [Nigam 68, Ebeling 97, Kelly 69]

9.6 Problems

9.6.1 Problem 1

Look at the coupled pendulum shown in Fig. 9.3 [Hairer 92]. All initial values are equal to zero at $t = 0$. The first pendulum is pushed into movement by a force $f(t)$ (N).

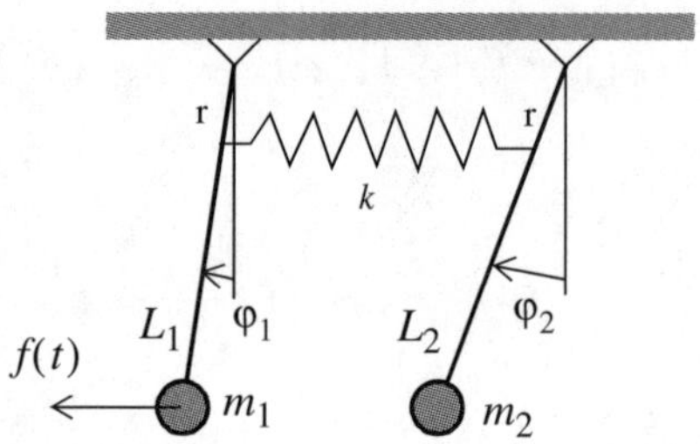

Fig. 9.3. Coupled pendulum

$$f(t) = \begin{cases} 0.01\sqrt{1-(1-t)^2} & |t-1| \le 1 \\ 0 & \text{otherwise} \end{cases} .$$

The nonlinear equations of motion for both masses are:

$$\ddot{\varphi}_1 = -\frac{g\sin\varphi_1}{L_1} - \frac{kr^2}{m_1 L_1^2}(\sin\varphi_1 - \sin\varphi_2)\cos\varphi_1 + \frac{f(t)}{m_1 L_1}$$

$$\ddot{\varphi}_2 = -\frac{g\sin\varphi_2}{L_2} - \frac{kr^2}{m_2 L_2^2}(\sin\varphi_2 - \sin\varphi_1)\cos\varphi_2$$

$$\begin{Bmatrix} \ddot{\varphi}_1 \\ \ddot{\varphi}_2 \end{Bmatrix} = -\begin{bmatrix} \dfrac{g\sin\varphi_1}{L_1} + \dfrac{kr^2}{m_1 L_1^2}(\sin\varphi_1 - \sin\varphi_2)\cos\varphi_1 \\ \dfrac{g\sin\varphi_2}{L_2} + \dfrac{kr^2}{m_2 L_2^2}(\sin\varphi_2 - \sin\varphi_1)\cos\varphi_2 \end{bmatrix} + \begin{Bmatrix} \dfrac{f(t)}{m_1 L_1} \\ 0 \end{Bmatrix} .$$

The linearised equations of motion around $\varphi_1 = \varphi_2 = 0$ become

$$\ddot{\varphi}_1 = -\frac{g\varphi_1}{L_1} - \frac{kr^2}{m_1 L_1^2}(\varphi_1 - \varphi_2) + \frac{f(t)}{m_1 L_1}$$

$$\ddot{\varphi}_2 = -\frac{g\varphi_2}{L_2} - \frac{kr^2}{m_2 L_2^2}(\varphi_2 - \varphi_1)$$

$$\begin{bmatrix} m_1 L_1^2 & 0 \\ 0 & m_2 L_2^2 \end{bmatrix} \begin{Bmatrix} \ddot{\varphi}_1 \\ \ddot{\varphi}_2 \end{Bmatrix} + \begin{bmatrix} g m_1 L_1 + kr^2 & -kr^2 \\ -kr^2 & g m_2 L_2 + kr^2 \end{bmatrix} \begin{Bmatrix} \varphi_1 \\ \varphi_2 \end{Bmatrix} = \begin{Bmatrix} L_1 f(t) \\ 0 \end{Bmatrix} . \quad (9.112)$$

We choose the following somewhat modified parameters, [Hairer 93], $L_1 = L_2 = 1$ m, $m_1 = 1$ kg, $m_2 = 0.99$ kg, $r = 0.1$ m, $k = 2$ N/m, $g = 1$ m/s^2, $t_{\text{end}} = 600$ s.

The linearised equations of motion are solved using the Wilson-θ method, with $\theta = 1.4$ and $\Delta t = 0.05$ s.

The solutions for φ_1 and φ_2 are shown in Fig. 9.4 and in Fig. 9.5.

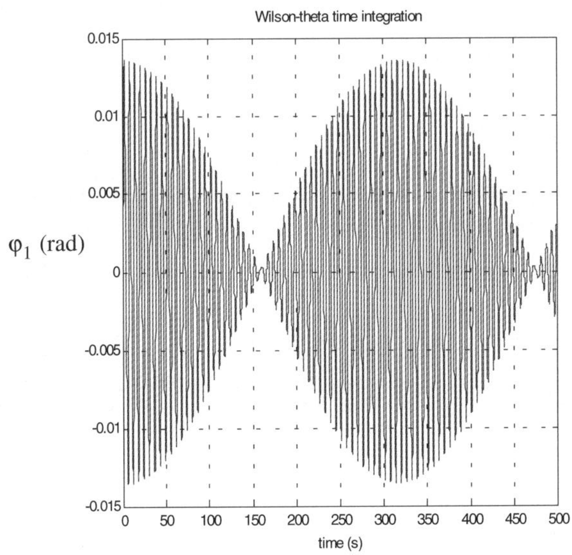

Fig. 9.4. Response φ_1

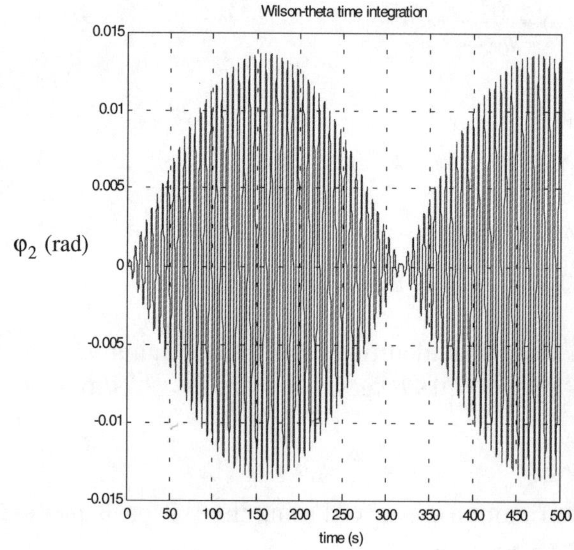

Fig. 9.5. Response φ_2

Solve (9.112) with the central difference method and the generalised α algorithm as described in previous sections.

9.6.2 Problem 2

Solve (9.112) numerically using the Runge–Kutta methods for the first-order and the second-order differential equation.

9.6.3 Problem 3

Solve (9.112) numerically using the Houbolt method.

10 Shock-Response Spectrum

10.1 Introduction

Separation of stages, the separation of the spacecraft from the last stage of the launch vehicle will induce very short duration loads in the internal structure of the spacecraft, the so-called shockloads. The duration of the shockload is, in general, very short with respect to the duration associated with the fundamental natural frequencies of the loaded dynamic mechanical system.

The effects of the shock loads are generally depicted in a shock-response spectrum (SRS). The SRS is essentially a plot that shows the responses of a number of single degree of freedom (sdof) systems to an excitation. The excitation is usually an acceleration–time history.

A SRS is generated by calculating the maximum response of a sdof system to a particular base transient excitation. Many sdof systems tuned to a range of natural frequencies are assessed using the same input-time history. A damping value must be selected in the analysis. A damping ratio of $\zeta = 0.05$, $Q = 10$, is commonly used. The final plot, the SRS, looks like a frequency-domain plot. It shows the largest response encountered for a particular sdof system anywhere within the analysed time. Thus the SRS provides an estimate of the response of an actual product and its various components to a given transient input (i.e. shock pulse) [Grygier 97].

A typical example of a time-history acceleration and associated SRS as illustrated in Fig. 10.1 and Fig. 10.2, are extracted from NASA-STD-7003 [Mulville 99].

In this chapter, the response of an sdof system, due to enforced acceleration, will be reviewed. Furthermore the calculation of SRSs will be discussed in detail. The maximum values occurring in time histories will be compared with the SRS approach and finally it will be shown how an existing SRS can be matched (with synthesised decaying sinusoids).

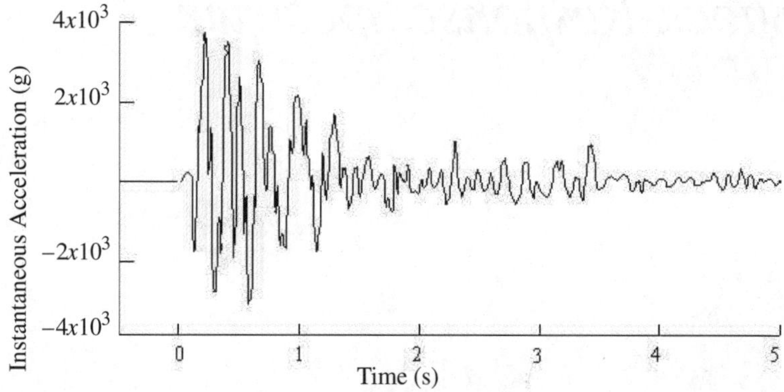

Fig. 10.1. Typical pyroshock acceleration time-history [Mulville 99]

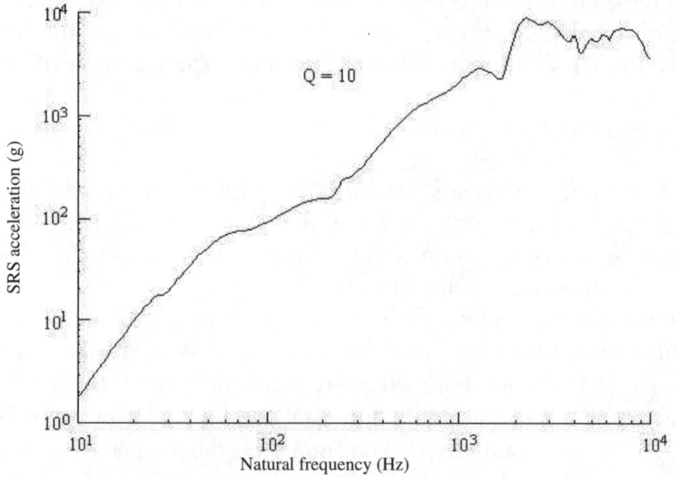

Fig. 10.2. Typical pyroschock maximum shock response spectrum (SRS) [Mulville 99]

10.2 Enforced Acceleration

An sdof system with a discrete mass m, a damper element c and a spring element k is placed on a moving base that is accelerated with an acceleration $\ddot{u}(t)$. The resulting displacement of the mass is $x(t)$. We introduce the natural (circular) frequency $\omega_n = \sqrt{\frac{k}{m}}$, the damped circular frequency $\omega_d = \omega_n\sqrt{1-\zeta^2}$, the critical damping constant $c_{crit} = 2\sqrt{km}$ and the damping ratio $\zeta = \frac{c}{c_{crit}}$. The amplification factor is defined as $Q = \frac{1}{2\zeta}$ where $Q = 10$ is generally assumed.

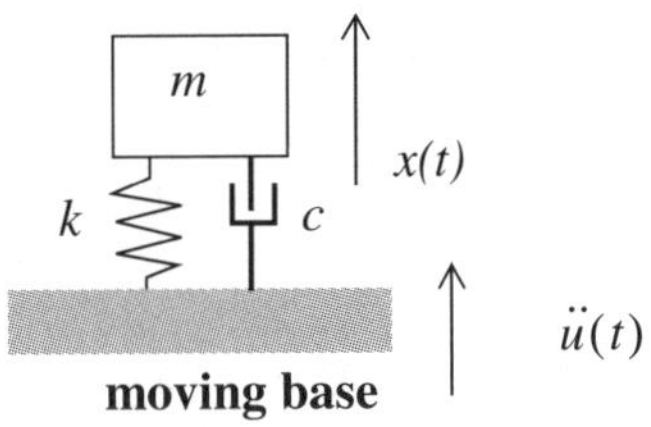

Fig. 10.3. Enforced acceleration on a damped sdof system

We introduce a relative motion $z(t)$, which is the displacement of the mass with respect to the base. The relative displacement is

$$z(t) = x(t) - u(t). \tag{10.1}$$

The equation of motion for the relative motion $z(t)$ is

$$\ddot{z}(t) + 2\zeta\omega_n\dot{z}(t) + \omega_n^2 z(t) = -\ddot{u}(t). \tag{10.2}$$

The enforced acceleration of the sdof system is transformed into an external force. The absolute displacement $x(t)$ can be calculated with

$$\ddot{x}(t) = \ddot{z}(t) + \ddot{u}(t) = -2\zeta\omega_n\dot{z}(t) - \omega_n^2 z(t). \tag{10.3}$$

The solution of (10.2), taking the initial condition with respect to displacement $z(0)$ and velocity $\dot{z}(0)$ into account, is

$$z(t) = z(0)e^{-\zeta\omega_n t}\left(\cos\omega_d t + \frac{\zeta}{\sqrt{1-\zeta^2}}\sin\omega_d t\right)$$

$$+ \dot{z}(0)e^{-\zeta\omega_n t}\frac{\sin\omega_d t}{\omega_d} - \int_0^t e^{-\zeta\omega_n\tau}\frac{\sin\omega_d\tau}{\omega_d}\ddot{u}(t-\tau)d\tau. \tag{10.4}$$

For SRS calculations $z(0) = \dot{z}(0) = 0$, hence

$$z(t)= -\int_0^t e^{-\zeta\omega_n\tau}\frac{\sin\omega_d\tau}{\omega_d}\ddot{u}(t-\tau)d\tau = -\int_0^t e^{-\zeta\omega_n(t-\tau)}\frac{\sin\omega_d(t-\tau)}{\omega_d}\ddot{u}(\tau)d\tau\,. \quad (10.5)$$

After differentiation of (10.5) with respect to time [Kelly 69] the relative velocity $\dot{z}(t)$ becomes

$$\dot{z}(t)= -\int_0^t e^{-\zeta\omega_d(t-\tau)}\cos(\omega_d(t-\tau))\ddot{u}(\tau)d\tau - \zeta\omega_n z(t)\,. \quad (10.6)$$

The absolute acceleration $\ddot{x}(t)$ can be obtained by applying (10.3) [Kelly 69]

$$\ddot{x}(t) = 2\zeta\omega_n\int_0^t e^{-\zeta\omega_n(t-\tau)}\cos(\omega_d(t-\tau))\ddot{u}(\tau)d\tau + \omega_n(2\zeta^2-1)z(t)\,. \quad (10.7)$$

The maximum acceleration $\ddot{x}(t)$ can be calculated by inserting the natural frequency $\omega_n = 2\pi f_n$ (rad/s) of the sdof system for every natural frequency. The maximum acceleration $\ddot{x}(t)$ will be plotted against the number of cycles per second f_n (Hz). This plot is called the shock-response spectrum (SRS) of the base excitation $\ddot{u}(t)$.

10.3 Numerical Calculation of the SRS, the Piecewise Exact Method

In this section, two similar methods of calculating numerically transient responses of sdof dynamic systems, are discussed:

1. A method as discussed in [Nigam 68, Ebeling 97]
2. A method as discussed by [Kelly 69]

In both methods the forcing function is assumed to vary linearly in a piecewise fashion and, based upon this assumption, an exact solution is determined.

The equation of relative $z(t)$ motion of the sdof dynamic system exposed to a base acceleration $\ddot{u}(t)$ is given by (10.2)

$$\ddot{z}(t) + 2\zeta\omega_n\dot{z}(t) + \omega_n^2 z(t) = -\ddot{u}(t)\,.$$

The base acceleration $\ddot{u}(t)$ is mostly given in a discrete form in a table; acceleration versus time. We assume a linear variation of the acceleration between two time steps t_j and t_{j+1}. The acceleration $\ddot{u}(t_{j+1})$ is expressed in terms of the acceleration $\ddot{u}(t_j)$. The time increment is $\Delta t_j = t_{j+1} - t_j$ and the increment of the acceleration is $\Delta\ddot{u}(t_j) = \ddot{u}(t_{j+1}) - \ddot{u}(t_j)$.

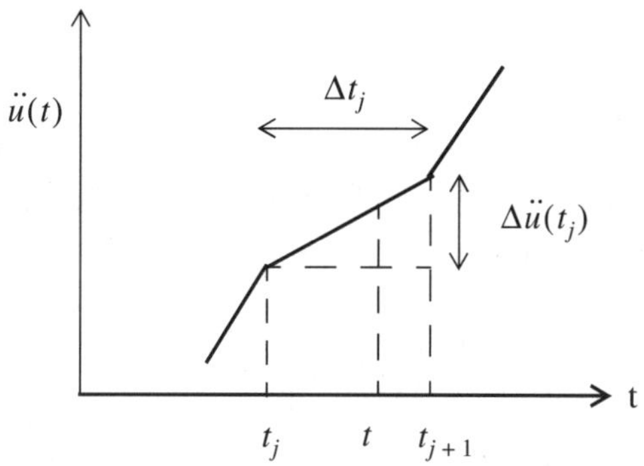

Fig. 10.4. Linearisation numerical scheme of acceleration $\ddot{u}(t)$

The acceleration $\ddot{u}(t_{j+1})$ at the time t_{j+1} becomes

$$\ddot{u}(t) = \ddot{u}(t_j) + \frac{\Delta\ddot{u}(t_j)}{\Delta t_j}(t - t_j), \quad t_j \le t \le t_{j+1}. \tag{10.8}$$

Equation (10.2) will be rewritten

$$\ddot{z}(t) + 2\zeta\omega_n\dot{z}(t) + \omega_n^2 z(t) = -\ddot{u}(t_j) - \frac{\Delta\ddot{u}(t_j)}{\Delta t_j}(t - t_j), \quad t_j \le t \le t_{j+1}. \tag{10.9}$$

The solution of (10.9) is

$$z(t) = z(t_j)e^{-\zeta\omega_n(t-t_j)}\left(\cos\omega_d(t-t_j) + \frac{\zeta}{\sqrt{1-\zeta^2}}\sin\omega_d(t-t_j)\right)$$

$$+ \dot{z}(t_j)e^{-\zeta\omega_n(t-t_j)}\frac{\sin\omega_d(t-t_j)}{\omega_d} - \int_{t_j}^{t} e^{-\zeta\omega_n(t-\tau)}\frac{\sin\omega_d(t-\tau)}{\omega_d}\ddot{u}(\tau)d\tau \tag{10.10}$$

The integral in (10.10) is given by [Kelly 69].

$$\int_{t_j}^{t} e^{-\zeta\omega_n(t-\tau)}\frac{\sin\omega_d(t-\tau)}{\omega_d}\ddot{u}(\tau)d\tau =$$

$$= \frac{-\ddot{u}(t_j)}{\omega_n^2}\left[1 - e^{-\zeta\omega_n(t-t_j)}\left(\cos\omega_d(t-t_j) + \frac{\zeta}{\sqrt{1-\zeta^2}}\sin\omega_d(t-t_j)\right)\right]$$

$$-\frac{\Delta\ddot{u}(t_j)}{\omega_n^2}\left[1 - \frac{2\zeta}{\omega_n(t-t_j)}(1 - e^{-\zeta\omega_n(t-t_j)}\cos\omega_d(t-t_j))\right.$$

$$\left.+\frac{\Delta\ddot{u}(t_j)}{\omega_n^2}\left[\frac{(1-2\zeta^2)}{\omega_d(t-t_j)}e^{-\zeta\omega_n(t-t_j)}\sin\omega_d(t-t_j)\right]\right. \quad (10.11)$$

The state vector $(z, \dot{z})^T$ at time t_{j+1} can be expressed in the state vector at time t_j and the piecewise linear given base acceleration $\ddot{u}$ at both t_j and t_{j+1} [Nigam 68, Gupta 92, Ebeling 97],

$$\begin{Bmatrix} z(t_{j+1}) \\ \dot{z}(t_{j+1}) \end{Bmatrix} = [A]\begin{Bmatrix} z(t_j) \\ \dot{z}(t_j) \end{Bmatrix} + [B]\begin{Bmatrix} \ddot{u}(t_j) \\ \ddot{u}(t_{j+1}) \end{Bmatrix}, \quad (10.12)$$

with

- $[A] = \begin{bmatrix} a_{11} & a_{12} \\ a_{21} & a_{22} \end{bmatrix}$

and

- $[B] = \begin{bmatrix} b_{11} & b_{12} \\ b_{21} & b_{22} \end{bmatrix}$

The absolute acceleration $\ddot{x}(t_{j+1})$ is given by (10.3)

$$\ddot{x}(t_{j+1}) = -2\zeta\omega_n\dot{z}(t_{j+1}) - \omega_n^2 z(t_{j+1}). \quad (10.13)$$

With $\Delta t_j = t_{j+1} - t_j$, the elements of the matrix $[A]$ are

- $a_{11} = e^{-\zeta\omega_n\Delta t_j}\left(\frac{\zeta}{\sqrt{1-\zeta^2}}\sin\omega_d\Delta t_j + \cos\omega_d\Delta t_j\right)$
- $a_{12} = e^{-\zeta\omega_n\Delta t_j}\frac{\sin\omega_d\Delta t_j}{\omega_d}$
- $a_{21} = -\frac{\omega_n}{\sqrt{1-\zeta^2}}e^{-\zeta\omega_n\Delta t_j}\sin\omega_d\Delta t_j$

- $a_{22} = e^{-\zeta\omega_n\Delta t_j}\left(\cos\omega_d\Delta t_j - \frac{\zeta}{\sqrt{1-\zeta^2}}\sin\omega_d\Delta t_j\right)$

With $\Delta t_j = t_{j+1} - t_j$ the elements of the matrix $[B]$ are

- $b_{11} = e^{-\zeta\omega_n\Delta t_j}\left(\left[\frac{2\zeta^2-1}{\omega_n^2\Delta t_j} + \frac{\zeta}{\omega_n}\right]\frac{\sin\omega_d\Delta t_j}{\omega_d} + \left[\frac{2\zeta}{\omega_n^3\Delta t_j} + \frac{1}{\omega_n^2}\right]\cos\omega_d\Delta t_j\right) - \frac{2\zeta}{\omega_n^3\Delta t_j}$

- $b_{12} = -e^{-\zeta\omega_n\Delta t_j}\left(\left(\frac{2\zeta^2-1}{\omega_n^2\Delta t_j}\right)\frac{\sin\omega_d\Delta t_j}{\omega_d} + \frac{2\zeta}{\omega_n^3\Delta t_j}\cos\omega_d\Delta t_j\right) - \frac{1}{\omega_n^2} + \frac{2\zeta}{\omega_n^3\Delta t_j}$

- $b_{21} = e^{-\zeta\omega_n\Delta t_j}\left[\frac{2\zeta^2-1}{\omega_n^2\Delta t_j} + \frac{\zeta}{\omega_n}\right]\left(\cos(\omega_d\Delta t_j) - \frac{\zeta}{\sqrt{1-\zeta^2}}\sin\omega_d\Delta t_j\right)$

  $-e^{-\zeta\omega_n\Delta t_j}\left[\frac{2\zeta}{\omega_n^3\Delta t_j} + \frac{1}{\omega_n^2}\right](\omega_d\sin\omega_d\Delta t_j + \zeta\omega_n\cos\omega_d\Delta t_j) + \frac{1}{\omega_n^2\Delta t_j}$

- $b_{22} = -e^{-\zeta\omega_n\Delta t_j}\left(\frac{2\zeta^2-1}{\omega_n^2\Delta t_j}\left(\cos\omega_d\Delta t_j - \frac{\zeta}{\sqrt{1-\zeta^2}}\sin\omega_d\Delta t_j\right)\right)$

  $-e^{-\zeta\omega_n\Delta t_j}\left(\frac{-2\zeta}{\omega_n^3\Delta t_j}(\omega_d\sin\omega_d\Delta t_j + \zeta\omega_n\cos\omega_d\Delta t_j)\right) - \frac{1}{\omega_n^2\Delta t_j}$

In [Gupta 92] the following expressions for b_{21} and b_{22} are given:

- $b_{21} = -\frac{a_{11}-1}{\omega_n^2\Delta t_j} - a_{12}$
- $b_{22} = -b_{21} - a_{12}$

In [Kelly 69] a very similar numerical approach, as discussed by [Nigam 68, Gupta 92, Ebeling 97], is proposed

$$z(t_{j+1}) = B_1 z(t_j) + B_2\dot{z}(t_j) + B_3\ddot{u}(t_j) + B_4\Delta\ddot{u}(t_j) \tag{10.14}$$

$$\frac{\dot{z}(t_{j+1})}{\omega_n} = B_6 z(t_j) + B_7\dot{z}(t_j) + B_8\ddot{u}(t_j) + B_9\Delta\ddot{u}(t_j), \tag{10.15}$$

with (10.13)

$$\ddot{x}(t_{j+1}) = -2\zeta\omega_n\dot{z}(t_{j+1}) - \omega_n^2 z(t_{j+1})$$

where

- $\Delta t_j = t_{j+1} - t_j$

- $\Delta\ddot{u}(t_j) = \ddot{u}(t_{j+1}) - \ddot{u}(t_j)$
- $B_1 = e^{-\zeta\omega_n\Delta t_j}\left(\frac{\zeta}{\sqrt{1-\zeta^2}}\sin\omega_d\Delta t_j + \cos\omega_d\Delta t_j\right)$
- $B_2 = e^{-\zeta\omega_n\Delta t_j}\frac{\sin\omega_d\Delta t_j}{\omega_d}$
- $B_3 = \frac{1}{\omega_n^2}(1 - B_1)$
- $B_4 = \frac{1}{\omega_n^2}\left[1 - \frac{2\zeta}{\omega_n\Delta t_j}(1 - e^{-\zeta\omega_n\Delta t_j}\cos\omega_d\Delta t_j) - (1-2\zeta^2)\left(e^{-\zeta\omega_n\Delta t_j}\frac{\sin\omega_d\Delta t_j}{\omega_d\Delta t_j}\right)\right]$
- $B_6 = -\omega_n B_2$
- $B_7 = \frac{e^{-\zeta\omega_n\Delta t_j}}{\omega_n}\left(\cos\omega_d\Delta t_j - \frac{\zeta}{\sqrt{1-\zeta^2}}\sin\omega_d\Delta t_j\right)$
- $B_8 = -\frac{B_2}{\omega_n}$
- $B_9 = \frac{B_1 - 1}{\omega_n^3\Delta t_j}$

For the calculation of the SRS the following parameters are important [Assink 95]:

1. The damping ratio ζ of the sdof dynamic system.
2. The number of sdof systems for which the maximum response is calculated.
3. The minimum time frame of the transient T_{min} (s). The minimum time frame is the maximum of either $T_{min} \geq \frac{1}{f_{min}}$ or twice the maximum shock time $T_{min} \geq 2t_{shock}$.
4. The time increment Δt must be less than 10% of the reciprocal value of the maximum frequency f_{max} (Hz) involved in the calculation of the SRS, $\Delta t \leq \frac{0.1}{f_{max}}$. The minimum number of time steps n within the time frame T_{min} is $n = \frac{T_{min}}{\Delta t} = 10\frac{f_{max}}{f_{min}}$.

A half-sine pulse $\ddot{u}_{\text{base}} = 200\sin\frac{\pi t}{\tau}$, ($0 \le t \le \tau$) $= 0.0005$ s and $\ddot{u}_{\text{base}} = 0$, $t < 0, t > \tau$ is applied to the base of series of sdof dynamic systems to calculate the SRS of the HSP. The total time is $t_{\text{end}} = 0.05$ s and $\Delta t = 0.00001 \le \frac{0.1}{f_{\max}} = \frac{0.1}{3000} = 0.00003$ s. The damping ratio $\zeta = 0.05$, $Q = 10$. The Kelly method is applied to obtain the SRS.

The calculated SRS (absolute acceleration) is illustrated in Fig. 10.5.

10.4 Response Analysis in Combination with Shock-Response Spectra

A multi-dof linear system, excited with an acceleration $\ddot{u}_{\text{base}}$ at the base, is represented by the equation

$$[M]\{\ddot{x}\} + [C]\{\dot{x}\} + [K]\{x\} = \{0\}. \tag{10.16}$$

The matrix equation for the relative displacement vector $\{z\} = \{x\} - \{u\}$, the relative velocities $\{\dot{z}\} = \{\dot{x}\} - \{\dot{u}\}$ and the relative acceleration $\{\ddot{z}\} = \{\ddot{x}\} - \{\ddot{u}\}$, with respect to the base, can be written as

$$[M]\{\ddot{z}\} + [C]\{\dot{z}\} + [K]\{z\} = -[M]\{T\}\ddot{u}_{\text{base}}, \tag{10.17}$$

with $\{T\}$ the rigid-body vector with respect to the base.
From the undamped eigenvalue problem (10.17)

$$([K] - \lambda_i[M])\{\phi_i\} = \{0\}, \tag{10.18}$$

the eigenvalues λ_i and associated modes $\{\phi_i\}$ can be obtained and used for the modal analysis (modal displacement method (MDM)) approach. We assume

$$\{z\} = [\phi_1, \phi_2, \phi_3, ...]\begin{Bmatrix} \eta_1 \\ \eta_2 \\ \eta_3 \\ . \end{Bmatrix} = [\Phi]\{\eta\}, \tag{10.19}$$

where

- $[\Phi]$ the modal matrix.
- $\{\eta\}$ the vector of generalised coordinates.

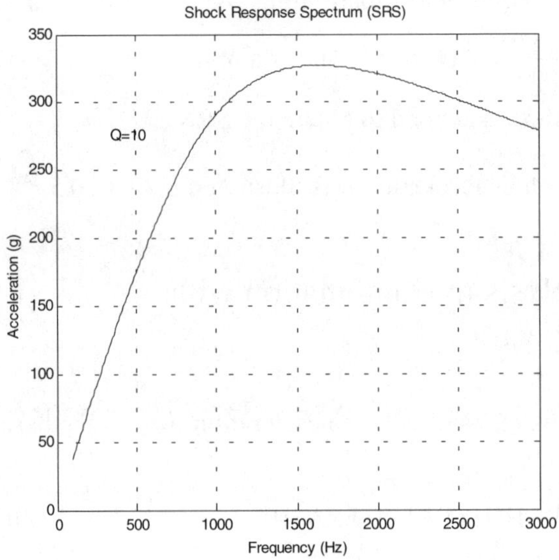

Fig. 10.5. The SRS of a half-sine pulse (HSP) with amplitude A=200 g and a time duration $\tau = 0.0005$ s

The modes are orthogonal with respect to the mass matrix $[M]$ and the stiffness matrix $[K]$,

$$[\Phi]^T[M][\Phi] = \langle m \rangle, \ \ [\Phi]^T[K][\Phi] = \langle \lambda_i m_i \rangle = \langle \omega_i^2 m_i \rangle . \tag{10.20}$$

If we introduce the modal damping $c_i = 2\zeta_i\omega_i$ the equations of motion expressed in the generalised coordinates become

$$\ddot{\eta}_i + 2\zeta_i\omega_i\dot{\eta}_i + \omega_i^2\eta_i = \frac{-[\phi_i]^T[M]\{T\}}{[\phi_i]^T[M][\phi_i]}\ddot{u}_{\text{base}} = -\Gamma_i\ddot{u}_{\text{base}} , \tag{10.21}$$

with

- ζ_i the modal damping ratio with respect to mode 'i'.
- ω_i the natural frequency corresponding with mode $\{\phi_i\}$.
- Γ_i the modal participation factor.

Equation (10.21) is similar to (10.2). Equation (10.2) is applied to calculate the maximum (peak) response of the sdof system to obtain the SRS corresponding with the base acceleration $\ddot{u}_{\text{base}}$. The peak responses, or SRS of the sdof dynamic systems as described by (10.21), will be a fraction Γ_i of the base acceleration SRS of $\ddot{u}_{\text{base}}$. The acceleration SRS of the generalised coordinate $\ddot{\eta}_i$ is given by

$$SRS(\ddot{\eta}_i) = \Gamma_i SRS(\ddot{u}_{\text{base}}) . \tag{10.22}$$

The modal contribution to the SRS of the physical degrees of freedom; $\{z\}, \{\dot{z}\}, \{\ddot{z}\}$ become

$$SRS(\ddot{z}, \omega_i) = \{\phi_i\} SRS(\ddot{\eta}_i) , \tag{10.23}$$

and the contribution to the SRS of the velocities

$$SRS(\dot{z}, \omega_i) = \frac{\{\phi_i\} SRS(\ddot{\eta}_i)}{\omega_i} , \tag{10.24}$$

and the contribution to the SRS of the displacement

$$SRS(z, \omega_i) = \frac{\{\phi_i\} SRS(\ddot{\eta}_i)}{\omega_i^2} . \tag{10.25}$$

The SRS of the absolute acceleration $SRS(\ddot{x})$ can be obtained in a similar way as for the SRS for the relative acceleration $SRS(\ddot{z})$.

The total SRS for the absolute acceleration is a particular summation over all the modal contributions $SRS(\ddot{x}, \omega_i)$. In [Gupta 92, Haelsig 72] two summation methods were discussed. The first one is an absolute summation taking all modes into account

$$SRS(\ddot{x}) = \sum_{i=1}^{n} \left| SRS(\ddot{x}, \omega_i) \right| , \tag{10.26}$$

and the second one is the square root of the summation of the squared values, the SRSS value

$$SRS(\ddot{x}) = \sqrt{\sum_{i=1}^{n} \{SRS(\ddot{x}, \omega_i)\}^2} . \tag{10.27}$$

A four mass-spring system with the discrete mass $m = 5$ kg, the spring stiffness $k = 1000000$ N/m and the damping ratio is $\zeta = 0.05$ ($Q = 10$) is illustrated in Fig. 10.6.

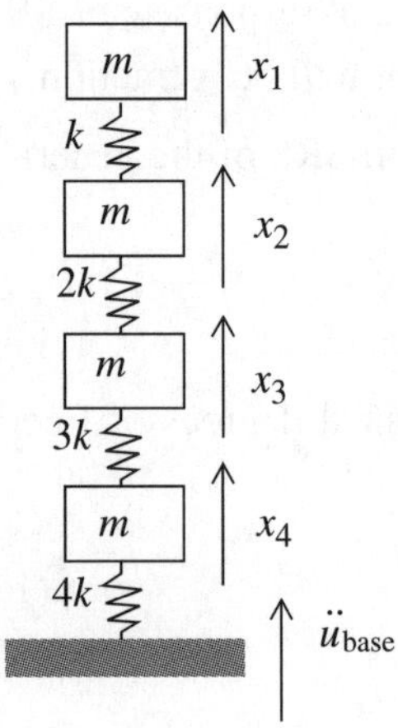

Fig. 10.6. Four mass–spring–system

The base acceleration is:

1. A half-sine pulse $\ddot{u}_{\text{base}} = 200\sin\frac{\pi t}{\tau}$, $0 = t \le \tau$ g and $\ddot{u}_{\text{base}} = 0$, $t < 0, t > \tau$ with $\tau = 0.0005$ s.
2. A Shock Response Spectrum based upon a HSP with an amplitude $A = 200$ g, a time duration $\tau = 0.0005$ s and $Q = 10$.

For both cases the acceleration transient responses $\{\ddot{x}\}$ and the $SRS(\ddot{x}, f)$ will be calculated and compared.

We will solve (10.17)

$$[M]\{\ddot{z}\} + [C]\{\dot{z}\} + [K]\{z\} = -[M]\{T\}\ddot{u}_{\text{base}}$$

applying the modal displacement method (MDM) and taking all 4 modes into account. The absolute displacement vector is $\{x\} = \{z\} + \{u\}$, the absolute velocity vector $\{\dot{x}\} = \{\dot{z}\} + \{\dot{u}\}$ and the absolute acceleration vector $\{\ddot{x}\} = \{\ddot{z}\} + \{\ddot{u}\}$. To calculate the spring forces only the relative displacement vector $\{z\}$ is required. The force matrix $[S]$ is defined as

$$[S] = k\begin{bmatrix} 1 & -1 & 0 & 0 \\ 0 & 2 & -2 & 0 \\ 0 & 0 & 3 & -3 \\ 0 & 0 & 0 & 4 \end{bmatrix}.$$

The forces in the springs are $\{F\} = [S]\{z\}$.

The physical relative degrees of freedom $\{z\}$ are transformed into the generalised coordinates $\{\eta\}$ using the modal matrix $[\Phi]$, thus $\{z\} = [\Phi]\{\eta\}$.

The stress modes can be calculated with

$$[\Phi_\sigma] = [S][\Phi].$$

The decoupled equations of motion expressed in the generalised coordinates $\{\eta\}$ and adding the "ad hoc" modal damping ratio ζ become

$$\ddot{\eta}_i + 2\zeta_i\omega_i\dot{\eta} + \omega_i^2\eta_i = \frac{-\{\phi_i\}^T[M]\{T\}}{\{\phi_i\}^T[M]\{\phi_i\}}\ddot{u}_{base} = -\Gamma_i\ddot{u}_{\text{base}} = f.$$

To solve the acceleration in the time domain, the Newmark-β method [Wood 90] will be applied with $\beta = 0.25$ and $\gamma = 0.5$.

$$\{\eta\}_{n+1} = \{\eta\}_n + \Delta t\{\dot{\eta}\}_n + \frac{\Delta t^2}{2}(1-2\beta)\{\ddot{\eta}\}_n + \Delta t^2\beta\{\ddot{\eta}\}_{n+1}$$

$$\{\dot{\eta}\}_{n+1} = \{\dot{\eta}\}_n + \Delta t(1-\gamma)\{\ddot{\eta}\}_n + \Delta t\gamma\{\ddot{\eta}\}_{n+1}$$

$$[D] = [I] + \gamma\Delta t\langle 2\zeta_i\omega_i\rangle + \beta\Delta t^2\langle\omega_i^2\rangle$$

$$[D]\{\ddot{\eta}\}_{n+1} = \{f\}_{n+1} - \langle 2\zeta_i\omega_i\rangle(\{\dot{\eta}\}_n + \Delta t(1-\gamma)\{\ddot{\eta}\}_n)$$
$$-\langle\omega_i^2\rangle\left(\{\eta\}_n + \Delta t\{\dot{\eta}\}_n + \frac{\Delta t^2}{2}(1-2\beta)\{\ddot{\eta}\}_n\right).$$

At $t = 0$, $\{\ddot{\eta}\}_1 = \{\ddot{\eta}(0)\}$

$$\{\ddot{\eta}\}_1 = \{f\}_1 - \langle 2\zeta_i\omega_i\rangle\{\dot{\eta}\}_1 - \langle\omega_i^2\rangle\{\eta\}_1,$$

with the initial conditions

- $\{\dot{\eta}\}_1 = \{\dot{\eta}(0)\} = ([\Phi]^T[\Phi])^{-1}[\Phi]^T\{\dot{z}(0)\}$
- $\{\eta\}_1 = \{\eta(0)\} = ([\Phi]^T[\Phi])^{-1}[\Phi]^T\{z(0)\}$

The natural frequencies $\{f_n\}$ (Hz) and associated mode shapes $[\Phi]$ of the dynamic system illustrated in Fig. 10.6 are

$$\{f_n\} = \begin{Bmatrix} 40.4 \\ 94.0 \\ 151.6 \\ 218.2 \end{Bmatrix}\text{Hz},\ [\Phi] = \begin{bmatrix} 0.7766 & -0.5978 & 0.1972 & -0.0232 \\ 0.5261 & 0.4458 & -0.6974 & 0.1950 \\ 0.3160 & 0.5785 & 0.4372 & -0.6118 \\ 0.1420 & 0.3303 & 0.5325 & 0.7663 \end{bmatrix}.$$

The stress modes become $[\Phi_\sigma] = [S][\Phi]$

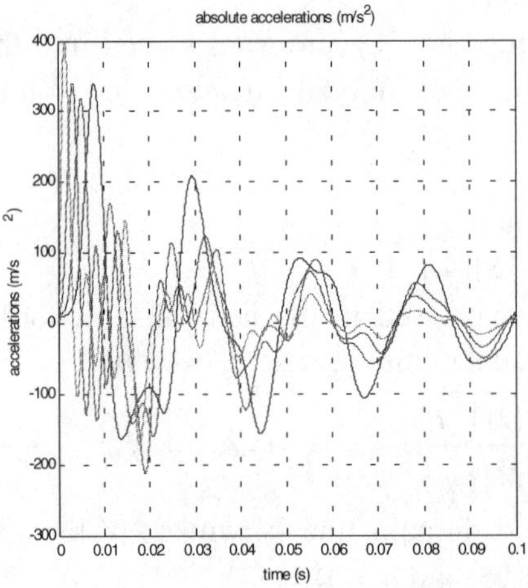

Fig. 10.7. Accelerations

$$[\Phi_\sigma] = 10^6 \begin{bmatrix} 0.2505 & -1.0437 & 0.8946 & -0.2182 \\ 0.4202 & -0.2653 & -2.2693 & 1.6134 \\ 0.5221 & 0.7446 & -0.2857 & -4.1342 \\ 0.5679 & 1.3213 & 2.1299 & 3.0651 \end{bmatrix}.$$

The vector of modal participation factors $\{\Gamma\}$ is

$$\{\Gamma\} = \begin{Bmatrix} 1.7608 \\ 0.7568 \\ 0.4695 \\ 0.3262 \end{Bmatrix}.$$

Question 1

With the initial conditions $z(0) = 0$ and $\dot{z}(0) = 0$ the time acceleration $\{\ddot{x}(t)\}$ are calculated and illustrated in Fig. .

The relative displacement $\{z(t)\}$ is calculated and shown in Fig. 10.8. The time histories of the spring forces $\{F(t)\}$ are illustrated in Fig. 10.9.

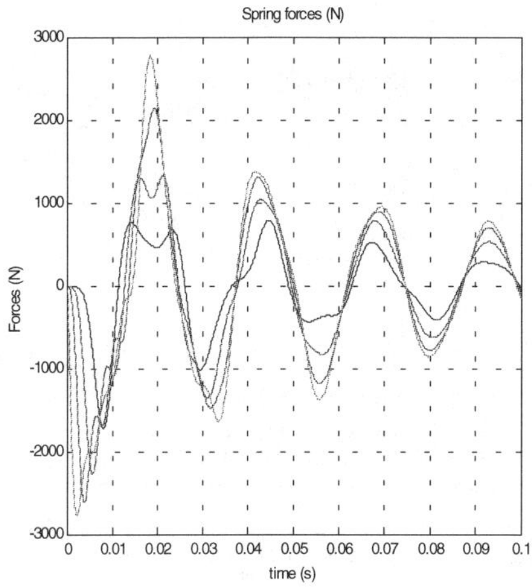

Fig. 10.8. Relative displacements

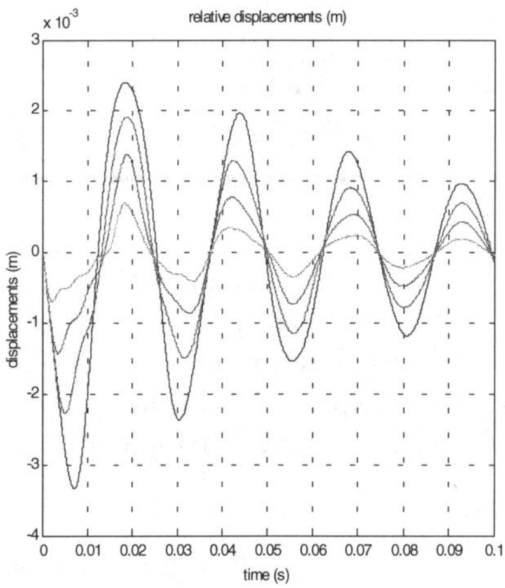

Fig. 10.9. Spring forces

The maximum values of the time histories of the absolute acceleration $\{\ddot{x}(t)\}$ and the spring forces $\{F(t)\}$ are given in Table 10.1.

Table 10.1. Maximum values of the accelerations and the spring forces

Node #	Absolute acceleration (m/s^2)	Spring #	Spring force (N)
x_1	340	x_1–x_2	786
x_2	319	x_2–x_3	1331
x_3	342	x_3–x_4	2145
x_4	400	x_4	2774

Table 10.2. SRS acceleration/displacement generalised coordinates

Mode #	Natural frequency f_i (Hz)	SRS (f_i) (m/s^2)	Modal participation factor Γ_i	Γ_i SRS (f_i) (m/s^2)	$\dfrac{\Gamma_i \mathrm{SRS}(f_1)}{(2\pi f_1)^2}$ (m)
1	40.4233	147.6260	1.7608	259.9357	4.0294e-3
2	94.0432	342.8672	0.7568	259.4977	0.7432e-3
3	151.6007	550.8845	0.4695	258.6392	$0.2851x10^{-3}$
4	218.1650	788.2119	0.3262	257.1538	$0.1369x10^{-3}$

Question 2

The Shock spectra for the acceleration and the displacement per mode are given in Table 10.2. The shock spectra per mode are given in Table 10.3, and the absolute and SRSS values of the acceleration in Table 10.4. The calculation of the shock spectra values of the spring force can be found in Table 10.5, and the absolute and SRSS values are given in Table 10.6.

Observations

The SRS approach, absolute and SRSS values, bounds the maximum values of the time histories.

Table 10.3. Acceleration

Node #	Mode 1 $\{\phi_i\}\Gamma_i \mathrm{SRS}(f_i)$ (m/s^2)	Mode 2 $\{\phi_i\}\Gamma_i \mathrm{SRS}(f_i)$ (m/s^2)	Mode 3 $\{\phi_i\}\Gamma_i \mathrm{SRS}(f_i)$ (m/s^2)	Mode 4 $\{\phi_i\}\Gamma_i \mathrm{SRS}(f_i)$ (m/s^2)
x_1	201.8738	–155.1394	51.0037	–5.9718
x_2	136.7599	115.6969	–180.3807	50.1338

Table 10.3. Acceleration (Continued)

Node #	Mode 1 $\{\phi_i\}\Gamma_i SRS(f_i)$ (m/s^2)	Mode 2 $\{\phi_i\}\Gamma_i SRS(f_i)$ (m/s^2)	Mode 3 $\{\phi_i\}\Gamma_i SRS(f_i)$ (m/s^2)	Mode 4 $\{\phi_i\}\Gamma_i SRS(f_i)$ (m/s^2)
x_3	82.1471	150.1255	113.0865	−157.3186
x_4	36.9065	85.7168	137.7211	197.0530

Table 10.4. SRS($\ddot{x}$) acceleration

Node #	$\sum_1^4 \lvert\{\phi_i\}\Gamma_i SRS(f_i)\rvert$ (m/s^2)	$\sqrt{\sum_1^4 [\{\phi_i\}\Gamma_i SRS(f_i)]^2}$ (m/s^2)
x_1	414	260
x_2	483	259
x_3	503	259
x_4	457	258

Table 10.5. Forces per mode

Spring #	Mode 1 (N)	Mode 2 (N)	Mode 3 (N)	Mode 4 (N)
x_1–x_2	1009.4	−775.7	255.0	−29.9
x_2–x_3	1693.2	−197.2	−646.9	220.8
x_3–x_4	2103.9	553.4	−81.5	−565.8
x_4	2288.4	982.0	607.2	419.5

Table 10.6. Spring forces

Spring #	Sum Absolute values (N)	SRSS (N)
x_1-x_2	2070	1299
x_2-x_3	2758	1837
x_3-x_4	3305	2249
x_4	4297	2597

10.5 Matching Shock Spectra with Synthesised Time Histories

It is not possible to run an SRS on a shaker table, because it has no time history. The calculation of a time history from a given or specified SRS (time-history synthesis) is not unique and the recalculation of a time history is a process of trial and error [Smallwood 74a]. It is assumed that a time history that results in a SRS in accordance with the given or specified SRS will cause the same damage in the structure under test. However, the time-history synthesis is very much dependent on the physical limitations of the exciter. These limitations are illustrated in Table 10.7 [Smallwood 74a]:

Table 10.7. Exciter limitations

Limitation #	Initial	Final	Maximum
1	$\ddot{u}_{\text{base}}(0) = 0$	$\ddot{u}_{\text{base}}(T) = 0$	limited
2	$\dot{u}_{\text{base}}(0) = 0$	$\dot{u}_{\text{base}}(T) = 0$	limited
3	$u_{\text{base}}(0) = 0$	$u_{\text{base}}(T) = 0$	limited

The acceleration is actually limited by the force capabilities of the exciter and the start and the final acceleration, velocity and displacement of the applied transient must be zero.

Smallwood [Smallwood 74a] lists a number of possible transients that meet the limitations indicated in Table 10.7:

- Sums of decaying sinusoids
- Sum of waveforms
- Shaker optimised cosines

- Fast sine sweeps
- Modulated random noise
- Classical pulses

A discussion of all possible techniques for time-history synthesis is beyond the scope of this book. Only sums of decaying sinusoids will be discussed.

Decaying sinusoids

The equation of motion for the relative motion $z(t)$ is (10.2)

$$\ddot{z}(t) + 2\zeta\omega_n\dot{z}(t) + \omega_n^2 z(t) = -\ddot{u}(t) .$$

The solution of (10.2), with initial condition with respect to displacement $z(0)$ and velocity $\dot{z}(0)$ is (10.4)

$$z(t) = z(0)e^{-\zeta\omega_n t}\left(\cos\omega_d t + \frac{\zeta}{\sqrt{1-\zeta^2}}\sin\omega_d t\right)$$

$$+ \dot{z}(0)e^{-\zeta\omega_n t}\frac{\sin\omega_d t}{\omega_d} - \int_0^t e^{-\zeta\omega_n\tau}\frac{\sin\omega_d\tau}{\omega_d}\ddot{u}(t-\tau)d\tau .$$

For SRS calculations $z(0) = \dot{z}(0) = 0$, hence

$$z(t) = -\int_0^t e^{-\zeta\omega_n\tau}\frac{\sin\omega_d\tau}{\omega_d}\ddot{u}(t-\tau)d\tau = -\int_0^t e^{-\zeta\omega_n(t-\tau)}\frac{\sin\omega_d(t-\tau)}{\omega_d}\ddot{u}(\tau)d\tau$$

If the base excitation $\ddot{u}$ is equal to the Dirac delta function $\delta(\tau)$ then

$$\ddot{z}(t) = -e^{-\zeta\omega_n t}\frac{\sin\omega_d t}{\omega_d} = -h(t) ,\ t \geq 0 . \tag{10.28}$$

Assuming $\ddot{z}(t)$ is the acceleration response of one of the generalised coordinates of the uncoupled equations of motion, the transient response of a structure consists of the superposition of decaying sinusoids that have been exposed to shocks (delta functions). Thus excitation consisting of sums of decaying sinusoids appears to be a natural choice [Nelson 74] as a transient vibration test of substructures and components exposed to shocks environments. The usual basic decaying sinusoid is given by [Smallwood 74b]

$$g_i(t) = \begin{bmatrix} A_i e^{-\zeta_i\omega_i t}\sin(\omega_i t) \ ,\ t \geq 0 \\ 0 \ ,\ t < 0 \end{bmatrix} . \tag{10.29}$$

The associated velocity $v_i(t) = \int_0^t g_i(\tau)d\tau$, with $v_i(0) = 0$ becomes

$$v_i(t) = \frac{A_i\{e^{-\zeta_i\omega_i t}[\cos(\omega_i t) + \zeta_i \sin(\omega_i t)] - 1\}}{\omega_i(\zeta_i^2 + 1)}, \tag{10.30}$$

and the corresponding displacement $s_i(t) = \int_0^t v_i(\tau)d\tau$, with $s_i(0) = 0$ is

$$s_i(t) = \frac{A_i\left\{e^{-\zeta_i\omega_i t}[2\zeta_i\cos(\omega_i t) - \sin(\omega_i t) + \zeta_i^2\sin(\omega_i t)] + \zeta_i^2\omega_i t + \zeta_i\omega_i t - 2\zeta_i\right\}}{\omega_i^2(\zeta_i^2 + 1)^2}, \tag{10.31}$$

with

- $A_i = 1$ m/s^2, amplitude
- $\zeta_i = 0.05$, decay rate
- $\omega_i = 25$ rad/s, circular frequency

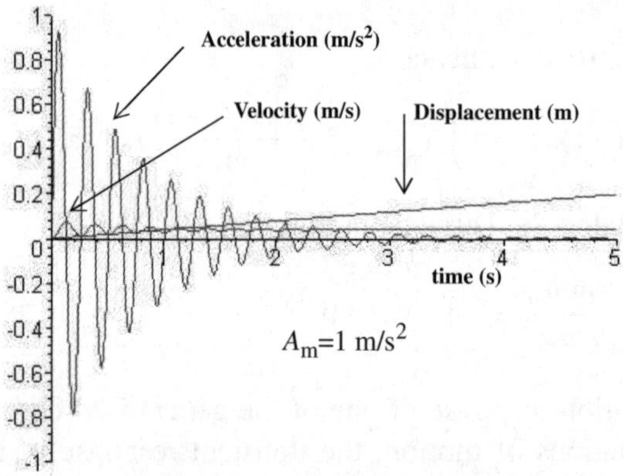

Fig. 10.10. Acceleration, velocity and displacement associated with the associated decaying sinusoids.

The plot in Fig. 10.10 shows that the velocity and the displacement transients do not converge to zero with increasing time, i.e. the constraints as stated in Table 10.7 are thus violated when basis decaying sinusoids are applied.

Velocity and displacement compensation is needed to achieve zero velocity and displacement with increasing time. Smallwood and Nord [Smallwood 74b] and Nelson and Prasthofer [Nelson 74] suggested compensation methods to obtain zero velocity and displacement with increasing time.

Smallwood and Nord

$$\ddot{u}_{\text{base}} = \sum_{i=1}^{n} g_i(t) + U(t+\tau) A_{\text{m}} e^{-\zeta_{\text{m}}\omega_{\text{m}}(t+\tau)} \sin\omega_{\text{m}}(t+\tau) \tag{10.32}$$

$$g_i(t) = A_i U(t-\tau_i) e^{-\zeta_i\omega_i(t-\tau_i)} \sin\omega_i(t-\tau_i), \tag{10.33}$$

with

- $A_{\text{m}} = -\omega_{\text{m}}(1+\zeta_{\text{m}}^2)\sum_{i=1}^{n}\frac{A_i}{\omega_i(1+\zeta_i^2)}$.
- $\tau = \frac{\omega_{\text{m}}(1+\zeta_{\text{m}}^2)}{A_{\text{m}}}\left\{\frac{2\zeta_{\text{m}}A_{\text{m}}}{\omega_{\text{m}}^2(1+\zeta_{\text{m}}^2)^2} + \sum_{i=1}^{n}\left[\frac{A_i\tau_i}{\omega_i(1+\zeta_i^2)} + \frac{2\zeta_i A_i}{\omega_i^2(1+\zeta_i^2)^2}\right]\right\}$.
- $U(t)$ is the unit step function, $U(t) = 0, t<0$ and $U(t) = 1, t \ge 0$.
- The decaying sinusoids $g_i(t)$ are in fact $g_i(t-\tau)$ and start after τ (s), and the correcting time history $A_{\text{m}} e^{-\zeta_{\text{m}}\omega_{\text{m}}(t+\tau)} \sin\omega_{\text{m}}(t+\tau)$ is, in fact, $A_{\text{m}} e^{-\zeta_{\text{m}}\omega_{\text{m}}t} \sin\omega_{\text{m}}(t)$ and starts at t=0.

The magnitude A_{m} and the shift τ of the velocity and the displacement compensating pulse are fixed by the other parameters.

Nelson and Prasthofer

$$\ddot{u}_{\text{base}} = \sum_{i=1}^{n} g_i(t) \tag{10.34}$$

$$g_i(t) = A_i\{(K_1 e^{-at} - K_2 e^{-bt}) + K_3 e^{-ct}\sin(\omega_i t + \theta)\}, \tag{10.35}$$

with

- $K_1 = \frac{\omega_{\text{d}} a^2}{(a-b)[(c-a)^2 + \omega_{\text{d}}^2]}$
- $K_2 = \frac{\omega_{\text{d}} b^2}{(a-b)[(c-b)^2 + \omega_{\text{d}}^2]}$

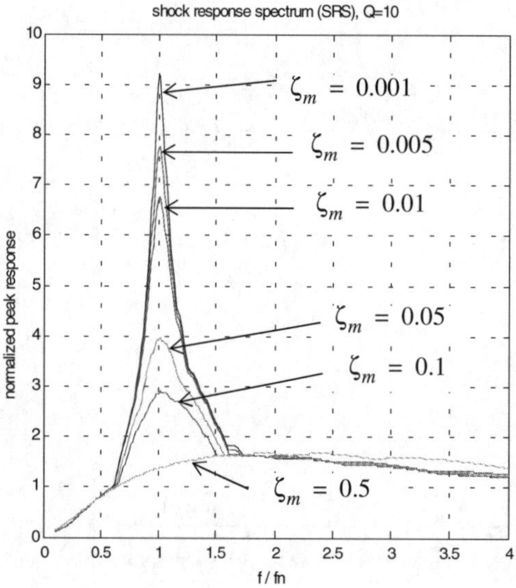

Fig. 10.11. Normalised peak response decaying sinusoid

- $K_3 = \sqrt{\frac{(c^2 - \omega_d^2)^2 + 4c^2\omega_d^2}{[(b-c)^2 + \omega_d^2][(a-c)^2 + \omega_d^2]}}$
- $\theta = \mathrm{atan}\left(\frac{-2c\omega_d}{c^2 - \omega_d^2}\right) - \mathrm{atan}\left(\frac{\omega_d}{a-c}\right) - \mathrm{atan}\left(\frac{\omega_d}{b-c}\right)$
- $a = \frac{\omega_i}{2\pi}$
- $b = 2\zeta_i\omega_i$
- $c = \zeta_i\omega_i$
- $\omega_d = \omega_i\sqrt{1-\zeta_i^2}$

The normalised peak acceleration response of decaying sinusoids is the SRS divided by the maximum value of the decaying sinusoid $g_i(t)$. The maximum value of the decaying sinusoid (g) is

$$g_{\max} = A_i e^{-\zeta_i\omega_i \mathrm{atan}\left(-\frac{1}{\zeta_i}\right)} \sin\left\{\omega_i \mathrm{atan}\left(-\frac{1}{\zeta_i}\right)\right\} \tag{10.36}$$

The normalised peak acceleration response of a decaying sinusoid is illustrated in Fig. •. Fig. [Smallwood 74a] proposes a flow diagram that can be applied to select decaying sinusoids to match a given shock response-spectrum to estimate the amplitude A_i in conjunction with the frequency ω_i and the decay rate ζ_i.

In this example the decaying sinusoids of Smallwood and Nord will be applied to match the SRS as illustrated in Fig. 10.5.

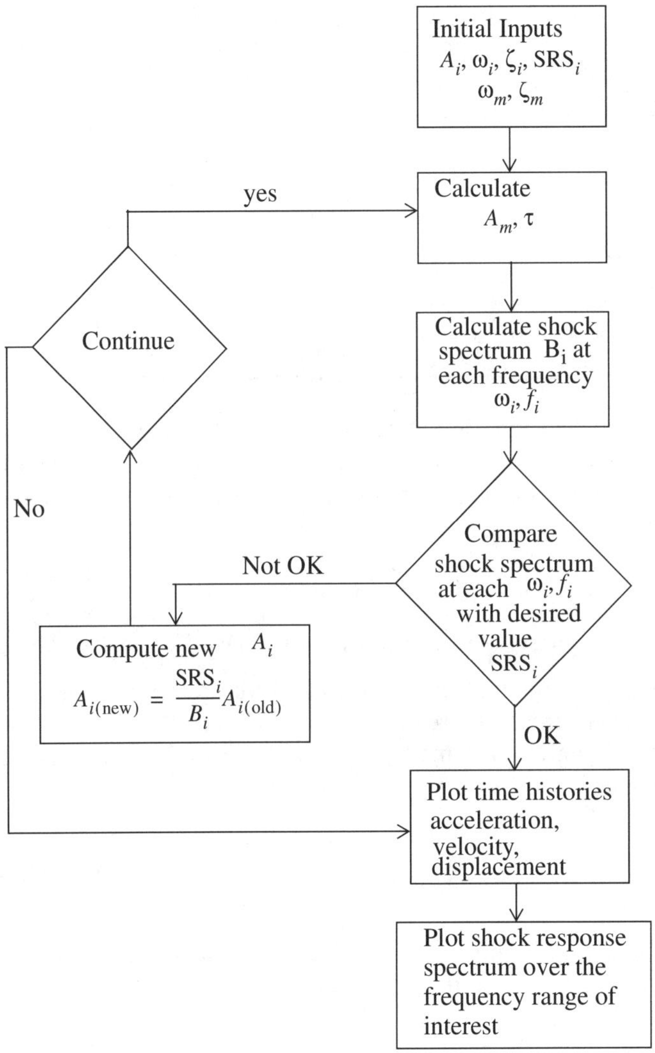

Fig. 10.12. Flow diagram for selecting decaying sinusoids to match a gives shock-response spectrum [Smallwood 74b]

This SRS is based upon the half-sine pulse (HSP), $\ddot{u}_{base} = 200 \sin\frac{\pi t}{\tau}$ g, $0 \le t \le \tau = 0.0005$ s. The procedure as illustrated in Fig. will be used to match the SRS with decaying sinusoids.
The SRS will be matched with the following values, see Table 10.8.

Table 10.8. Components of decaying sinusoids [Smallwood 74b]

Frequency #	f_i ((Hz)	ζ_i (%)	A_i (g)	τ_i (s)
1	250	20	35	0
2	500	10	50	0
3	750	10	68	0
4	1000	5	47	0
5	1250	5	47	0
6	1500	5	46	0
	f_m (Hz)	ζ_m (%)	A_m (g)	τ (s)
7	100	100	-87.7	0.0015

The matched acceleration time history of the combined Smallwood decaying sinusoids is illustrated in Fig. 10.13. The trapezoidal rule [Schwarz 89] was applied to calculate the velocity $v(t)$ and displacement $s(t)$ time histories.

$$v(t+\Delta t) = v(t) + 0.5\Delta t\{\ddot{u}(t) + \ddot{u}(t+\Delta t)\} \quad (10.37)$$

$$s(t+\Delta t) = s(t) + 0.5\Delta t\{v(t) + v(t+\Delta t)\}\,. \quad (10.38)$$

Several numerical integration methods, trapezoidal rule, Simpson's rule and the Newton–Cotes method, are described in [Hairer 96]. A very popular numerical integration method is Simpson's rule, with $x_k = a + kh,\ x_n = b$,

$$\int_a^b f(x)dx \approx h[f(x_0) + 4f(x_1) + 2f(x_2) + 4f(x_3) + 2f(x_4) + \ldots\ldots + f(x_n)]\,. \quad (10.39)$$

The velocity time history is shown in Fig. 10.14, and the displacement time history in Fig. 10.15. The original SRS and the matched SRS are both shown in Fig. 10.16.

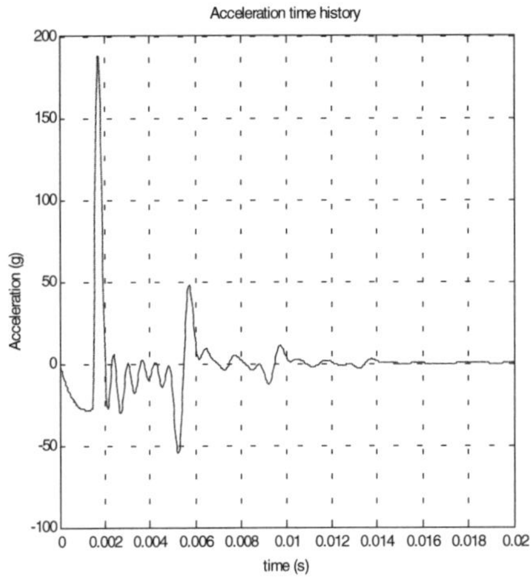

Fig. 10.13. Matched acceleration time history of combined Smallwood decaying sinusoids

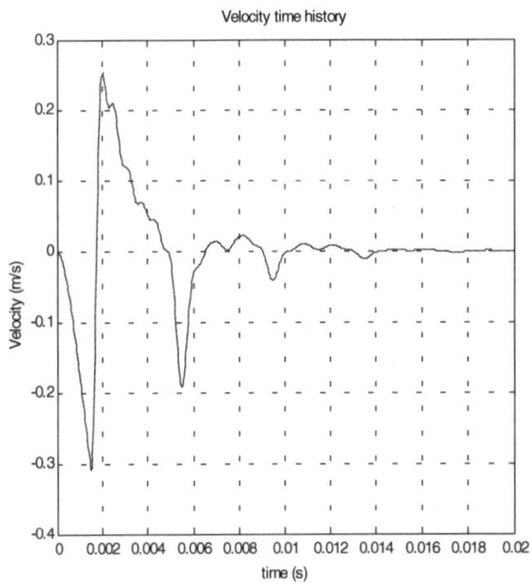

Fig. 10.14. Velocity time history of combined Smallwood decaying sinusoids

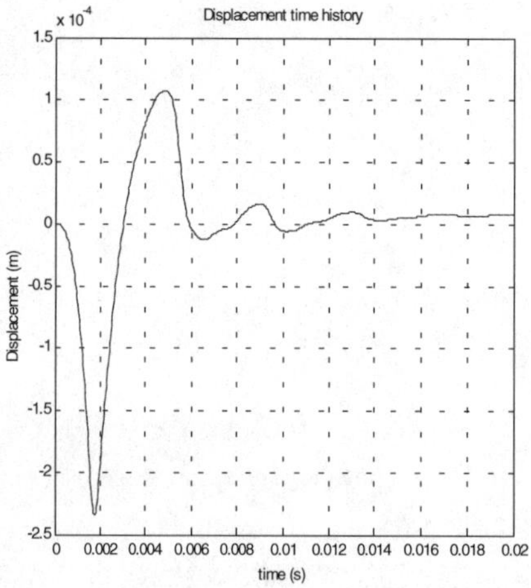

Fig. 10.15. Displacement time history of combined Smallwood decaying sinusoids

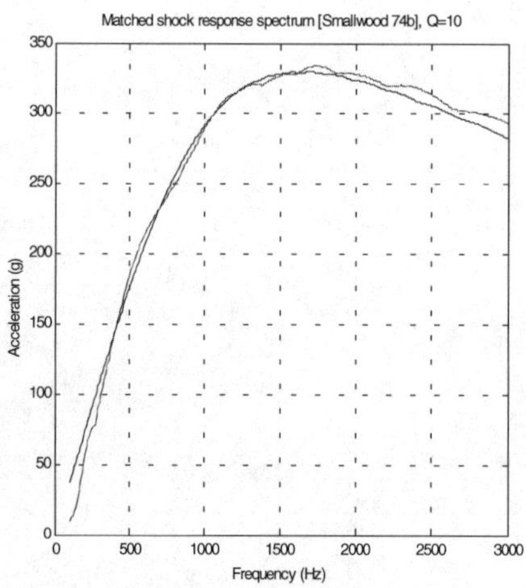

Fig. 10.16. Original and matched SRS

10.6 Problems

10.6.1 Problem 1

Calculate, using the Wilson-θ method, the absolute acceleration SRS of the following 4 pulses:

1. Rising triangular pulse
2. Decaying triangular pulse
3. Rectangular pulse
4. Half-sine pulse

and illustrate the SRSs in one figure. The pulses start at $t = \tau = 0.0005$ s and the duration $t_{\text{duration}} = \tau$. The amplitudes of the pulses are unity. The results obtained with Kelly's numerical approach are illustrated in Fig. 10.17.

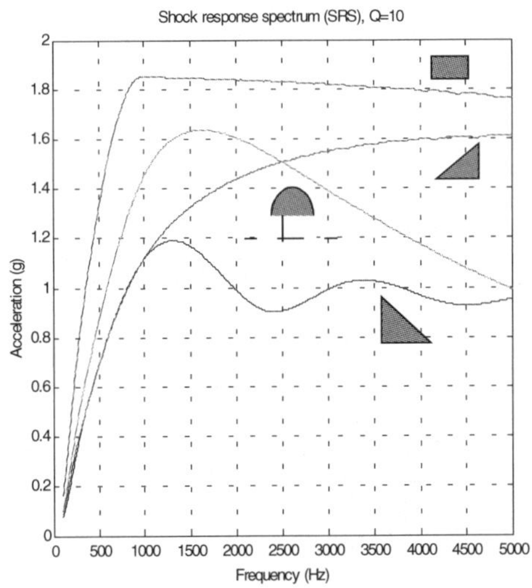

Fig. 10.17. SRS of rising and decaying triangular pulse, rectangular pulse and the half sine pulse

The equation of relative motion $z(t)$ of an sdof dynamic system, exposed to a base acceleration $\ddot{u}(t)$, is given by (10.2)

$$\ddot{z}(t) + 2\zeta\omega_n\dot{z}(t) + \omega_n^2 z(t) = -\ddot{u}(t).$$

The constant time increment is $\Delta t = t_{j+1} - t_j$.

The Wilson-θ method is defined as follows [Wood 90]

$\{Z_{j+1}\} = [A_\theta]\{Z_j\} + \{F_\theta\}$,

with

- $\{Z_j\} = \begin{bmatrix} z_j \\ \Delta t \dot{z}_j \end{bmatrix}$
- $[A_\theta] = D_\theta^{-1} \begin{bmatrix} 1 + 2\Delta t\theta\zeta\omega_n - \theta(1-\theta)\Delta t^2\omega_n^2 & 1 \\ -\Delta t^2\omega_n^2 & 1 - (1-\theta)2\Delta t\zeta\omega_n - \theta(1-\theta)\Delta t^2\omega_n^2 \end{bmatrix}$
- $D_\theta = 1 + 2\Delta t\theta\zeta\omega_n + \theta^2\Delta t^2\omega_n^2$
- $\{F_\theta\} = -\Delta t^2 D_\theta^{-1} \begin{bmatrix} \theta\{\theta\ddot{u}_{j+1} + (1-\theta)\ddot{u}_j\} \\ \{\theta\ddot{u}_{j+1} + (1-\theta)\ddot{u}_j\} \end{bmatrix}$
- $\theta = 0.5$
- j is the time step

and the absolute acceleration $\ddot{x}$ of the sdof system is

$$\ddot{x}_{j+1} = -2\zeta\omega_n\dot{z}_{j+1} - \omega_n^2 z_{j+1}.$$

10.6.2 Problem 2

Match the SRS of a half-sine pulse (Fig. 10.5) with decaying sinusoids using the proposed method of Nelson and Prasthofer [Nelson 74], $\ddot{u}_{\text{base}} = 200\sin\frac{\pi t}{\tau}$ g, $0 \le t \le \tau = 0.0005$ s, elsewhere $\ddot{u}_{\text{base}} = 0$.

11 Random Vibration of Linear Dynamic Systems

11.1 Introduction

By random vibration of linear dynamic systems the vibration of deterministic linear systems exposed to random (stochastic) loads is meant.

Random processes are characterised by the fact that their behaviour cannot be predicted in advance and therefore can be treated only in a statistical manner.

A microstochastic process is for example, the "the Brownian motion" of particles and molecules, [Wax 54], a macrostochastic process is, for example, the motion of the earth during a earthquake.

During the launch of a spacecraft with a launch vehicle the spacecraft will be exposed to random loads both of mechanical and acoustic nature. The mechanical random loads are the base acceleration excitation at the interface between the launch vehicle and the spacecraft. The random loads are caused by several sources, i.e., the interaction between the launch-vehicle structure and the engines; exhaust noise, combustion. Also, turbulent boundary layers will introduce random loads.

In this chapter the theory of random vibrations of linear systems will be briefly reviewed.

For further study on the theory of random vibration the following references [Bismark-Nasr 99, Bolotin 81, Crandall 63, Crandall 73, Elishakoff 83, Elishakoff 94, Heinrich 78, Lin 76, Lutes 96, Maymon 98, Newland 94, Robson 71, Schueller 87, Wirsching 95] are recommended.

11.2 Random Process

A random process is random in time. With the aid of probabilistic theory [Papoulis 65] of random processes the probability can be described. The mean value and the mean-square values are of great importance for random processes. We can make a distinction between ensemble and time averages. In this section we will briefly restate the properties of random processes.

The ensemble average of a collection of sampled records $x_1(t)$, $x_2(t)$, $x_3(t)$,...,$x_n(t)$ at a certain time t_1 is defined as

$$E\{x(t_1)\} = \frac{1}{n}\sum_{j=1}^{n} x_j(t_1). \tag{11.1}$$

The time average (temporal mean, [McConnel 95]) value of a record $x(t)$, over a very long sampling time T, is given by

$$\langle x \rangle = \lim_{T\to\infty}\frac{1}{T}\int_0^T x(t)dt. \tag{11.2}$$

First we make in Table 11.1 a qualification of random processes.

Table 11.1. Qualification of random process

Random process		
	Stationary	Ergodic
	Stationary	Non ergodic
	Non stationary	Non ergodic

A random process $x(t)$ is stationary if the ensemble statistics are independent of a time-shift τ (s), which means, for example, for ensemble averages of $x(t)$ and $x(t+\tau)$

$$E\{x(t_1)\} = \frac{1}{n}\sum_{j=1}^{n} x_j(t_1) = \frac{1}{n}\sum_{j=1}^{n} x_j(t_1+\tau). \tag{11.3}$$

For an ergodic random process the ensemble statistics are equal to the time averages, i.e. for the average values

$$E\{x(t_1)\} = \frac{1}{n}\sum_{j=1}^{n} x_j(t_1) = \lim_{T\to\infty}\frac{1}{T}\int_0^T x(t)dt = \langle x \rangle. \tag{11.4}$$

An ergodic process, is by definition, a stationary process.

For our purposes we will assume a stationary and ergodic random process $x(t)$.

The cumulative probability $F(X)$, that $x(t) \le X$, is given by

$$F(X) = \int_{-\infty}^{X} f(x)dx\,, \tag{11.5}$$

with

- $f(x)$ the probability density function with the following properties
- $f(x) \geq 0$
- $\int_{-\infty}^{\infty} f(x)dx = 1$
- $F(X+dx) - F(X) = \int_{X}^{X+dx} f(x)dx = f(X)dx \quad X \leq x(t) \leq X + dx$

The cumulative probability function $F(x)$ has the following properties:

- $F(-\infty) = 0$
- $0 \leq F(x) \leq 1$
- $F(\infty) = 1$
- $f(x) = \frac{dF(x)}{dx}$

Examples of probability density functions (p.d.f.) are [Scheidt 94]:

- The constant distribution $G(a, b)$, X is called equally distributed over the interval [a,b], $X \sim G(a, b)$, $f(x) = \frac{1}{b-a}$, $a \leq x \leq b$, $f(x) = 0$ elsewhere.
- The normal distribution $N(\mu, \sigma)$, $\sigma > 0$. X is normal distributed with the parameters μ and σ, $X \sim N(\mu, \sigma)$ when $f(x) = \frac{1}{\sigma\sqrt{2\pi}} e^{-\frac{(x-\mu)^2}{2\sigma^2}}$.
- The log normal distribution $LN(\mu, \sigma)$, $\sigma > 0$. X is log normal distributed with the parameters μ and σ, $X \sim LN(\mu, \sigma)$, $x > 0$, when $f(x) = \frac{1}{\sigma\sqrt{2\pi}} e^{-\frac{(\ln(x)-\mu)^2}{2\sigma^2}}$.
- The Rayleigh distribution $R(\sigma)$, $\sigma > 0$. X is Rayleigh distributed with the parameter σ, $X \sim R(\sigma)$, $x > 0$, when $f(x) = \frac{2x}{\sigma^2} e^{-\frac{x^2}{\sigma^2}}$.

For an ergodic random process the term $f(x)dx$ may be approximated by

$$f(x)dx \approx \lim_{T \to \infty} \frac{1}{T} \sum_i \delta t_i\,, \tag{11.6}$$

with

- δt_i the lingering period of $x(t)$ between $\alpha \le x \le \beta$

The mode is defined as the peak of the p.d.f. $f(x)$, and the mean value μ has an equal moment to the left and to the right of it:

$$\int_{-\infty}^{\infty} (x - \mu) f(x) dx = 0\,. \tag{11.7}$$

This means that the averaged value of x can be calculated with

$$E(x) = \mu = \frac{\int_{-\infty}^{\infty} x f(x) dx}{\int_{-\infty}^{\infty} f(x) dx} = \int_{-\infty}^{\infty} x f(x) dx\,. \tag{11.8}$$

We can also define the n-the moment about the mean μ is follows

$$\mu_n = \int_{-\infty}^{\infty} (x - \mu)^n f(x) dx\,. \tag{11.9}$$

The second moment is called the variance of a signal.

$$\sigma^2 = \mu_2 = \int_{-\infty}^{\infty} (x - \mu)^2 f(x) dx\,, \tag{11.10}$$

and σ is called the standard deviation.

Suppose a sinusoidal signal $x(t) = A\sin\omega t$. Over one period T (s) the signal $x(t)$ will cross a certain area twice when $x(t) = X + \delta x \le A$, with a total time $2\delta t$. The p.d.f. can be estimated with $f(x)dx = \frac{2\delta t}{T} = \frac{\omega\delta t}{\pi}$ and with $\delta x = \omega A\cos\omega t\delta t$ the p.d.f. becomes

$$f(x) = \frac{1}{\pi A\cos\omega t} = \frac{1}{\pi A\sqrt{1 - \left(\frac{x(t)}{A}\right)^2}}\,.$$

The mean value μ is, in accordance with (11.8)

$$E(x) = \mu = \int_{-\infty}^{\infty} xf(x)dx = 0,$$

and the variance σ^2 becomes

$$\sigma^2 = \mu_2 = \int_{-\infty}^{\infty} (x-\mu)^2 f(x)dx = \frac{A^2}{2}.$$

In general, within the framework of linear vibrations it may be assumed that the averaged (mean) values of the response of linear systems exposed to dynamic loads will be equal to zero. So the second moments about the mean μ, the variance, are equal to the mean square values, or $E(x^2) = \sigma^2 = \int_{-\infty}^{\infty} x^2 f(x)dx$.

A random process x is randomly distributed, $0 \le x \le 1$, with a p.d.f.

- $f(x) = 1, \ 0 \le x \le 1$
- $f(x) = 0, \ x < 0$ and $x > 1$

Calculate the mean value, the mean square and the variance of x:

- The mean value $E(x) = \mu = \int_{-\infty}^{\infty} xf(x)dx = \int_0^1 xdx = \frac{1}{2}$
- The mean square $E(x^2) = \int_{-\infty}^{\infty} x^2 f(x)dx = \int_0^1 x^2 dx = \frac{1}{3}$
- The variance $\sigma^2 = E(x^2) - \mu^2 = \int_{-\infty}^{\infty} (x-\mu)^2 f(x)dx = E(x^2) - \mu^2 = \frac{1}{12}$
- The standard deviation $\sigma = 0.289$.

For a stationary and ergodic random process $x(t)$ we have established the following relations for the mean value

$$\mu_x = \langle x \rangle = E(x) = \int_{-\infty}^{\infty} xf(x)dx = \lim_{T \to \infty} \frac{1}{T} \int_0^T x(t)dt, \tag{11.11}$$

and for the mean-square value

$$\langle x^2 \rangle = E(x^2) = \int_{-\infty}^{\infty} x^2 f(x)dx = \lim_{T \to \infty} \frac{1}{T} \int_0^T x^2(t)dt = \sigma_x^2 + \mu_x^2. \tag{11.12}$$

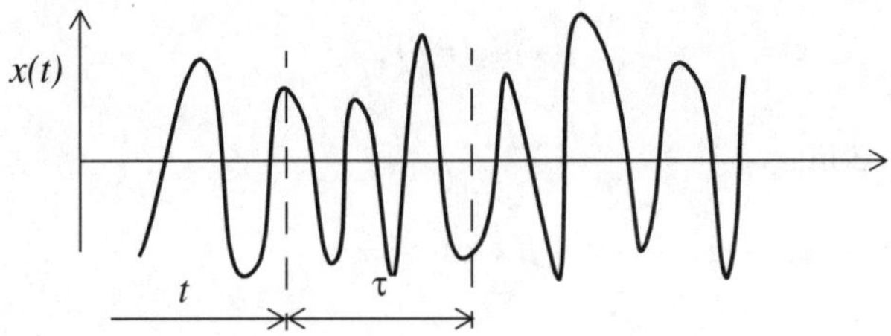

Fig. 11.1. Autocorrelation

The autocorrelation function of a stationary and an ergodic random process $x(t)$ is defined as (Fig. 11.1)

$$R_{xx}(\tau) = E\{x(t)x(t+\tau)\} = \lim_{T\to\infty}\frac{1}{T}\int_0^T x(t)x(t+\tau)dt \tag{11.13}$$

with the following general properties

- $\lim_{\tau\to\infty} R_{xx}(\tau) = 0$
- $R_{xx}(\tau)$ is a real function
- $R_{xx}(\tau)$ is a symmetric function, $R_{xx}(\tau) = R_{xx}(-\tau)$, $R_{xx}(-\tau) = E\{x(t-\tau)x(t)\}$
- $R_{xx}(0) = E(x^2) = \lim_{T\to\infty}\frac{1}{T}\int_0^T x^2(t)dt$
- $R_{xx}(0) \geq |R_{xx}(\tau)|$, which can be proven with the relation $\lim_{T\to\infty}\frac{1}{T}\int_0^T [x(t)\pm x(t+\tau)]^2 dt \geq 0$
- The correlation between $x(t)$ and $\dot{x}(t)$ is $R_{x\dot{x}}(\tau) = -R_{\dot{x}x}(\tau)$, thus $R_{x\dot{x}}(0) = -R_{\dot{x}x}(0) = 0$
- If $x(t) = \alpha y(t) + \beta z(t)$ then $R_{xx}(\tau) = \alpha^2 R_{yy}(\tau) + \beta^2 R_{zz}(\tau) + \alpha\beta R_{yz}(\tau) + \alpha\beta R_{zy}(\tau)$
- The Fourier transform requirement is satisfied for the autocorrelation function $\int_{-\infty}^{\infty} |R_{xx}(\tau)| d\tau < \infty$

The crosscorrelation function is defined as

$$R_{xy}(\tau) = E\{x(t)y(t+\tau)\} = \lim_{T\to\infty}\frac{1}{T}\int_0^T x(t)y(t+\tau)dt. \tag{11.14}$$

The autocorrelation $R_{xx}(\tau)$ of $x(t) = A\sin\omega t$ is

$$R_{xx}(\tau) = \frac{1}{T}\int_0^T x(t)x(t+\tau)dt = \frac{\omega A^2}{2\pi}\int_0^{\frac{2\pi}{\omega}} \sin\omega t \sin\omega(t+\tau)dt = \frac{A^2}{2}\cos\omega\tau.$$

The mean square of $x(t)$ is

$$E(x^2) = R_{xx}(0) = \frac{A^2}{2}.$$

11.3 Power-Spectral Density

The Fourier transform of a function $x(t)$ is defined as [James 93]

$$F\{x(t)\} = X(\omega) = \int_{-\infty}^{\infty} x(t)e^{-j\omega t}dt, \tag{11.15}$$

and the inverse of the Fourier transform

$$F^{-1}\{X(\omega)\} = x(t) = \frac{1}{2\pi}\int_{-\infty}^{\infty} X(\omega)e^{j\omega t}d\omega, \tag{11.16}$$

assuming that $\int_{-\infty}^{\infty} |x(t)|dt < \infty$.

We may write for the autocorrelation function $R_{xx}(\tau)$

$$R_{xx}(\tau) = \lim_{T\to\infty}\frac{1}{T}\int_0^T x(t)x(t+\tau)dt = \lim_{T\to\infty}\frac{1}{2T}\int_{-T}^T x(t)x(t+\tau)dt. \tag{11.17}$$

The Fourier transform of the autocorrelation function $R_{xx}(\tau)$ is called the power spectral density function $S_{xx}(\omega)$ (also called autospectral density [McConnel 95])

$$S_{xx}(\omega) = \int_{-\infty}^{\infty} R_{xx}(\tau)e^{-j\omega\tau}d\tau = 2\int_0^{\infty} R_{xx}(\tau)\cos(\omega\tau)d\tau, \tag{11.18}$$

and

$$R_{xx}(\tau) = \frac{1}{2\pi}\int_{-\infty}^{\infty} S_{xx}(\omega)e^{j\omega\tau}d\omega = \frac{1}{\pi}\int_{-\infty}^{\infty} S_{xx}(\omega)\cos(\omega\tau)d\omega\,. \tag{11.19}$$

Both the autocorrelation function $R_{xx}(\tau)$ and the power spectral density $S_{xx}(\omega)$ are symmetric functions about $\tau = 0$ and $\omega = 0$.

The pair of (11.18) and (11.19) is called the Wiener–Khintine relationship [Harris 74].

For the correlation function $R(\tau) = \sigma^2 e^{-b|\tau|}$ will give the following power- spectral density function $S(\omega)$

$$S(\omega) = 2\sigma^2\int_0^{\infty} e^{-b\tau}\cos(\omega\tau)d\tau = \frac{2\sigma^2 b}{(b^2+\omega^2)}\,.$$

The correlation function $R(\tau)$ is given by (see Fig. 11.2)

$$R(\tau) = \sigma^2\begin{cases} 1-\frac{|\tau|}{\varepsilon} & |\tau| \le \varepsilon \\ 0 & |\tau| \ge \varepsilon \end{cases}\,.$$

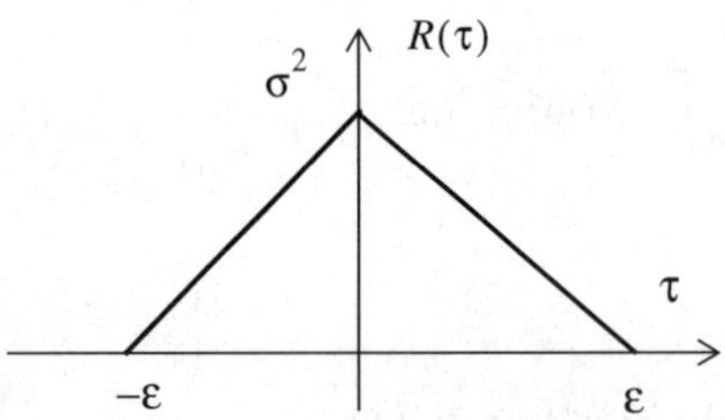

Fig. 11.2. Correlation function

The power-spectral density function $S(\omega)$ becomes

$$S(\omega) = 2\int_0^{\infty} R(\tau)\cos(\omega\tau)d\tau = \frac{2\sigma^2}{\varepsilon\omega^2}\{1-\cos(\omega\varepsilon)\}\,.$$

The total energy E of the signal $x(t)$ is given by [James 93]

$$E = \int_{-\infty}^{\infty} [x(t)]^2 dt\,. \tag{11.20}$$

The (average) power P of the signal $x(t)$ is given by [James 93]

$$P = \lim_{T \to \infty} \frac{1}{2T} \int_{-T}^{T} [x(t)]^2 dt . \qquad (11.21)$$

With use of (11.16) we can rewrite (11.20)

$$E = \int_{-\infty}^{\infty} x(t)x(t)dt = \int_{-\infty}^{\infty} x(t)\left[\frac{1}{2\pi}\int_{-\infty}^{\infty} X(\omega)e^{j\omega t}d\omega\right]dt , \qquad (11.22)$$

or by changing the order of integration (11.22) becomes

$$E = \frac{1}{2\pi}\int_{-\infty}^{\infty} X(\omega)\left[\int_{-\infty}^{\infty} x(t)e^{j\omega t}dt\right]d\omega = \frac{1}{2\pi}\int_{-\infty}^{\infty} X(\omega)X^*(\omega)d\omega , \qquad (11.23)$$

hence

$$E = \int_{-\infty}^{\infty} [x(t)]^2 dt = \frac{1}{2\pi}\int_{-\infty}^{\infty} X(\omega)X^*(\omega)d\omega = \frac{1}{2\pi}\int_{-\infty}^{\infty} |X(\omega)|^2 d\omega , \qquad (11.24)$$

The resulting equation (11.24) is called Parseval's theorem [James 93], with

- $|X(\omega)|^2$ the energy spectral density, if $z = x + jy$, and $z^* = x - jy$ resulting in $zz^* = x^2 + y^2 = |z^2|$.

Equation (11.21), using Parseval's theorem, can be written as

$$P = \lim_{T \to \infty} \frac{1}{2T} \int_{-T}^{T} [x(t)]^2 dt = \frac{1}{2\pi}\int_{-\infty}^{\infty} \lim_{T \to \infty} \frac{1}{2T}|X(\omega)|^2 d\omega , \qquad (11.25)$$

with

- $\lim_{T \to \infty} \frac{1}{2T}|X(\omega)|^2$ the power spectral density (PSD) function of $x(t)$

Using (11.17), (11.18) after multiplication by $e^{j\omega t}e^{-j\omega t}$, can be written

$$S_{xx}(\omega) = \int_{-\infty}^{\infty} \lim_{T \to \infty} \frac{1}{2T}\left[\int_{-T}^{T} x(t)x(t+\tau)dt\right]e^{j\omega t}e^{-j\omega(t+\tau)} d\tau = \lim_{T \to \infty} \frac{1}{2T}|X(\omega)|^2 . \qquad (11.26)$$

The average power P, (11.25), can be expresses as follows

$$P = \frac{1}{2\pi}\int_{-\infty}^{\infty} \lim_{T \to \infty} \frac{1}{2T}|X(\omega)|^2 d\omega = \frac{1}{2\pi}\int_{-\infty}^{\infty} S_{xx}(\omega)d\omega = R_{xx}(0) , \qquad (11.27)$$

hence

$$R_{xx}(0) = E(x^2) = \mu_x^2 = \frac{1}{2\pi}\int_{-\infty}^{\infty} S_{xx}(\omega)d\omega\,. \tag{11.28}$$

$S_{xx}(\omega)$ has the following properties

- $S_{xx}(\omega) = S_{xx}(-\omega)$
- $S_{xx}(\omega) \geq 0$

The PSD function $S_{xx}(\omega)$ is two-sided. It is more practical to replace ω (rad/s) with f (Hz, cycles/s) and to replace the two-sided PSD function $S_{xx}(\omega)$ with a one-sided PSD function $W_{xx}(f)$ and than (11.28) now becomes

$$R_{xx}(0) = E(x^2) = \mu_x^2 = \int_{0}^{\infty} W_{xx}(f)df\,, \tag{11.29}$$

with

- $W_{xx}(f) = 2S_{xx}(\omega)$

White Noise

If the power-spectral density function of a signal $x(t)$ is constant over the complete frequency range, $W_{xx}(f) = W_o$, $0 \leq f \leq \infty$ we talk about white noise.

The power-spectral density function $S_{xx}(\omega) = \frac{W_o}{2}$, $-\infty \leq \omega \leq \infty$. The autocorrelation function $R_{xx}(\tau)$ can now be calculated using (11.19).

$$R_{xx}(\tau) = \frac{1}{2\pi}\int_{-\infty}^{\infty} S_{xx}(\omega)e^{j\omega\tau}d\omega = \frac{W_o}{2}\frac{1}{2\pi}\int_{-\infty}^{\infty} e^{j\omega\tau}d\omega = \frac{W_o}{2}\delta(\tau)\,,$$

with

- $\int_{-\infty}^{\infty} \delta(\tau)d\tau = 1$, $\delta(\tau)$ the Dirac delta function.

We have a random process with a constant (white noise) PSD function between two frequencies (band-limited). Calculate the associated auto correlation function $R_{xx}(\tau)$

- $W(f) = W_o, f_1 \leq f \leq f_2$
- $W(f) = 0, f < f_1$ and $f > f_2$

The autocorrelation function $R_{xx}(\tau)$ is

$$R_{xx}(\tau) = \frac{1}{2\pi}\int_{-\infty}^{\infty} S_{xx}(\omega)e^{j\omega\tau}d\omega = \frac{1}{2\pi}\int_{-\infty}^{\infty} S_{xx}(\omega)\cos\omega\tau d\omega . \qquad (11.30)$$

Equation (11.30) can be easily proved because

$$\frac{1}{2\pi}\int_{-\infty}^{\infty} S_{xx}(\omega)\sin\omega\tau d\omega = 0 , \qquad (11.31)$$

and therefore

$$R_{xx}(\tau) = \frac{1}{2\pi}\int_{-\infty}^{\infty} S_{xx}(\omega)\cos\omega\tau d\omega = \frac{1}{\pi}\int_{0}^{\infty} S_{xx}(\omega)\cos\omega\tau d\omega . \qquad (11.32)$$

Hence

$$R_{xx}(\tau) = \frac{W_o}{2\pi}\int_{2\pi f_1}^{2\pi f_2} \cos\omega\tau d\omega = \frac{W_o}{2\pi\tau}[\sin 2\pi f_2\tau - \sin 2\pi f_1\tau] , \qquad (11.33)$$

resulting in

$$R_{xx}(0) = \lim_{\tau \to 0}\frac{W_o}{2\pi\tau}[\sin 2\pi f_2\tau - \sin 2\pi f_1\tau] = W_o[f_2 - f_1] . \qquad (11.34)$$

The result of (11.34) corresponds with the result we had if we used (11.29).

Assume a very narrow bandwidth $[f_2 - f_1] = \delta f$. Then (11.33) then becomes

$$R_{xx}(\tau) = \frac{W_o}{2\pi\tau}[\sin 2\pi(f_1 + \delta f)\tau - \sin 2\pi f_1\tau] . \qquad (11.35)$$

Using the Taylor series $f(x + \delta x) = f(x) + \frac{f'(x)}{1!}\delta x + \frac{f''(x)}{2!}\delta x^2 + ...$

$$\sin 2\pi(f_1 + \delta f)\tau \approx \sin 2\pi f_1\tau + \frac{\cos 2\pi f_1\tau}{1!}2\pi\delta f\tau . \qquad (11.36)$$

The auto correlation function $R_{xx}(\tau)$, with (11.35), can now be calculated

$$R_{xx}(\tau) = \frac{W_o}{2\pi\tau}[(\cos 2\pi f_1\tau)2\pi\delta f\tau] = W_o\delta f\cos 2\pi f_1\tau \qquad (11.37)$$

$$R_{xx}(0) = W_o\delta f . \qquad (11.38)$$

In the following Table 11.2 some useful relations between the autocorrelation function $R_{xx}(\tau)$ and the power-spectral density function $S_{xx}(\omega)$ are given.

Table 11.2. Properties of auto correlation and power-spectral density function

$x(t)$	$R_{xx}(\tau)$	$S_{xx}(\omega)$
$\alpha x(t)$	$\alpha^2 R_{xx}(\tau)$	$\alpha^2 S_{xx}(\omega)$
$\dfrac{dx(t)}{dt}$	$-\dfrac{d^2 R_{xx}(\tau)}{d\tau^2}$	$\omega^2 S_{xx}(\omega)$
$\dfrac{dx^n(t)}{dt^n}$	$(-1)^n \dfrac{d^{2n} R_{xx}(\tau)}{d\tau^{2n}}$	$\omega^{2n} S_{xx}(\omega)$
$x(t)e^{\pm j\omega_o t}$	$R_{xx}(\tau)e^{\pm j\omega_o t}$	$S_{xx}(\omega \mp \omega_o)$

11.4 Deterministic Linear Dynamic System

The linear deterministic system is illustrated in Fig. 11.3. The system shows no random properties and the properties will not change with time. The system is excited with a random load $f(t)$ and the response (output) of the system is denoted by $x(t)$. The random responses $x(t)$ are very generalised and may be: displacements, velocities, accelerations, forces, stresses. The linear system will be characterised using the impulse response function. Linear means that doubling the loads $f(t)$ will result in twice as much response $x(t)$. Besides presenting the forces and responses in the time domain the forces and responses are transformed in the frequency domain.

A linear system may be represented either simply, as a single degree of freedom system (sdof), or more complex with multi-degrees of freedom (mdof), or even as a continuum. But in the solution of the responses the modal superposition will be applied many times and the problem will be reduced in solving many uncoupled sdof dynamic systems.

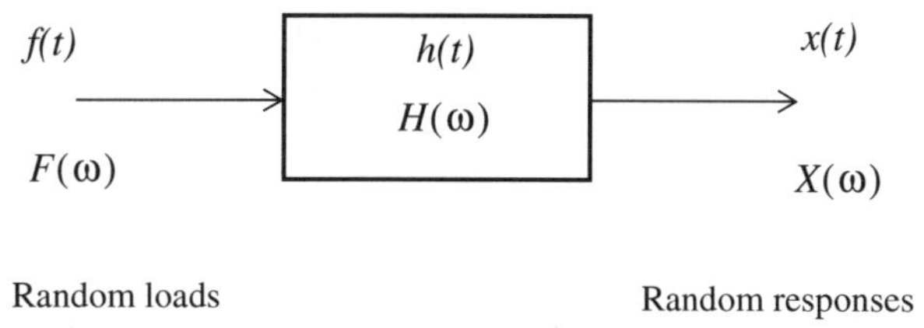

Fig. 11.3. Deterministic dynamic system

The response $x(t)$ of the linear system, due to the force $f(t)$, and the impulse response function $h(t)$ is given by the convolution integral

$$x(t) = \int_{-\infty}^{\infty} h(\tau)f(t-\tau)d\tau, \tag{11.39}$$

or in the frequency domain [Harris 74] by

$$X(\omega) = H(\omega)F(\omega), \tag{11.40}$$

with

- $f(t)$ the generalised external force in the time domain.
- $F(\omega) = \int_{-\infty}^{\infty} f(t)e^{-j\omega t}dt$, the Fourier transform of $f(t)$.
- $h(t)$, the damped impulse response function $h(t) = e^{-\zeta\omega_n t}\dfrac{\sin\omega_n\sqrt{1-\zeta^2}}{\omega_n\sqrt{1-\zeta^2}}$.

  This impulse response function can be derived from the sdof dynamic system $\ddot{x}(t) + 2\zeta\omega_n\dot{x}(t) + \omega_n^2 x(t) = f(t)$.
- $H(\omega)$ the frequency response function, the Fourier transform of $h(t)$,

  $$H(\omega) = \int_{-\infty}^{\infty} h(t)e^{-j\omega t}dt.$$
- $x(t)$ the response of the sdof system.
- $X(\omega)$ the Fourier transform of $x(t)$, $X(\omega) = \int_{-\infty}^{\infty} x(t)e^{-j\omega t}dt$ and

  $$x(t) = \frac{1}{2\pi}\int_{-\infty}^{\infty} X(\omega)e^{j\omega t}d\omega.$$

The PSD function of $x(t)$, in accordance with (11.26) , is

$$S_{xx}(\omega) = \lim_{T \to \infty} \frac{1}{2T} X(\omega) X^*(\omega) = \lim_{T \to \infty} \frac{1}{2T} |X(\omega)|^2, \tag{11.41}$$

with

- $X^*(\omega) = \int_{-\infty}^{\infty} x(t) e^{j\omega t} dt$ the conjugate of $X(\omega)$.

The PSD function of the random response $x(t)$ can be expressed in the PSD function of the random loads, applying (11.40)

$$S_{xx}(\omega) = \lim_{T \to \infty} \frac{1}{2T} X(\omega) X^*(\omega) = \lim_{T \to \infty} \frac{1}{2T} H(\omega) H^*(\omega) F(\omega) F^*(\omega), \tag{11.42}$$

or

$$S_{xx}(\omega) = |H(\omega)|^2 S_{FF}(\omega). \tag{11.43}$$

Equation (11.43) is very important to analyse the response characteristics of linear dynamic systems.

The cross-PSD function $S_{xF}(\omega)$ is

$$S_{xF}(\omega) = \lim_{T \to \infty} \frac{1}{2T} X(\omega) F^*(\omega) \tag{11.44}$$

$$S_{xF}(\omega) = S^*_{Fx}(\omega), \tag{11.45}$$

with

- $\Re(S_{xF}(\omega))$ the real part of the cross-PSD function is called the co-spectral density (CSD) function
- $\Im(S_{xF}(\omega))$ the imaginary part of the cross-PSD function is called the quad- spectral density (QSD) function

$$S_{xF}(\omega) = \lim_{T \to \infty} \frac{1}{2T} X(\omega) F^*(\omega) = \lim_{T \to \infty} \frac{1}{2T} H(\omega) F(\omega) F^*(\omega) = H(\omega) S_{FF}(\omega). \tag{11.46}$$

The cross-PSD function is generally a complex-valued function.

11.4.1 Force-Loaded Sdof System

The mass–spring–damper system is loaded by the force $F(t)$ at the mass m. The mass is suspended by a linear spring with spring stiffness k and a damper with damping constant c.

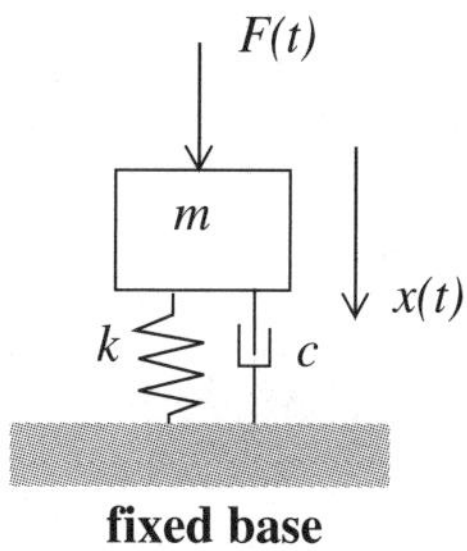

Fig. 11.4. Sdof system loaded with force $F(t)$

The equation of motion is

$$m\ddot{x}(t) + c\dot{x}(t) + kx(t) = F(t). \tag{11.47}$$

Dividing (11.47) by the mass m the equation of motion of the damped sdof system becomes

$$\ddot{x}(t) + 2\zeta\omega_n\dot{x}(t) + \omega_n^2 x(t) = \frac{F(t)}{m} = f(t), \tag{11.48}$$

with

- $\omega_n = \sqrt{\frac{k}{m}}$ the natural frequency (rad/s)
- $\zeta = \frac{c}{2\sqrt{km}}$ the damping ratio

For harmonic motions, we write

$$x(t) = X(\omega)e^{j\omega t} \text{ and } f(t) = F(\omega)e^{j\omega t}. \tag{11.49}$$

Substituting (11.49) into (11.48) the following result is obtained

$$X(\omega) = H(\omega)F(\omega), \tag{11.50}$$

with

- $H(\omega) = \frac{1}{\omega_n^2 - \omega^2 + 2j\zeta\omega\omega_n}$ the frequency response function.

The square of the modulus of $H(\omega)$ is

$$|H(\omega)|^2 = \frac{1}{(\omega_n^2 - \omega^2)^2 + (2\zeta\omega\omega_n)^2}. \tag{11.51}$$

The mean-square response of $x(t)$, due to the random load $f(t)$ with the PSD function $S_{FF}(\omega)$, using (11.43), becomes

$$E\{x(t)^2\} = R_{xx}(0) = \frac{1}{2\pi}\int_{-\infty}^{\infty} |H(\omega)|^2 S_{FF}(\omega)d\omega . \tag{11.52}$$

For the time being the forcing function $f(t)$ has a constant PSD function (white noise) $S_{FF}(\omega) = \frac{W_{FF}(f)}{2} = S_F$, thus

$$E\{x(t)^2\} = R_{xx}(0) = \frac{S_F}{2\pi}\int_{-\infty}^{\infty} |H(\omega)|^2 d\omega . \tag{11.53}$$

The integral $\int_{-\infty}^{\infty} |H(\omega)|^2 d\omega$ has a closed-form solution [Newland 94]. If $H_2(\omega) = \frac{B_0 + j\omega B_1}{A_0 + j\omega A_1 - \omega^2 A_2}$ then $I_2 = \int_{-\infty}^{\infty} |H_2(\omega)|^2 d\omega = \frac{\pi\{A_0 B_1^2 + A_2 B_0^2\}}{A_0 A_1 A_2}$.

The modulus of the frequency-response function $|H(\omega)|^2$ is given by (11.51). Thus

- $A_0 = \omega_n^2$
- $A_1 = 2\zeta\omega_n$
- $A_2 = 1$
- $B_0 = 1$
- $B_1 = 0$

The integral (11.53) will be

$$E\{x(t)^2\} = R_{xx}(0) = \frac{S_F}{2\pi}\int_{-\infty}^{\infty} |H(\omega)|^2 d\omega = \frac{S_F}{4\zeta\omega_n^3} . \tag{11.54}$$

The mean square of the acceleration $\ddot{x}(t)$, $E\{\ddot{x}^2(t)\}$, with a white-noise forcing function does not exist [Piszczek 86].

11.4.2 Enforced Acceleration

An sdof system with a discrete mass m, a damper element c and a spring element k is placed on a moving base which is accelerated with an acceleration $\ddot{u}(t)$. The resulting displacement of the mass is $x(t)$. We introduce a relative motion $z(t)$ which is the displacement of the mass with respect to the base. The relative displacement is

$$z(t) = x(t) - u(t). \tag{11.55}$$

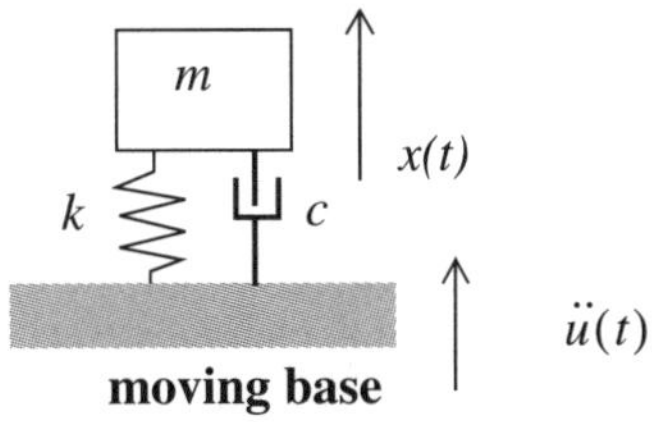

Fig. 11.5. Enforced acceleration of a damped sdof system

The equation of motion of the sdof system, illustrated in Fig. 11.5, is

$$\ddot{z}(t) + 2\zeta\omega_n\dot{z}(t) + \omega_n^2 z(t) = -\ddot{u}(t). \tag{11.56}$$

The enforced acceleration of the sdof system is transformed into an external force.

The absolute displacement $x(t)$ can be calculated with

$$\ddot{x}(t) = \ddot{z}(t) + \ddot{u}(t) = -2\zeta\omega_n\dot{z}(t) - \omega_n^2 z(t). \tag{11.57}$$

The PSD function $S_{zz}(\omega)$, of the relative motion $z(t)$, becomes (using (11.43))

$$S_{zz}(\omega) = |H(\omega)|^2 S_{\ddot{u}\ddot{u}}(\omega) \tag{11.58}$$

with

- $S_{\ddot{u}\ddot{u}}(\omega) = \lim\limits_{T\to\infty}\frac{1}{2T}\ddot{U}(\omega)\ddot{U}^*(\omega)$, with a dimensions of (m/s^2)^2/(Rad/s). Using $W_{\ddot{u}\ddot{u}}(f) = 2S_{\ddot{u}\ddot{u}}(\omega)$ the dimensions of the PSD function are (m/s^2)^2/Hz

We will derive the mean-square values both in the time and frequency domain starting with derivation of the autocorrelation function $R_{zz}(\tau)$. Using (11.19)

$$R_{zz}(\tau) = \frac{1}{2\pi}\int_{-\infty}^{\infty} S_{zz}(\omega)e^{j\omega\tau}d\omega,$$

and inserting (11.58) we obtain the following expression

$$R_{zz}(\tau) = \frac{1}{2\pi}\int_{-\infty}^{\infty} |H(\omega)|^2 S_{\ddot{u}\ddot{u}}(\omega)e^{j\omega\tau}d\omega,$$

with the frequency response function $H(\omega)$

$$H(\omega) = \frac{1}{\omega_n^2\left(1 - \frac{\omega^2}{\omega_n^2} + 2j\frac{\omega}{\omega_n}\right)}, \tag{11.59}$$

and $|H(\omega)|^2$

$$|H(\omega)|^2 = \frac{1}{\omega_n^4\left\{\left(1 - \frac{\omega^2}{\omega_n^2}\right)^2 + \left(2\frac{\omega}{\omega_n}\right)^2\right\}}. \tag{11.60}$$

For the time being the forcing function $f(t) = -\ddot{u}(t)$ has a constant PSD function (white noise) $S_{\ddot{u}\ddot{u}}(\omega) = \frac{W_{\ddot{u}\ddot{u}}(f)}{2} = S_{\ddot{u}} = \frac{W_{\ddot{u}}}{2}$.

Now the auto correlation function $R_{zz}(\tau)$ can be written as

$$R_{zz}(\tau) = \frac{S_{\ddot{u}}}{2\pi}\int_{-\infty}^{\infty} |H(\omega)|^2 e^{j\omega\tau}d\omega = \frac{S_{\ddot{u}}}{2\pi\omega_n^4}\int_{-\infty}^{\infty} \frac{1}{\left\{\left(1 - \frac{\omega^2}{\omega_n^2}\right)^2 + \left(2\frac{\omega}{\omega_n}\right)^2\right\}} e^{j\omega\tau}d\omega$$

$$. \tag{11.61}$$

Finally we obtain for $R_{zz}(\tau)$, [Bismark-Nasr 99]

$$R_{zz}(\tau) = \frac{S_{\ddot{u}}e^{-\zeta\omega_n\tau}}{4\zeta\omega_n^3}\left(\left[\cos(\omega_d\tau) + \frac{\zeta}{\sqrt{1-\zeta^2}}\sin(\omega_d\tau)\right]\right), \ (\tau \geq 0). \tag{11.62}$$

The mean square of the response $z(t)$ becomes

$$E\{z(t)^2\} = R_{zz}(0) = \frac{S_{\ddot{u}}}{4\zeta\omega_n^3} = \frac{W_{\ddot{u}}}{8\zeta\omega_n^3}. \tag{11.63}$$

We know that

$$R_{\ddot{z}\ddot{z}}(\tau) = -\frac{d^2R_{zz}(\tau)}{d\tau^2}, \tag{11.64}$$

and

$$R_{\ddot{z}\ddot{z}}(\tau) = \frac{d^4 R_{zz}(\tau)}{d\tau^4} . \tag{11.65}$$

We will repeat the mean-square calculations of $z(t)$ in the frequency domain.

Using (11.43) the mean-square response of $z(t)$, due to the random load $f(t)$ with the PSD function $S_{\ddot{u}\ddot{u}}(\omega)$, becomes

$$E\{z(t)^2\} = R_{zz}(0) = \frac{1}{2\pi}\int_{-\infty}^{\infty} |H(\omega)|^2 S_{\ddot{u}\ddot{u}} d\omega . \tag{11.66}$$

For the time being the forcing function $f(t)$ has a constant PSD function (white noise) $S_{\ddot{u}\ddot{u}}(\omega) = \frac{W_{\ddot{u}\ddot{u}}(f)}{2} = S_{\ddot{u}} = \frac{W_{\ddot{u}}}{2}$, thus

$$E\{x(t)^2\} = R_{xx}(0) = \frac{S_{\ddot{u}}}{2\pi}\int_{-\infty}^{\infty} |H(\omega)|^2 d\omega . \tag{11.67}$$

The modulus of the frequency response function $|H(\omega)|^2$ is given by (11.51). Thus the integral (11.67) becomes

$$E\{z(t)^2\} = \frac{S_{\ddot{u}}}{2\pi}\int_{-\infty}^{\infty} |H(\omega)|^2 d\omega = \frac{S_{\ddot{u}}}{4\zeta\omega_n^3} . \tag{11.68}$$

This is analogous to (11.54).

The mean-square value of the relative velocity $\dot{z}(t)$ is given by

$$E\{\dot{z}^2(t)\} = \frac{S_{\ddot{u}}}{2\pi}\int_{-\infty}^{\infty} \omega^2 |H(\omega)|^2 d\omega = \frac{S_{\ddot{u}}}{4\zeta\omega_n} . \tag{11.69}$$

Now we want to calculate the mean-square value of the acceleration $\ddot{x}(t)$, which can be calculated using (11.57), $\ddot{x}(t) = -2\zeta\omega_n\dot{z}(t) - \omega_n^2 z(t)$. The autocorrelation function for the acceleration is

$$R_{\ddot{x}\ddot{x}}(\tau) = (2\zeta\omega_n)^2 R_{\dot{z}\dot{z}}(\tau) + \omega_n^4 R_{zz}(\tau) + 2\zeta\omega_n^3 R_{\dot{z}z}(\tau) + 2\zeta\omega_n^3 R_{z\dot{z}}(\tau) . \tag{11.70}$$

The autocorrelation $R_{\dot{z}z}(\tau) = -R_{z\dot{z}}(\tau)$ between the velocity $\dot{z}(t)$ and the displacement $z(t)$. Therefore

$$R_{\ddot{x}\ddot{x}}(0) = (2\zeta\omega_n)^2 R_{\dot{z}\dot{z}}(0) + \omega_n^4 R_{zz}(0) , \tag{11.71}$$

or

$$E\{\ddot{x}^2(t)\} = (2\zeta\omega_n)^2 E\{\dot{z}^2(t)\} + \omega_n^4 E\{z^2(t)\} = \frac{\omega_n S_{\ddot{u}}}{4\zeta}(1+4\zeta^2), \qquad (11.72)$$

or

$$E\{\ddot{x}^2(t)\} = \frac{\omega_n S_{\ddot{u}}}{4\zeta}(1+4\zeta^2) = \frac{\pi}{2} f_n Q W_{\ddot{u}}(1+4\zeta^2) \approx \frac{\pi}{2} f_n Q W_{\ddot{u}}, \qquad (11.73)$$

where

- $Q = \frac{1}{2\zeta}$ the amplification factor
- f_n the natural frequency (Hz)

In general, the mean value of the acceleration $\ddot{x}(t)$ is zero, $\mu_{\ddot{x}} = 0$. The variance of the acceleration $\ddot{x}(t)$,

$$\sigma_{\ddot{x}}^2 = E\{\ddot{x}^2(t)\} - \mu_{\ddot{x}}^2 = E\{\ddot{x}^2(t)\} = \ddot{x}_{rms}^2 \qquad (11.74)$$

where

- rms the root mean square

The modulus of the frequency transfer function $|H(\omega)|^2$, (11.51), shows a maximum value at the natural frequency $\omega = \omega_n$

$$|H(\omega_n)|^2 = \frac{Q^2}{\omega_n^2}.$$

The bandwidth $\Delta\omega$ at half-power, i.e. $|H(\omega)|^2 = \frac{Q^2}{2\omega_n^2}$, is $\Delta\omega \approx 2\zeta\omega_n$ or $\Delta f \approx 2\zeta f_n$.

The mean-square value of the acceleration $\ddot{x}(t)$ can now be written as

$$E\{\ddot{x}^2(t)\} = \frac{\pi}{2} f_n Q W_{\ddot{u}} = \frac{\pi}{2}\Delta f_n Q^2 W_{\ddot{u}}. \qquad (11.75)$$

Most of the contribution of the power to the mean-square value of $E\{\ddot{x}^2(t)\}$ is stored in a very peaked area with a bandwidth $\frac{\pi}{2}\Delta f_n$ and a height Q^2. The contribution to the power outside the bandwidth $\frac{\pi}{2}\Delta f_n$ is much less. If the PSD function of the forcing function $W_{\dddot{u}\dddot{u}}(f)$ is rather constant in the bandwidth $\frac{\pi}{2}\Delta f_\mathrm{n}$ and in the neighbourhood of the natural frequency f_n (Hz), (11.73) may be approximated by

$$E\{\ddot{x}^2(t)\} = \frac{\pi}{2} f_\mathrm{n} Q W_{\ddot{u}\ddot{u}}(f_\mathrm{n}) . \tag{11.76}$$

Equation (11.76) is called Miles' equation, [Miles 54], and is normally written as

$$\ddot{x}_{\mathrm{rms}} = \sqrt{\frac{\pi}{2} f_\mathrm{n} Q W_{\ddot{u}\ddot{u}}(f_\mathrm{n})} . \tag{11.77}$$

If the PSD function $W_{\ddot{u}}(f_\mathrm{n})$ has the dimension (g^2/Hz), the rms value of the acceleration $\ddot{x}_{\mathrm{rms}}$ has the dimension (g), in practice often denoted by Grms.

The equation of motion of the sdof system, illustrated in Fig. 11.5, expressed in the absolute responses is

$$m\ddot{x}(t) + c\{\dot{x}(t) - \dot{u}(t)\} + k\{x(t) - u(t)\} = 0 . \tag{11.78}$$

Equation (11.78) divided by the mass m will result in

$$\ddot{x}(t) + 2\zeta\omega_\mathrm{n}\dot{x}(t) + \omega_\mathrm{n}^2 x(t) = 2\zeta\omega_\mathrm{n}\dot{u}(t) + \omega_\mathrm{n}^2 u(t) = f(t) . \tag{11.79}$$

The PSD function of the forcing function $f(t)$ is defined as

$$S_{FF}(\omega) = \lim_{T\to\infty}\frac{1}{2T}F(\omega)F^*(\omega) = \lim_{T\to\infty}\frac{1}{2T}\left|[2\zeta\omega_\mathrm{n} j\omega U(\omega) + \omega_\mathrm{n}^2 U(\omega)]\right|^2 \tag{11.80}$$

or

$$S_{FF}(\omega) = \lim_{T\to\infty}\frac{1}{2T}[2j\zeta\omega_n\omega U(\omega) + \omega_n^2 U(\omega)][-2j\zeta\omega_\mathrm{n}\omega U^*(\omega) + \omega_\mathrm{n}^2 U^*(\omega)] . \tag{11.81}$$

Further expanded this gives

$$S_{FF}(\omega) = \lim_{T\to\infty}\frac{1}{2T}(2\zeta\omega_\mathrm{n}\omega)^2 U(\omega)U^*(\omega) + \omega_\mathrm{n}^4 U(\omega)U^*(\omega)$$

$$\lim_{T\to\infty}\frac{1}{2T}[\ +j\omega_\mathrm{n}^3\omega U(\omega)U^*(\omega) - j\omega_\mathrm{n}^3\omega U(\omega)U^*(\omega)] . \tag{11.82}$$

Thus, finally,

$$S_{FF}(\omega) = [(2\zeta\omega_\mathrm{n}\omega)^2 + \omega_\mathrm{n}^4](S_{uu}(\omega)) , \tag{11.83}$$

or expressed in the PSD function of the enforced acceleration $\ddot{u}(t)$, (11.83) becomes

$$S_{FF}(\omega) = \left[\left(\frac{2\zeta\omega_n}{\omega}\right)^2 + \left(\frac{\omega_n}{\omega}\right)^2\right] S_{\ddot{u}\ddot{u}}(\omega). \tag{11.84}$$

The PSD function of the absolute acceleration $\ddot{x}(t)$ will be

$$S_{\ddot{x}\ddot{x}}(\omega) = \omega^4 |H(\omega)|^2 S_{FF}(\omega) = \omega^4 |H(\omega)|^2 \left[\left(\frac{2\zeta\omega_n}{\omega}\right)^2 + \left(\frac{\omega_n}{\omega}\right)^2\right] S_{\ddot{u}\ddot{u}}(\omega). \tag{11.85}$$

Equation (11.85) can now be rewritten as

$$S_{\ddot{x}\ddot{x}}(\omega) = |\hat{H}(\omega)|^2 S_{FF}(\omega) = \left|\frac{\omega_n + 2j\zeta\omega}{\omega_n^2 - \omega^2 + 2j\zeta\omega\omega_n}\right|^2 S_{\ddot{u}\ddot{u}}(\omega). \tag{11.86}$$

The mean-square response of $\ddot{x}(t)$, due to the random load $f(t)$ with the constant PSD function $S_{\ddot{u}\ddot{u}}(\omega) = S_{\ddot{u}} = \frac{W_{\ddot{u}}}{2}$ becomes

$$E\{\ddot{x}(t)^2\} = R_{\ddot{x}\ddot{x}}(0) = \frac{S_{\ddot{u}}}{2\pi}\int_{-\infty}^{\infty} |\hat{H}(\omega)|^2 d\omega, \tag{11.87}$$

with

- $A_0 = \omega_n^2$
- $A_1 = 2\zeta\omega_n$
- $A_2 = 1$
- $B_0 = \omega_n$
- $B_1 = 2\zeta$

The integral in (11.87) now becomes

$$E\{\ddot{x}(t)^2\} = R_{\ddot{x}\ddot{x}}(0) = \frac{S_{\ddot{u}}}{2\pi}\int_{-\infty}^{\infty} |\hat{H}(\omega)|^2 d\omega = \frac{\omega_n S_{\ddot{u}}}{4\zeta}(1 + 4\zeta^2) \approx \frac{\pi}{2} f_n Q W_{\ddot{u}}. \tag{11.88}$$

The result of (11.88) is the same as that obtained in (11.73).

We have an sdof dynamic system with a natural frequency $f_n = 100$ Hz and a damping ratio $\zeta = 0.05$, $Q = \frac{1}{2\zeta}$. The white-noise PSD function of the base acceleration is $W_{\ddot{u}} = 0.1\ \mathrm{g^2/Hz}$. Calculate the rms acceleration of the sdof system.

The rms acceleration response of the sdof system can be calculated using either (11.73) or (11.88), the Miles's equation, (11.77)

$$\ddot{x}_{rms} = \sqrt{\frac{\pi}{2} f_n Q W_{\ddot{u}\ddot{u}}(f_n)} = \sqrt{\frac{\pi}{2} 100\ 10\ 0.1} = 12.53\ \text{Grms}.$$

11.4.3 Multi-Inputs and Single Output (MISO)

In Fig. 11.6 an sdof system is shown with both an enforced acceleration $\ddot{u}(t)$ at the base and a direct force $F(t)$. We will now calculate the PSD function of the force and acceleration. This will result in PSD functions and cross-PSD functions.

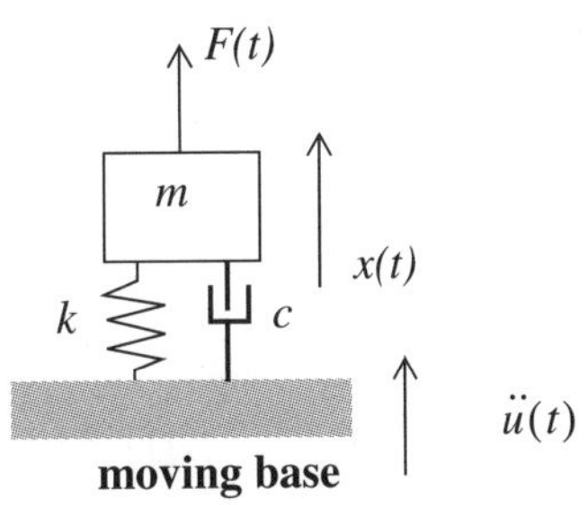

Fig. 11.6. Multi-inputs ($F(t)$ and $\ddot{u}(t)$) single output ($x(t)$)

The equation of motion of the sdof system, illustrated in Fig. 11.6, is a summation of (11.48) and (11.79).

$$\ddot{x}(t) + 2\zeta\omega_n\dot{x}(t) + \omega_n^2 x(t) = f(t) + 2\zeta\omega_n\dot{u}(t) + \omega_n^2 u(t) = q(t). \tag{11.89}$$

The force $f(t)$ is $f(t) = \dfrac{F(t)}{m}$.

The PSD function of $q(t)$ is

$$\left[\left(\frac{2\zeta\omega_n}{\omega}\right)^2 + \left(\frac{\omega_n}{\omega}\right)^2\right] S_{\ddot{u}\ddot{u}}(\omega) + S_{FF}(\omega) + 2\left(\frac{\omega_n}{\omega}\right)^2 S_{F\ddot{u}}(\omega), \tag{11.90}$$

with

- $S_{FF}(\omega)$ the PSD function of $f(t)$

- $S_{\ddot{u}\ddot{u}}(\omega)$ the PSD function of the enforced acceleration
- $S_{F\ddot{u}}(\omega) = S^{*}_{\ddot{u}F}(\omega)$ the cross-PSD function of $f(t)$, $\ddot{u}(t)$ and is, in general, a complex-valued function

11.5 Deterministic Mdof Linear Dynamic System

Dynamic systems may be exposed to random forces and or random enforced motions (i.e. acceleration at the base). Both kinds of random loads will be discussed in the following sections.

11.5.1 Random Forces

In general, the equations of motion of a discrete mdof dynamic system can be written as

$$[M]\{\ddot{x}(t)\} + [C]\{\dot{x}(t)\} + [K]\{x(t)\} = \{F(t)\}, \tag{11.91}$$

and consists of the following matrices and vectors:

- the mass matrix $[M]$
- the stiffness matrix $[K]$
- the damping matrix $[C]$
- The force vector $\{F(t)\}$
- the displacement, velocity and acceleration vectors $\{x(t)\}$, $\{\dot{x}(t)\}$ and $\{\ddot{x}(t)\}$

For linear mdof systems the mass, stiffness and damping matrix do not vary with time and are deterministic, however, the displacement, velocity, acceleration and force vector do usually change with time and are random.

Using the modal displacement (superposition) method (MDM) the physical displacement vector $x(t)$ will be depicted on the independent set of vectors; the modal matrix $[\Phi]$

$$x(t) = [\Phi]\{\eta(t)\}, \tag{11.92}$$

with

- $\{\eta(t)\}$ the vector of generalised coordinates

The modal matrix $[\Phi] = [\phi_1, \phi_2, ...,\phi_n]$ has the following orthogonality properties with respect to the mass matrix and the stiffness matrix

$$[\Phi]^T[M][\Phi] = \langle m\rangle \text{ and } [\Phi]^T[K][\Phi] = \langle\lambda m\rangle \tag{11.93}$$

- $\langle m\rangle$ the diagonal matrix of generalised or modal masses
- $\langle\lambda m\rangle$ the diagonal matrix of the eigenvalues multiplied by the generalised masses

Using (11.92) and premultiplying (11.91) by the transpose of the modal matrix $[\Phi]^T$ the result will be

$$[\Phi]^T[M][\Phi]\{\ddot{\eta}(t)\} + [\Phi]^T[C][\Phi]\{\dot{\eta}(t)\} + [\Phi]^T[K][\Phi]\{\eta(t)\} = [\Phi]^T\{F(t)\} \tag{11.94}$$

Making use of the orthogonality properties of the modal matrix, (11.93), the equations of motion are expressed in generalised coordinates, generalised masses, eigenvalues and generalised forces

$$\langle m\rangle\{\ddot{\eta}(t)\} + \langle c\rangle\{\dot{\eta}(t)\} + \langle\lambda m\rangle\{\eta(t)\} = f(t), \tag{11.95}$$

with

- $f(t)$ the vector of generalised forces
- $\langle c\rangle$ the diagonal matrix of the generalised damping. This means the damping matrix $[C]$ consists of proportional damping. Generally, we will add on an ad hoc basis modal viscous damping to the uncoupled equations of motion of the generalised coordinates, $\frac{c_i}{m_i} = 2\zeta_i\omega_i$, $c_i = \{\phi_i\}^T[C]\{\phi_i\}$.
- $m_i = \{\phi_i\}^T[M]\{\phi_i\}$ the generalised mass associated with mode $\{\phi_i\}$.
- $\zeta_i = \dfrac{c_i}{2\sqrt{k_i m_i}}$ the modal damping ratio.
- $k_i = \{\phi_i\}^T[K]\{\phi_i\} = \lambda_i m_i$ the generalised stiffness.
- $\lambda_i = \omega_i^2$ the eigenvalues of the eigenvalue problem $([K] - \lambda_i[M])\{\phi_i\} = \{0\}$.

Finally, the uncoupled equation of motion of the generalised coordinates enforced with the generalised forces becomes

$$\ddot{\eta}_i(t) + 2\zeta_i\omega_i\dot{\eta}(t) + \omega_i^2\eta_i(t) = \frac{\{\phi_i\}^T\{F(t)\}}{m_i} = \frac{q_i(t)}{m_i},\ i=1,2,\dots. \tag{11.96}$$

The PSD function of the external generalised forces $q_i(t)$ is defined as

$$S_{q_iq_j}(\omega) = \lim_{T\to\infty}\frac{1}{2T}Q_i(\omega)Q_j^*(\omega) = \lim_{T\to\infty}\frac{1}{2T}\{\phi_i\}^T\{F_i(\omega)\}\{F_j^*(\omega)\}^T\{\phi_j\}, \quad (11.97)$$

or

$$S_{q_iq_j}(\omega) = \{\phi_i\}^T[S_{F_iF_j}(\omega)]\{\phi_i\}, \quad (11.98)$$

with

- $[S_{F_iF_j}(\omega)]$ the matrix of PSD and cross-PSD functions

$$[S_{F_iF_j}(\omega)] = \begin{bmatrix} S_{F_1F_1}(\omega) & S_{F_1F_2}(\omega) & . & S_{F_1F_n}(\omega) \\ S_{F_2F_1}(\omega) & S_{F_2F_2}(\omega) & . & . \\ . & . & . & S_{F_{n-1}F_n}(\omega) \\ S_{F_nF_1}(\omega) & . & S_{F_nF_{n-1}}(\omega) & S_{F_nF_n}(\omega) \end{bmatrix}$$

The PSD function of the generalised coordinate $\eta_i(t)$ is

$$S_{\eta_i\eta_i}(\omega) = \left|\frac{H_i(\omega)}{m_i}\right|^2 S_{QQ}(\omega) = \frac{H_i(\omega)}{m_i}\{\phi_i\}^T[S_{F_iF_j}(\omega)]\{\phi_i\}\frac{H_j^*(\omega)}{m_j}, \quad (11.99)$$

with

- $H_i(\omega) = \dfrac{1}{\omega_i^2 - \omega^2 + 2j\zeta\omega\omega_i}$

The matrix of PSD and cross-PSD functions of the generalised coordinates becomes

$$[S_{\eta_i\eta_i}(\omega)] = \langle\frac{H_i(\omega)}{m_i}\rangle[\Phi]^T[S_{F_iF_j}(\omega)][\Phi]\langle\frac{H_j^*(\omega)}{m_j}\rangle. \quad (11.100)$$

- $[S_{\eta_i\eta_i}(\omega)]$ the matrix of PSD and cross PSD functions of the generalised coordinates

$$[S_{\eta_i\eta_i}(\omega)] = \begin{bmatrix} S_{\eta_1\eta_1}(\omega) & S_{\eta_1\eta_2}(\omega) & . & S_{\eta_1\eta_n}(\omega) \\ S_{\eta_2\eta_1}(\omega) & S_{\eta_2\eta_2}(\omega) & . & . \\ . & . & . & S_{\eta_{n-1}\eta_n}(\omega) \\ S_{\eta_n\eta_1}(\omega) & . & S_{\eta_n\eta_{n-1}}(\omega) & S_{\eta_n\eta_n}(\omega) \end{bmatrix}$$

The matrix of PSD functions and cross-PSD functions of the physical displacements $\{x(t)\} = [\Phi]\{\eta(t)\}$ with $\eta(t) = \Pi(\omega)e^{j\omega t}$ is

$$[S_{x_ix_i}(\omega)] = \lim_{T\to\infty}\frac{1}{2T}\{X_i(\omega)\}\{X_j(\omega)^*\}^T = \lim_{T\to\infty}\frac{1}{2T}[\Phi]\{\Pi_i(\omega)\}\{\Pi_i^*(\omega)\}^T[\Phi]^T, \tag{11.101}$$

or

$$[S_{x_ix_i}(\omega)] = [\Phi][[S_{\eta_i\eta_j}(\omega)]][\Phi]^T. \tag{11.102}$$

Finally, we end with

$$[S_{x_ix_j}(\omega)] = [\Phi]\left[\langle\frac{H_i(\omega)}{m_i}\rangle[\Phi]^T[S_{F_iF_j}(\omega)][\Phi]\langle\frac{H_j^*(\omega)}{m_j}\rangle\right][\Phi]^T. \tag{11.103}$$

The matrix of mean squares of $x(t)$ can be calculated

$$[E(x_ix_j)] = R_{x_ix_j}(0) = \frac{1}{2\pi}\int_{-\infty}^{\infty}[S_{x_ix_j}(\omega)]d\omega = \int_0^{\infty}[W_{x_ix_j}(f)]df. \tag{11.104}$$

Equation (11.103) can also be written as

$$[W_{x_ix_j}(\omega)] = [\Phi]\left[\langle\frac{H_i(2\pi f)}{m_i}\rangle[\Phi]^T[W_{F_iF_j}(f)][\Phi]\langle\frac{H_j^*(2\pi f)}{m_j}\rangle\right][\Phi]^T. \tag{11.105}$$

11.5.2 Random Base Excitation

An mdof dynamic system will have an acceleration base excitation at the base (Fig. 11.7)

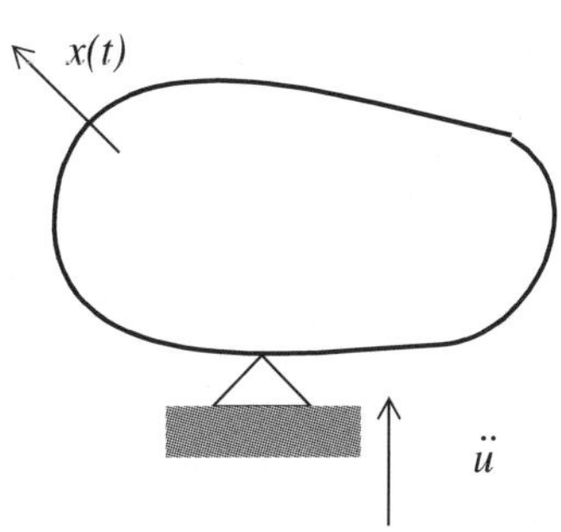

Fig. 11.7. Base excitation

The undamped equations of motion are

$$[M]\{\ddot{x}(t)\} + [K]\{x(t)\} = \{0\}. \tag{11.106}$$

If we introduce the relative motion vector $\{z(t)\} = \{x(t)\} - \{u(t)\}$, with $[K]\{T\}u(t) = \{0\}$, (11.106) becomes

$$[M]\{\ddot{z}(t)\} + [K]\{z(t)\} = -[M]\{T\}\ddot{u}(t) \tag{11.107}$$

$$[W_{z_i z_j}(f)] = [\Phi]\left[\langle\frac{H_i(2\pi f)}{m_i}\rangle[\Phi]^T[W_{F_i F_j}(f)][\Phi]\langle\frac{H_j^*(2\pi f)}{m_j}\rangle\right][\Phi]^T, \tag{11.108}$$

with

- $[W_{F_i F_j}(f)] = (-1)^2[M]\{T\}\{T\}^T[M]W_{\ddot{u}\ddot{u}}(f)$
- $H_i(2\pi f) = \dfrac{1}{(2\pi)^2(f_i^2 - f^2 + 2j\zeta f f_i)}$ the frequency-response function
- $m_i = \{\phi_i\}^T[M]\{\phi_i\}$ the generalised mass
- $[\Phi] = [\phi_1, \phi_1, ..., \phi_n]$ the modal matrix
- $(-(2\pi f_i)^2[M] + [K])\{\phi_i\} = \{0\}$ the eigenvalue problem

Equation (11.108) can be written as

$$[W_{z_i z_j}(f)] = |[H_{z\ddot{u}}(f)]|^2 W_{\ddot{u}\ddot{u}(f)} = [H_{z\ddot{u}}(f)][H^*_{z\ddot{u}}(f)]^T W_{\ddot{u}\ddot{u}(f)} \tag{11.109}$$

$$[W_{z_i z_j}(f)] = \left([\Phi]\left[\langle\frac{H_i(2\pi f)}{m_i}\rangle[\Phi]^T[M]\{T\}\{T\}^T[M][\Phi]\langle\frac{H_j^*(2\pi f)}{m_j}\rangle\right][\Phi]^T\right)W_{\ddot{u}\ddot{u}}(f). \tag{11.110}$$

Thus, the FRF $[H_{z\ddot{u}}(f)]$ is

$$[H_{z\ddot{u}}(f)] = -[\Phi]\left[\langle\frac{H_i(2\pi f)}{m_i}\rangle[\Phi]^T[M]\{T\}\right]. \tag{11.111}$$

The minus sign reflects the negative RHS of (11.107).

The FRF $[H_{zu}(\omega)]$ between the displacement $U(\omega)$ and the relative displacement $Z(\omega)$, $\{Z(\omega)\} = [H_{zu}(\omega)]U(\omega) = -\omega^2[H_{z\ddot{u}}(\omega)]U(\omega)$, becomes

$$[H_{zu}(\omega)] = \omega^2[\Phi]\left[\langle\frac{H_i(\omega)}{m_i}\rangle[\Phi]^T[M]\{T\}\right]. \tag{11.112}$$

The absolute displacement vector is $\{x(t)\} = \{z(t)\} + \{T\}u$. The PSD function of the absolute displacement $\{x(t)\}$ is defined as

$$S_{x_ix_j}(\omega)= \lim_{T\to\infty}\frac{1}{2T}\{X(\omega)\}\{X_j^*(\omega)\}^T, \quad (11.113)$$

or

$$S_{x_ix_j}(\omega)= \lim_{T\to\infty}\frac{1}{2T}[\{Z(\omega)\}+\{T\}(U(\omega))][\{Z^*(\omega)\}+\{T\}(U^*(\omega))]^T, \quad (11.114)$$

or

$$S_{x_ix_j}(\omega)= \lim_{T\to\infty}\frac{1}{2T}(([H_{zu}(\omega)]U(\omega)+\{T\}U(\omega)])[H_{zu}^*(\omega)]^T U^*(\omega)+\{T\}^T U^*(\omega)), \quad (11.115)$$

and, finally,

$$S_{x_ix_j}(\omega)= ([H_{zu}(\omega)][H_{zu}^*(\omega)]^T+\{T\}[H_{zu}^*(\omega)]^T+[H_{zu}(\omega)]\{T\}^T+\{T\}\{T\}^T)S_{uu}(\omega). \quad (11.116)$$

The PSD function for the acceleration $\ddot{x}(t)$ becomes

$$S_{\ddot{x}_i\ddot{x}_j}(\omega)= ([H_{zu}(\omega)][H_{zu}^*(\omega)]^T+\{T\}[H_{zu}^*(\omega)]^T+[H_{zu}(\omega)]\{T\}^T+\{T\}\{T\}^T)S_{\ddot{u}\ddot{u}}(\omega), \quad (11.117)$$

or, in the frequency (Hz) domain

$$W_{\ddot{x}_i\ddot{x}_j}(f)= \left([H_{\ddot{x}\ddot{u}}(2\pi f)][H_{\ddot{x}\ddot{u}}^*(2\pi f)]^T\right)W_{\ddot{u}\ddot{u}}(f), \quad (11.118)$$

with

- $[H_{\ddot{x}\ddot{u}}(2\pi f)] = [H_{zu}(2\pi f)]+\{T\}$
- $[H_{zu}((2\pi f))] = (2\pi f)^2[\Phi]\left[\langle\frac{H_i(2\pi f)}{m_i}\rangle[\Phi]^T[M]\{T\}\right]$
- $H_i(2\pi f) = \frac{1}{(2\pi)^2(f_i^2-f^2+2j\zeta f f_i)}$
- $\{T\}$ the rigid-body motion in the excitation direction $\{\ddot{u}(t)\}$

11.5.3 Random Stresses and Forces

Besides the responses of the dofs, internal forces or stresses should be calculated to predict the strength characteristics of the structure itself.

The matrix of cross-power spectral densities of the physical stresses (forces) $S_{\sigma_i\sigma_j}(\omega)$ can also be calculated

$$[S_{\sigma_i\sigma_j}(\omega)] = [\Phi_\sigma][S_{\eta_i\eta_j}(\omega)][\Phi_\sigma]^T, \tag{11.119}$$

where

- $[\Phi_\sigma]$ is the matrix of the stress or force modes

The stress or force modes $[\Phi_\sigma]$ can be calculated using the mode shapes $[\Phi]$ and a so-called stress or force matrix $[D_\sigma]$. The stress mode is defined as

$$[\Phi_\sigma] = [D_\sigma][\Phi]. \tag{11.120}$$

When random loads are applied to the deterministic system, (11.119) can be written as

$$[W_{\sigma_i\sigma_j}(f)] = [D_\sigma][\Phi][H_{\sigma F}(2\pi f)][W_{FF}(f)][H^*{}_{\sigma F}(2\pi f)][\Phi]^T[D_\sigma]^T, \tag{11.121}$$

and when random base accelerations are applied to the deterministic system, (11.119) can be written as

$$[W_{\sigma_i\sigma_j}(f)] = [D_\sigma][\Phi][H_{\sigma\ddot{u}}(2\pi f)][W_{\ddot{u}\ddot{u}}(f)][H^*{}_{\sigma\ddot{u}}(2\pi f)][\Phi]^T[D_\sigma]^T, \tag{11.122}$$

with

- $[H_{\sigma F}(2\pi f)] = [D_\sigma][\Phi]\langle\frac{H_i(2\pi f)}{m_i}\rangle[\Phi]^T$
- $[H_{\sigma\ddot{u}}(2\pi f)] = -[D_\sigma][\Phi]\langle\frac{H_i(2\pi f)}{m_i}\rangle[\Phi]^T[M]\{T\}$
- $H_i(2\pi f) = \dfrac{1}{(2\pi)^2(f_i^2 - f^2 + 2j\zeta f f_i)}$
- $\{T\}$ the rigid-body motion in the excitation direction $\{\ddot{u}(t)\}$

A 3 mass–spring dynamic system, as shown in Fig. 11.8, is excited at the base with a constant band-limited random acceleration $W_{\ddot{u}\ddot{u}}(f) = 0.01$ $g^2/$Hz in a frequency range $5 \le f \le 500$ Hz. The mass distribution is $m_1 = 200$ kg, $m_2 = 150$ kg and $m_3 = 100$ kg. The stiffness distribution is $k_1 = 3x10^8$ N/m, $k_2 = 2x10^8$ N/m and $k_3 = 1x10^8$ N/m. The modal damping ratio for all modes is $\zeta = 0.05$ or $Q = 10$.

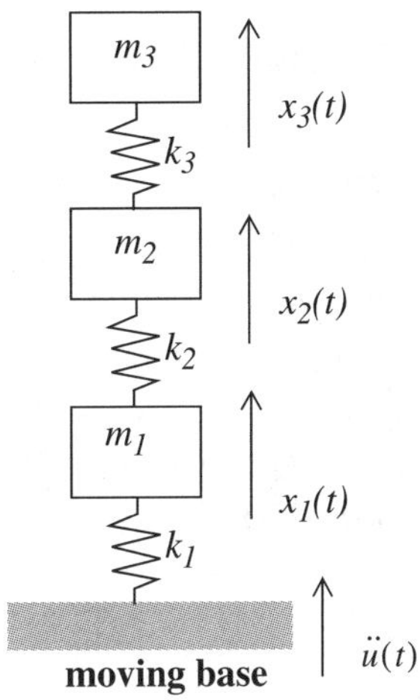

Fig. 11.8. 3 mass–spring system with enforced acceleration at the base

The undamped equations of motion are

$$\begin{bmatrix} m_1 & 0 & 0 \\ 0 & m_2 & 0 \\ 0 & 0 & m_3 \end{bmatrix} \begin{Bmatrix} \ddot{x}_1 \\ \ddot{x}_2 \\ \ddot{x}_3 \end{Bmatrix} + \begin{bmatrix} k_1 + k_2 & -k_2 & 0 \\ -k_2 & k_2 + k_3 & -k_3 \\ 0 & -k_3 & k_3 \end{bmatrix} \begin{Bmatrix} x_1 \\ x_2 \\ x_3 \end{Bmatrix} = \begin{Bmatrix} 0 \\ 0 \\ 0 \end{Bmatrix}, \tag{11.123}$$

or (11.123) expressed in the relative motion $\{z(t)\}$ will result in

$$\begin{bmatrix} m_1 & 0 & 0 \\ 0 & m_2 & 0 \\ 0 & 0 & m_3 \end{bmatrix} \begin{Bmatrix} \ddot{z}_1 \\ \ddot{z}_2 \\ \ddot{z}_3 \end{Bmatrix} + \begin{bmatrix} k_1 + k_2 & -k_2 & 0 \\ -k_2 & k_2 + k_3 & -k_3 \\ 0 & -k_3 & k_3 \end{bmatrix} \begin{Bmatrix} z_1 \\ z_2 \\ z_3 \end{Bmatrix} = -\begin{bmatrix} m_1 & 0 & 0 \\ 0 & m_2 & 0 \\ 0 & 0 & m_3 \end{bmatrix} \begin{Bmatrix} 1 \\ 1 \\ 1 \end{Bmatrix} \ddot{u}(t). \tag{11.124}$$

The natural frequencies and associate mode shapes of the mdof dynamic system are

$$\{f_n\} = \begin{Bmatrix} 94.35 \\ 201.73 \\ 299.53 \end{Bmatrix} \text{Hz}, [\Phi] = \begin{bmatrix} -0.0224 & 0.0432 & -0.0513 \\ -0.0482 & 0.0386 & 0.0535 \\ -0.0743 & -0.0636 & -0.0210 \end{bmatrix}$$

The effective masses, for, respectively modes shapes 1, 2 and 3, are

$$\{M_{eff}\} = \begin{Bmatrix} 366.13 \\ 64.97 \\ 18.90 \end{Bmatrix} \text{kg}.$$

The most important mode shape represents the maximum effective mass. This first mode will show maximum responses. A $3x3$ matrix $W_{\ddot{x}\ddot{x}}(f)$ of PSD and cross- PSD functions of the acceleration will be calculated for every frequency. The quadrature of diagonal terms of $W_{\ddot{x}\ddot{x}}(f)$ are the mean-square values of the acceleration. Taking the square root of the mean-square (auto-spectrum) values will result in the root mean square values of the acceleration of the dofs x_1, x_2 and x_3. The plots of the PSD functions of the acceleration $\ddot{x}_1$, $\ddot{x}_2$ and $\ddot{x}_3$ are shown in Fig. 11.9.

The integration, to obtain the mean-square values, is done with the trapezium rule with a frequency increment $\Delta f = 1$ Hz.

The rms values of the acceleration are

$$\{\ddot{x}\}_{rms} = \begin{Bmatrix} \ddot{x}_1 \\ \ddot{x}_2 \\ \ddot{x}_3 \end{Bmatrix} = \begin{Bmatrix} 30.27 \\ 41.72 \\ 60.52 \end{Bmatrix} \frac{m}{s^2}.$$

The force matrix of the dynamic system, shown in Fig. 11.8, is defined as

$$[D_\sigma] = \begin{bmatrix} k_1 & 0 & 0 \\ -k_2 & k_2 & 0 \\ 0 & -k_3 & k_3 \end{bmatrix} \text{N}.$$

The diagonal terms of the PSD function matrix $W_{\sigma\sigma}(f)$ are plotted in Fig. 11.10.

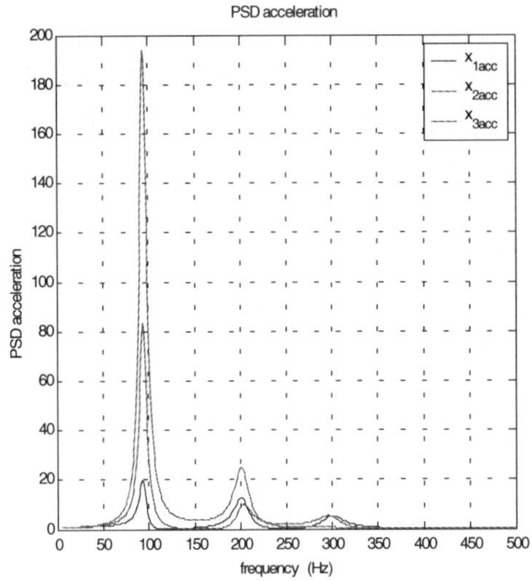

Fig. 11.9. PSD values of the acceleration $\ddot{x}_1$, $\ddot{x}_2$ and $\ddot{x}_3$ $(\mathrm{m/s^2})^2/\mathrm{Hz}$

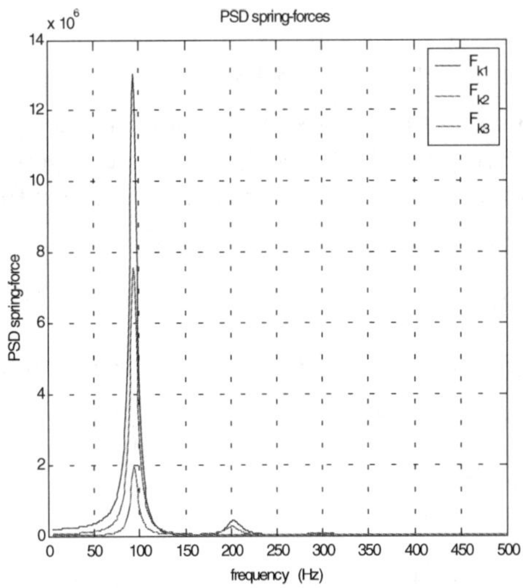

Fig. 11.10. PSD values of the spring forces ($\mathrm{N^2/Hz}$)

The square root of the diagonal terms of the auto-spectrum functions of the spring forces is

$$\sqrt{\langle R_{\sigma\sigma}(0)\rangle} = \langle F_{rms}\rangle = \sqrt{\int_5^{500} \langle W_{\sigma\sigma}(f)\rangle df} = \begin{bmatrix} 1.438x10^4 & 0 & 0 \\ 0 & 1.071x10^4 & 0 \\ 0 & 0 & 0.604x10^4 \end{bmatrix} \text{N.}$$

11.6 Analysis of Narrow-Band Processes

In this section some interesting properties of narrow-banded stationary processes will be discussed:

- Number of crossings per unit of time through a certain level
- Fatigue damage due to random excitation

Besides the rms value of the response of deterministic structures exposed to random forces, the above-mentioned properties are important properties for further investigation of the strength characteristics.

11.6.1 Crossings

Consider the event that a stationary process $x(t)$ will cross the level α from below with a certain positive velocity $\dot{x}(t) = v(t)$. One talks about a crossing with a positive slope (see Fig. 11.11). $N_\alpha(\tau)$ is the number of expected crossings for a time period τ. The random process $x(t)$ is a stationary process so the number of expected crossings does not depend on the time the process starts. The sum of numbers of zero crossing will be a linear function in time, hence:

$$N_\alpha(\tau_1 + \tau_2) = N_\alpha(\tau_1) + N_\alpha(\tau_2). \tag{11.125}$$

The number of positive crossings per unit of time ν_α^+, that the signal $x(t)$ will cross the level "α" with a positive slope (positive velocity), will be defined as

$$N_\alpha(\tau) = \nu_\alpha^+\tau. \tag{11.126}$$

The joint probability that the values of $x(t)$ and $v(t)$ are between certain values, for all times t, is defined as

$$f(\alpha, \beta)d\alpha d\beta = \text{Prob}(\alpha \le x(t) \le \alpha + d\alpha \text{ and } \beta \le v(t) \le \beta + d\beta). \tag{11.127}$$

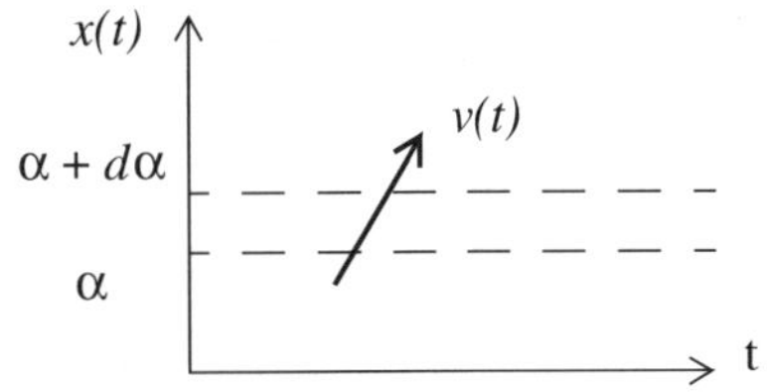

Fig. 11.11. Positive crossings

The cumulative joint probability is defined as

$$F(\alpha + \mathrm{d}\alpha, \beta + \mathrm{d}\beta) - F(\alpha, \beta) = \int_{\alpha}^{\alpha + \mathrm{d}\alpha} \int_{\beta}^{\beta + \mathrm{d}\beta} f(x, \dot{x}) dx d\dot{x}, \qquad (11.128)$$

or

$$F(\alpha + \mathrm{d}\alpha, \beta + \mathrm{d}\beta) - F(\alpha, \beta) \approx \frac{\partial^2 F(\alpha, \beta)}{\partial\alpha\partial\beta} d\alpha d\beta = f(\alpha, \beta) d\alpha d\beta, \qquad (11.129)$$

where
- $f(\alpha, \beta)$ the joint probability density function
- $F(\alpha, \beta)$ the joint cumulative joint probability function

Equations (11.127), (11.128) and (11.129) define the time per unit of time $x(t)$ in the interval $\alpha + \mathrm{d}\alpha$ with a velocity of approximately $v(t) \approx \beta$.

The number of expected crossings ν_{α}^{+} per unit time through the interval $[\alpha, \alpha + d\alpha]$ with a velocity $v(t) \approx \beta$ is estimated by dividing the amount of time per unit time spent inside this interval by the time required to pass this interval. The time τ to pass (up or down) the interval $[\alpha, \alpha + d\alpha]$ is

$$d\tau = \frac{d\alpha}{|\beta|}. \qquad (11.130)$$

The amount of time per unit of time the signal $x(t)$ is in the interval $[\alpha, \alpha + d\alpha]$ with a velocity in the interval $[\beta, \beta + d\beta]$ is $f(\alpha, \beta) d\alpha d\beta$. The expected number of crossings (up or down) ν_{α} per unit of time through the level α with all possible velocities $v(t) \approx \beta$ is

$$\frac{f(\alpha, \beta) d\alpha d\beta}{d\tau} = |\beta| f(\alpha, \beta) d\beta. \qquad (11.131)$$

The expected number of crossings (up or down) ν_α per unit of time through the level α with all possible velocities $v(t) \approx \beta$ is

$$\nu_\alpha = \int_{-\infty}^{\infty} |\beta| f(\alpha, \beta) d\beta \tag{11.132}$$

The expected number of positive zero crossings ν_α^+ can be easily derived from (11.132)

$$\nu_\alpha^+ = \frac{1}{2}\int_{-\infty}^{\infty} \beta f(\alpha, \beta) d\beta . \tag{11.133}$$

A special joint probability density function, as defined in (11.127), is the case of a Gaussian process, i.e. a process with a distribution [Gatti 99]

$$f(x, \dot{x}) = \frac{1}{2\pi\sigma_x\sigma_{\dot{x}}} e^{\left(-\frac{x^2}{2\sigma_x^2} - \frac{\dot{x}^2}{2\sigma_{\dot{x}}^2}\right)} . \tag{11.134}$$

Substituting (11.134) into (11.133) and performing the integration gives

$$\nu_\alpha^+ = \frac{1}{2}\int_{-\infty}^{\infty} \beta \frac{1}{2\pi\sigma_x\sigma_{\dot{x}}} e^{\left(-\frac{\alpha^2}{2\sigma_x^2} - \frac{\dot{\beta}^2}{2\sigma_{\dot{x}}^2}\right)} d\beta = \frac{\sigma_{\dot{x}}}{2\pi\sigma_x} e^{-\frac{\alpha^2}{2\sigma_x^2}} , \tag{11.135}$$

with

- $\sigma_x^2 = E(x^2) - \{E(x)\}^2 = E(x^2) = \int_0^{\infty} W_x(f)) df$
- $\sigma_{\dot{x}}^2 = E(\dot{x}^2) - \{E(\dot{x})\}^2 = E(\dot{x}^2) = \int_0^{\infty} (2\pi f)^2 W_x(f)) df$

In general, we assume zero mean values of the displacement $x(t)$ and velocity $\dot{x}(t)$. The number of zero positive zero crossings can be obtained from (11.135) with $\alpha = 0$

$$\nu_0^+ = \frac{\sigma_{\dot{x}}}{2\pi\sigma_x} = \frac{1}{2\pi}\sqrt{\frac{\int_0^{\infty} (2\pi f)^2 W_x(f)) df}{\int_0^{\infty} W_x(f)) df}} \tag{11.136}$$

The number of positive zero crossings ν_0^+ (Hz) is also called the equivalent or characteristic frequency of the random process. (11.135) can also written as

$$\nu_\alpha^+ = \frac{\sigma_{\dot{x}}}{2\pi\sigma_x} e^{-\frac{\alpha^2}{2\sigma_x^2}} = \nu_0^+ e^{-\frac{\alpha^2}{2\sigma_x^2}}. \tag{11.137}$$

For the mass–spring–damper system shown Fig. 11.5 the variance of the relative displacement σ_z^2 and the relative velocity $\sigma_{\dot{z}}^2$ has been calculated in (11.68) and (11.69).

The mean-square value of the relative displacement $z(t)$ is given by

$$E\{z(t)^2\} = \sigma_z^2 = \frac{S_{\ddot{u}}}{4\zeta\omega_n^3}. \tag{11.138}$$

The mean-square value of the relative velocity $\dot{z}(t)$ is given by

$$E\{\dot{z}^2(t)\} = \sigma_{\dot{z}}^2 = \frac{S_{\ddot{u}}}{4\zeta\omega_n} \tag{11.139}$$

With the aid of (11.136) the number of positive zero crossings per unit of time ν_0^+ (Hz) can be calculated,

$$\nu_0^+ = \frac{\sigma_{\dot{x}}}{2\pi\sigma_x} = \frac{1}{2\pi}\sqrt{\frac{\dfrac{S_{\ddot{u}}}{4\zeta\omega_n}}{\dfrac{S_{\ddot{u}}}{4\zeta\omega_n^3}}} = \frac{\omega_n}{2\pi} = f_n.$$

The number of positive zero crossings per unit of time ν_0^+ is equal to the natural frequency f_n of the sdof system, as illustrated in Fig. 11.5.

For the mdof dynamic system, as shown in Fig. 11.8, the number of positive zero crossings ν_0^+ for all dofs will be calculated, assuming a random base acceleration $W_{\ddot{u}\ddot{u}}(f) = 0.01\ \mathrm{g^2/Hz}$ in a frequency range $5 \le f \le 500$ Hz. The integration, to obtain the positive zero crossings per unit of time, is done using the trapezium rule with a frequency increment $\Delta f = 1$ Hz.

The positive zero crossings for the acceleration $\begin{Bmatrix} \ddot{x}_1 \\ \ddot{x}_2 \\ \ddot{x}_3 \end{Bmatrix}$ are

$$\{\nu_0^+\} = \begin{Bmatrix} 203 \\ 157 \\ 128 \end{Bmatrix} \text{Hz.}$$

The positive zero crossings for the forces in $\begin{Bmatrix} k_1 \\ k_2 \\ k_3 \end{Bmatrix}$ are

$$\{\nu_0^+\} = \begin{Bmatrix} 106 \\ 105 \\ 128 \end{Bmatrix} \text{Hz.}$$

11.6.2 Fatigue Damage due to Random Excitation

There are quite a number of failure modes, one of them is the failure of a structure due to fatigue behaviour of materials. Fatigue appears when the structure is exposed to oscillating loads (stresses). The material will crack and failure occurs. Fatigue damage is caused by microplastic deformations (strains) that will damage the structure of the material locally and accumulate to microcracks and ultimately to failure of the structure.

With the Palmgren–Miner rule one is able to predict the fatigue life of a structure or part of the structure caused by cumulative damage when the construction is exposed to oscillating loads or stresses.

At a certain stress level σ_i (in the case of random vibration the one-sigma value the stress) one can take the allowable number of oscillations N_i from a so-called s–N curve. In general, the relation between the stress level and the allowable number of oscillations, the s–N or Woehler fatigue curve, is

$$N(s)s^b = a, \tag{11.140}$$

where a and b are constants.

The model of cumulative damage, as formulated by Palmgren and Miner, the Palmgren–Miner damage function $D(t)$ [Gatti 99] is

$$D(t) = \sum_{i=1}^{N(t)} \Delta D_i . \qquad (11.141)$$

We assume that the damage function $D(t)$ is a nondecreasing function of time that starts at zero for a new structure and is normalised to unity when failure occurs; the instant of time t_{failure} at which $D(t_{\text{failure}}) = 1$.

The Palmgren–Miner rule can be formulated as follows:
if the i-th cycle occurs at the stress level σ_i at which, in accordance with the s–N curve, N_i cause failure, then the i-th increment of damage

$$\Delta D_i = \frac{1}{N_i} . \qquad (11.142)$$

If we group cycles of approximately equal amplitude together, we will have a situation in which we can identify n_i cycles at the stress level σ_i. Then each one of these groups i will produce $\frac{n_i}{N_i}$ incremental damage.
The failure condition becomes

$$D = \sum_i \frac{n_i}{N_i} = 1 . \qquad (11.143)$$

Nothing is stated about the sequence of the stress levels.
The number of positive crossings at the level a is given by (11.137)

$$\nu_\alpha^+ = \frac{\sigma_{\dot{x}}}{2\pi\sigma_x} e^{-\frac{\alpha^2}{2\sigma_x^2}} = \nu_0^+ e^{-\frac{\alpha^2}{2\sigma_x^2}}$$

With (11.137) we are able to calculate the number of peaks $n_{\text{p}}(\alpha)$ per unit of time of $x(t)$ in the range $\alpha \le x(t) \le \alpha + d\alpha$

$$n_{\text{p}}(\alpha) = \nu_\alpha^+ - \nu_{\alpha + d\alpha}^+ = -\frac{d\nu_\alpha^+}{d\alpha} d\alpha , \qquad (11.144)$$

thus

$$n(\alpha) = \nu_0^+ \frac{\alpha}{\sigma_x^2} e^{-\frac{\alpha^2}{2\sigma_x^2}} d\alpha . \qquad (11.145)$$

The total number of peaks during the time period T is given by $n(\alpha)T$.

We will introduce (11.145) into (11.143) and will replace the summation with an integration,

$$E\{D(T)\} = \sum_i \frac{n_i}{N_i} = T\int_0^\infty \frac{n(\alpha)}{N(\alpha)}d\alpha. \tag{11.146}$$

The cumulative damage function $D(t_{\text{failure}})$ has been replaced by the expected value of the cumulative damage function $E\{D(T)\}$ at the time T the structure will fail.

Substituting into (11.140) the number of allowable oscillations at stress level $s = \alpha$ is

$$N(\alpha) = a\alpha^{-b}. \tag{11.147}$$

Equations (11.145) and the (11.147) substituted into (11.146) will result in

$$E\{D(T)\} = T\int_0^\infty \frac{n(\alpha)}{N(\alpha)}d\alpha = \frac{\nu_0^+ T}{a\sigma_x^2}\int_0^\infty \alpha^{b+1} e^{-\frac{\alpha^2}{2\sigma_x^2}} d\alpha = \frac{\nu_0^+ T}{a}(\sqrt{2}\sigma_x)^b \Gamma\left(1+\frac{b}{2}\right). \tag{11.148}$$

Assuming failure at $E\{D(T)\} = 1$, the time to failure can now be calculated

$$T = \frac{a}{\nu_0^+(\sqrt{2}\sigma_x)^b \Gamma\left(1+\frac{b}{2}\right)}. \tag{11.149}$$

The Gamma function $\Gamma(z)$ is defined by $\Gamma(z) = \int_0^\infty t^{z-1}e^{-t}dt$.

An sdof dynamic system is excited at the base with random acceleration $\ddot{u}(t)$ with a PSD function $W_{\ddot{u}}(f)$. The sdof system is shown in Fig. 11.5. The relative rms displacement z_{rms} of the mass can be calculated with (11.68)

$$z_{\text{rms}} = \frac{S_{\ddot{u}}(\omega)}{4\zeta\omega_n^3} = \sqrt{\frac{QW_{\ddot{u}}(f_n)}{(2\pi f_n)^3}}.$$

The rms stress in the spring is

$$s_{\text{rms}} = \frac{kz_{\text{rms}}}{A},$$

with

- k the spring constant (N/m)
- A the cross section (m^2)

The s–N curve is given by $N(s)s^b = a$
with

- $a = 1.56x10^{39}$
- $b = 4$

The natural frequency of the sdof system is $f_n = 100$ Hz, the mass 100 kg and the modal damping ratio $\zeta = 0.05$. The cross section of the spring $A = 10^{-4}$ m^2. In a frequency range from 50–500 Hz the PSD of the base acceleration is $W_{\ddot{u}} = 0.1$ g/Hz^2.

The spring constant is $k = (2\pi f_n)^2 m = (2\pi 100)^2 100 = 3.948\ 10^7$ N/m.

The rms value of the relative displacement

$$z_{\mathrm{rms}} = \sqrt{\frac{QW_{\ddot{u}}(f_n)}{(2\pi f_n)^3}} = \sqrt{\frac{10x0.1x9.81^2}{(2\pi 100)^3}} = 6.229x10^{-4}\ \mathrm{m}.$$

The rms stress in the spring becomes

$$s_{\mathrm{rms}} = \frac{kz_{\mathrm{rms}}}{A} = \frac{3.948x10^7 x6.229x10^{-4}}{10^{-4}} = 2.459x10^8\ \mathrm{Pa}.$$

The number of positive zero crossing $\nu_0^+ = f_n = 100$ Hz.

The time to failure T, the fatigue lifetime, can now be calculated

$$T = \frac{a}{\nu_0^+(\sqrt{2}\sigma_s)^b\Gamma\left(1+\frac{b}{2}\right)} = \frac{1.56x10^{39}}{100(\sqrt{2}x2.459x10^8)^4\Gamma(3)} = 533\ \mathrm{s}.$$

11.7 Some Practical Aspects

In most cases the random mechanical loads for spacecraft and subsystems of spacecraft are as illustrated below. The PSD values of the acceleration depend on the frequency (Hz). In general, the frequency range is between 20–2000 Hz. The specification must be accompanied by the Grms value of the random acceleration in the frequency range. An example of a typical acceleration specification is given below.

- 20–150Hz 6dB/oct
- 150–700Hz $W_{\ddot{u}} = 0.04$ g^2/Hz

- 700–2000Hz –3 dB/oct.
- Grms=7.3 g

The graphical representation of the random acceleration specification is shown in Fig. 11.9.

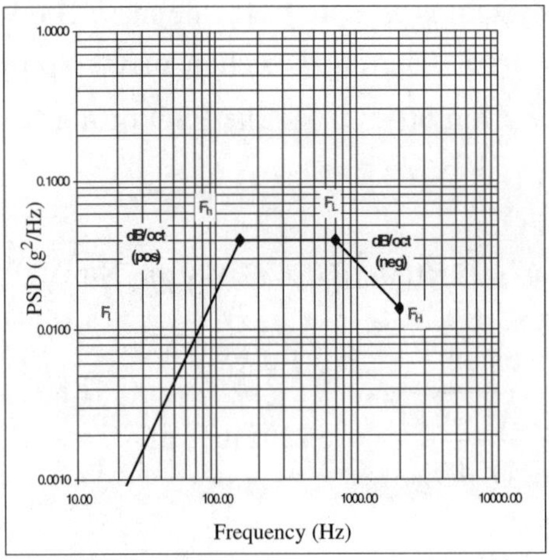

Fig. 11.12. Specification PSD acceleration, Grms=7.3 g (courtesy NASA GSFC, FEMCI).

The octave band is defined by

$$\frac{f_2}{f_1} = 2^1 . \tag{11.150}$$

The number of octaves x between two frequencies f_x and the reference frequency f_{ref} can be calculated by

$$\frac{f_x}{f_{\text{ref}}} = 2^x , \tag{11.151}$$

and x can simply be obtained with

$$x = \frac{\ln\left(\frac{f_x}{f_{\text{ref}}}\right)}{\ln(2)} = \frac{\log\left(\frac{f_x}{f_{\text{ref}}}\right)}{\log(2)} . \tag{11.152}$$

The relation between the PSD values depends on the number of dBs per octave n dB/oct and the number of octaves between the two frequencies f_x and f_{ref}. The relation in (dB) between the $W_{\ddot{u}}(f_x)$ and $W_{\ddot{u}}(f_{\text{ref}})$ is

$$10\log\left(\frac{W_{\ddot{u}}(f_x)}{W_{\ddot{u}}(f_{\text{ref}})}\right) = nx = \frac{n\log\left(\frac{f_x}{f_{\text{ref}}}\right)}{\log(2)}\ (\text{dB}) \tag{11.153}$$

or

$$\frac{W_{\ddot{u}}(f_x)}{W_{\ddot{u}}(f_{\text{ref}})} = nx = \frac{n\log\left(\frac{f_x}{f_{\text{ref}}}\right)}{\log(2)} = \left(\frac{f_x}{f_{\text{ref}}}\right)^{\frac{n\log(2)}{10}} \approx \left(\frac{f_x}{f_{\text{ref}}}\right)^{\frac{n}{3}}. \tag{11.154}$$

The angle m (dB/freq) can be obtained by

$$m = \frac{\log W_{\ddot{u}}(f_x) - \log W_{\ddot{u}}(f_{\text{ref}})}{\log f_x - \log f_{\text{ref}}} = \frac{\log\left(\frac{W_{\ddot{u}}(f_x)}{W_{\ddot{u}}(f_{\text{ref}})}\right)}{\log\left(\frac{f_x}{f_{\text{ref}}}\right)} = \frac{n}{3}. \tag{11.155}$$

Now we can calculate the Grms value of the acceleration specification.

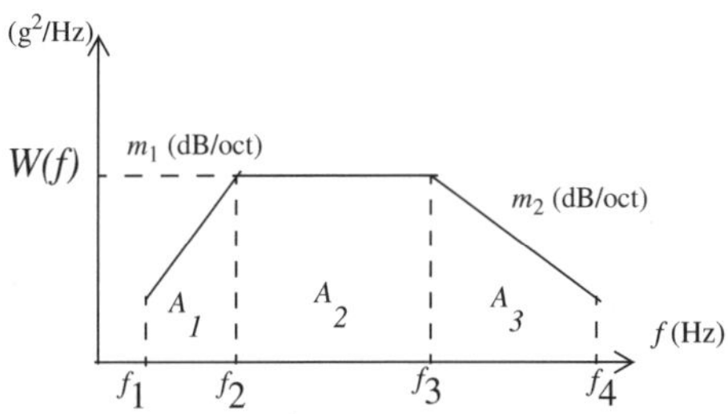

Fig. 11.13. Calculation Grms (g)

The G_{rms} value of the random acceleration specification can be calculated (see Fig. 11.13) using

$$G_{\text{rms}} = \sqrt{A_1 + A_2 + A_3}, \tag{11.156}$$

with

- $A_1 = \dfrac{Wf_2}{\frac{m_1}{3}+1}\left[1-\left(\dfrac{f_1}{f_2}\right)^{\frac{m_1}{3}+1}\right]\ (\text{g}^2)$
- $A_2 = W(f_3 - f_2)\ (\text{g}^2)$

- $A_3 = \frac{Wf_3}{\frac{m_2}{3}+1}\left[\left(\frac{f_4}{f_3}\right)^{\frac{m_2}{3}+1} - 1\right]$, (g^2), $\frac{m_2}{3} \neq -1$ and $m_2 < 0$ (dB/oct)

- For $\frac{m_2}{3} = -1$ we may use L'Hopital's rule and

$$A_3 = Wf_3 \ln\left(\frac{f_4}{f_3}\right) = 2.303\,Wf_3 \log\left(\frac{f_4}{f_3}\right) \; (g^2)$$

11.8 Problems

11.8.1 Problem 1

Calculate the Grms value of the following random acceleration specification.

- 20–150Hz 6dB/oct
- 150–700Hz $W_{\ddot{u}} = 0.04\ g^2/\text{Hz}$
- 700–2000 Hz –3 dB/oct.

Answer Grms=7.3 g.

11.8.2 Problem 2

A stationary random process has an autocorrelation function $R_{xx}(\tau)$, which is given by $R_{xx}(\tau) = Xe^{-a|\tau|}$, where $a > 0$, [Norton 98].

Determine

1. the mean-square value
2. power-spectral density function $S_{xx}(\omega)$ of the process

Answers: X, $\frac{aX}{(a^2+\omega^2)}$ $S_{xx}(\omega) = \int_{-\infty}^{\infty} R_{xx}(\tau)e^{-j\omega\tau}d\tau$.

11.8.3 Problem 3

The single-sided spectral density of the deflection $y(t)$ of an electric motor bearing is $W_{yy}(f) = W_o = 0.05$ mm²/Hz over the frequency band from 20–2000 Hz and is zero for all other frequencies. Determine the mean-square (m²), root mean-square deflection (m) and obtain an expression for the autocorrelation function $R_{yy}(\tau)$ (m²).

Answers: $9.9x10^{-5}$, $9.95x10^{-3}$, $\frac{0.05x10^{-6}}{2\pi\tau}[\sin(2\pi 2000\tau) - \sin(2\pi 20\tau)]$.

11.8.4 Problem 4

A structural component at its static equilibrium position is represented by a rigid slender bar of mass M (kg) and length L (m)], a spring k (N/m) and a damper c (Ns/m) as illustrated Fig. 11.14.

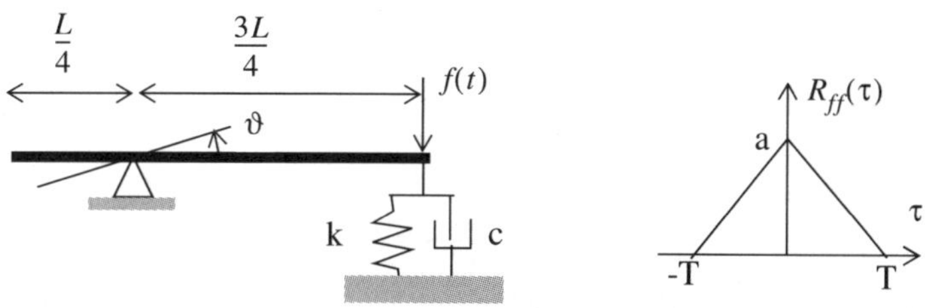

Fig. 11.14. Structural component + forcing function

The tip of the bar is subjected to a pulse $f(t)$ (N) with an autocorrelation function $R_{ff}(\tau)$ (N²) as shown in Fig. 11.14. By modelling the system as an sdof system, expressed by angle ϑ, with an equivalent mass, spring and damper, and derive an expression of the output spectral density $W_{\vartheta\vartheta}(f)$.

Answers: $\frac{28}{192}ML^2\ddot{\vartheta} + \frac{3}{4}L\dot{\vartheta}c + \frac{3}{4}L\vartheta k = f(t)$,

$$S_{\vartheta\vartheta}(\omega) = \frac{\frac{aT}{2\pi}\left\{\frac{\sin\left(\frac{\omega T}{2}\right)}{\frac{\omega T}{2}}\right\}^2}{\left(\frac{3}{4}Lk - \frac{28}{192}\omega^2 ML^2\right)^2 + \left(\frac{3}{4}Lc\omega\right)^2}, \quad W_{\vartheta\vartheta}(f) = 2S_{\vartheta\vartheta}(\omega).$$

11.8.5 Problem 5

A response $x(t)$ has an autocorrelation function $R_{xx}(\tau)$ that is given by

$R_{xx}(\tau) = \beta^2 e^{-\alpha\tau^2}$.

- Define the standard deviation σ_x of $x(t)$
- Define the standard deviation of $\sigma_{\dot{x}}$ and $\dot{x}(t)$
- Define the number of positive zero crossings $\nu_{0,x}^{+}$
- Define the number of zero crossings through $x(t) = \gamma$

Answer: β, $\beta\sqrt{2\alpha}$, $\nu_{0,x}^{+} = \frac{1}{\pi}\sqrt{\frac{\alpha}{2}}$, $\nu_{0,x=\gamma}^{+} = \frac{1}{\pi}\sqrt{\frac{\alpha}{2}}e^{-\frac{\gamma^2}{\sigma_x^2}}$ [Scheidt 94]

12 Low-Frequency Acoustic Loads, Structural Responses

12.1 Introduction

By acoustic vibration we mean the structural responses of structures exposed to acoustic loads or sound pressures. In this chapter we discuss the low-frequency acoustic vibrations because the equations of motion are solved using the modal approach, namely mode superposition [Madayag 69]. In the higher-frequency bands statistical energy analysis (SEA) is a good substitute for the classical modal approach.

In general, the modal characteristics of the dynamic system are calculated with the aid of the finite element method [Cook 89]. The accuracy is determined by the detail of the finite element model and the complexity of the structure. As stated above, the equations of motion will be solved using the classical modal approach and therefore linear structural behaviour is assumed.

The structure is assumed to be deterministic, however, the acoustic loads have a random nature. In this chapter the sound field will be assumed to be reverberant (diffuse). The sound intensity is the same in all directions.

Lightweight and large antenna structures and solar arrays, of spacecraft (Fig. 12.1) are very sensitive to acoustic loads during the launch phase. Spacecraft external structures are severely subjected to acoustic loads.

12.2 Acoustic Loads

In the dimensioning design specifications of spacecraft, solar arrays and antennae, acoustic loads are specified. These acoustic loads are generated during launch or in acoustic facilities for test purposes. It is very common to specify a reverberant sound field, which means that the intensity of the sound is the same for all directions.

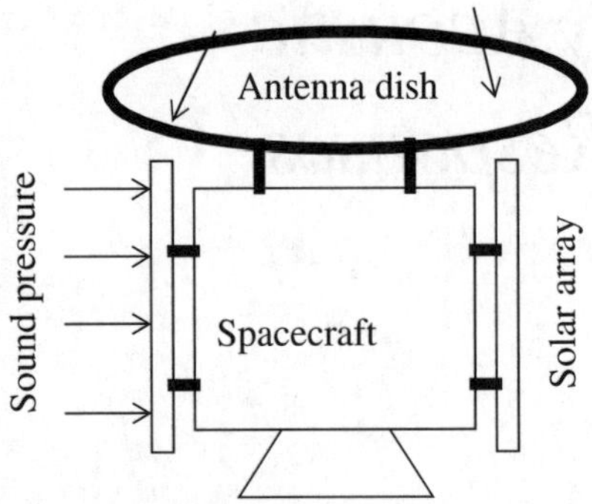

Fig. 12.1. Spacecraft structures exposed to acoustic loads

In general, the acoustic loads are specified as sound pressure levels (SPL) with units of deciBell (dB). The SPL is defined as

$$\mathrm{SPL} = 10\log\left(\frac{p^2}{p_{\mathrm{ref}}^2}\right), \tag{12.1}$$

where

- p is the rms pressure in a certain frequency band, in general one-octave of one-third octave band
- p_{ref} is the reference pressure 2×10^{-5} Pa

The x-th octave band of two sequential frequencies f_1 and f_2 is given by

$$\frac{f_2}{f_1} = 2^x, \tag{12.2}$$

in which:

- $x = 1$ the octave band applies and
- $x = \frac{1}{3}$ the one-third octave is applied, thus $\frac{f_2}{f_1} = 2^{\frac{1}{3}} = 1.260$

The centre frequency f_c (Hz) is defined as

$$f_c = \sqrt{f_{\min} f_{\max}}, \tag{12.3}$$

where

- $f_{\min}$ the minimum frequency in the frequency band (Hz)
- $f_{\max}$ the maximum frequency in the frequency band (Hz)

The frequency bandwidth Δf (Hz) is

$$\Delta f = f_{max} - f_{min} \,. \tag{12.4}$$

With $\frac{f_{max}}{f_{min}} = 2^x$ the bandwidth becomes

$$\Delta f = \left(2^{\frac{x}{2}} - 2^{-\frac{x}{2}}\right) f_c \,. \tag{12.5}$$

When:

- $x = 1$ the octave band $\Delta f = 0.7071 f_c$
- $x = \frac{1}{3}$ the one-third octave $\Delta f = 0.2316 f_c$

The power spectral density $W_p(f_c)$ (Pa^2/Hz) of the sound pressure is defined as

$$W_p(f_c) = \frac{p^2}{\Delta f} \,. \tag{12.6}$$

12.3 Equations of Motion

The equations of motion of a discrete number of coupled mass–spring systems with mass matrix $[M]$, stiffness matrix $[K]$, dynamic force vector $\{F(t)\}$, displacement vector $\{x\}$ and acceleration vector $\{\ddot{x}\}$ can be written as

$$[M]\{\ddot{x}\} + [K]\{x\} = \{F(t)\} \,. \tag{12.7}$$

The damping will be introduced later.

In general this discrete dynamic system is a finite element representation of a real structure.

The force vector $\{F(t)\}$ consists of a pressure (difference) applied to the area associated with a node. One force applied to the node is

$$F_i(t) = \int_{A_i} p(t) dA \approx p(t) A_i \,, \tag{12.8}$$

where

- $p(t)$ is the transient pressure exposed to the surface (Pa)
- A_i is the area associated with node i (m^2)

- $F_i(t)$ is the nodal force (N)

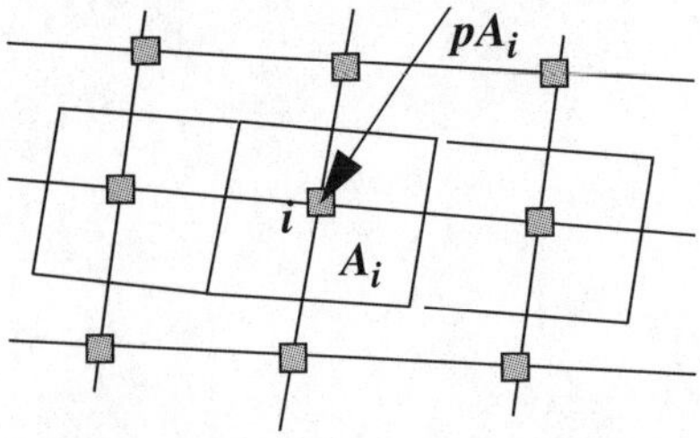

Fig. 12.2. Node i with associated area A_i exposed to pressure p

For a diffuse sound field a correlation exists between the nodal forces. Radiation effects are not taken into account.

The displacement vector $\{x(t)\}$ will be projected onto the independent mode shapes, the modal matrix $[\Phi]$, multiplied with the generalised coordinates $\{\eta(t)\}$

$$\{x(t)\} = [\Phi]\{\eta(t)\}. \tag{12.9}$$

Using the well-known orthogonality properties of the mode shapes with respect to the mass matrix $[M]$ and stiffness matrix $[K]$

$$[\Phi]^T[M][\Phi] = \langle\delta_{ij}m_i\rangle, [\Phi]^T[K][\Phi] = \langle\delta_{ij}m_i\lambda_i\rangle\ , \tag{12.10}$$

where

- δ_{ij} the Kronecker delta
- m_i generalised mass
- λ_i eigenvalue of dynamic system

The coupled equations of motion are decoupled and expressed in the generalised coordinates $\{\eta(t)\}$. We will also introduce the damping ratio ζ_i and the modal viscous damping term $2\zeta_i\sqrt{\lambda_i}\dot{\eta}_i(t) = 2\zeta_i\omega_i\dot{\eta}_i(t)$. The decoupled equations of motion become

$$\ddot{\eta}_i(t) + 2\zeta_i\omega_i\dot{\eta}_i(t) + \omega_i^2\eta_i(t) = \frac{\{\phi_i\}^T\{F(t)\}}{m_i} = f_i(t), i = 1, 2, ..., n\ . \tag{12.11}$$

The solution, with zero initial conditions, is

$$\eta_i(t)= \int_{-\infty}^{\infty} e^{-\zeta\omega_i\tau}\frac{\sin\omega_d\tau}{\omega_d}f_i(t-\tau)d\tau\ . \tag{12.12}$$

The damped natural (circular) frequency ω_d (rad/s) is defined as

$$\omega_d = \omega_i\sqrt{1-\zeta^2}. \tag{12.13}$$

In the frequency domain, with $\Pi(\omega) = \int_{-\infty}^{\infty} \eta(t)e^{-j\omega t}dt$

$$\Pi_i(\omega) = \frac{H_i(\omega)}{\omega_i^2}F_i(\omega), \tag{12.14}$$

with

- $H_i(\omega) = \dfrac{\omega_i^2}{(\omega_i^2 - \omega^2) + 2j\zeta_i\omega_i\omega}$ the frequency-response function
- $F_i(\omega) = \int_{-\infty}^{\infty} f_i(t)e^{-j\omega t}dt$ the Fourier transform of the force function $f_i(t)$

The ergodic stationary cross correlation function $R_{\eta_i\eta_j}(\tau)$ is given by

$$R_{\eta_i\eta_j}(\tau) = \lim_{T\to\infty}\frac{1}{2T}\int_{-T}^{T} \eta_i(t)\eta_j(t-\tau)dt. \tag{12.15}$$

The cross-power spectral density function $S_{\eta_i\eta_j}(\omega)$ is defined as [Harris 74]

$$S_{\eta_i\eta_j}(\omega) = \int_{-\infty}^{\infty} R_{\eta_i\eta_j}(\tau)e^{-j\omega\tau}d\tau, \tag{12.16}$$

and the reverse expression [Harris 74]

$$R_{\eta_i\eta_j}(\tau) = \frac{1}{2\pi}\int_{-\infty}^{\infty} S_{\eta_i\eta_j}(\omega)e^{j\omega\tau}d\omega. \tag{12.17}$$

Equation (12.16) and (12.17) form the Wiener–Khintchine relationships [Harris 74]. The crosscorrelation function of the generalised forces $f_i(t)$ and $f_j(t)$ is defined as

$$R_{f_if_j}(\tau) = \lim_{T\to\infty}\frac{1}{2T}\int_{-T}^{T} f_i(t)f_j(t-\tau)dt. \tag{12.18}$$

Equation (12.18), rewritten in matrix form, is

$$R_{f_if_j}(\tau) = \lim_{T \to \infty} \frac{1}{2T} \int_{-T}^{T} \frac{\{\phi_i\}^T \{F(t)\} \{F(t-\tau)\}^T \{\phi_j\}}{m_i m_j} dt \tag{12.19}$$

$$R_{f_if_j}(\tau) = \frac{1}{m_i m_j} \{\phi_i\}^T \langle A \rangle \left[\lim_{T \to \infty} \frac{1}{2T} \int_{-T}^{T} \{p_i(t)\} \{p_j(t-\tau)\}^T dt \right] \langle A \rangle \{\phi_j\}, \tag{12.20}$$

where

- $\langle A \rangle$ is the diagonal matrix of areas associated with the nodes of the finite element model (Fig. 12.2)

Finally, the crosscorrelation $R_{f_if_j}(\tau)$ can be related to the cross correlation matrix of pressures all over the surface of the structure.

$$R_{f_if_j}(\tau) = \frac{1}{m_i m_j} \{\phi_i\}^T \langle A \rangle [R_{p_ip_j}(\tau)] \langle A \rangle \{\phi_j\}, \tag{12.21}$$

where

- $[R_{p_ip_j}(\tau)]$ is the matrix of cross correlation functions of the pressures over the surface of the structure

Thus the cross-power spectral density becomes

$$S_{f_if_j}(\omega) = \frac{1}{m_i m_j} \{\phi_i\}^T \langle A \rangle [S_{p_ip_j}(\omega)] \langle A \rangle \{\phi_j\}, \tag{12.22}$$

where

- $[S_{p_ip_j}(\omega)]$ the matrix of cross power spectral densities of the pressures over the surface of the structure

The matrix of cross-power spectral densities of the pressures all over the surface of the structure and related to the nodes i=1,2,...,n is

$$[S_{p_ip_j}(\omega)] = \begin{bmatrix} S_{p_1p_1}(\omega) & S_{p_1p_2}(\omega) & \cdot\ \cdot & S_{p_1p_n}(\omega) \\ S_{p_2p_1}(\omega) & S_{p_2p_2}(\omega) & \cdot\ \cdot & S_{p_2p_n}(\omega) \\ \cdot & \cdot & \cdot\ \cdot & \cdot \\ \cdot & \cdot & \cdot\ \cdot & \cdot \\ S_{p_np_1}(\omega) & S_{p_np_2}(\omega) & \cdot\ \cdot & S_{p_np_n}(\omega) \end{bmatrix}. \tag{12.23}$$

Equation (12.16) can thus be written as

$$S_{\eta_i\eta_j}(\omega) =$$

$$\int_{-\infty}^{\infty} \lim_{T\to\infty}\frac{1}{2T}\left(\int_{-T}^{T}\left(\int_{-\infty}^{\infty} h_i(\alpha)f_i(t-\alpha)d\alpha\right)\left(\int_{-\infty}^{\infty} h_j(\beta-\tau)f_j(t-(\beta-\tau))d\beta\right)\right)e^{-j\omega\tau}d\tau dt \tag{12.24}$$

and

$$S_{\eta_i\eta_j}(\omega) = \int_{-\infty}^{\infty} R_{f_if_j}(\tau-\alpha+\beta)e^{-j\omega(\tau-\alpha+\beta)}d\tau\left(\int_{-\infty}^{\infty} h_i(\alpha)e^{-j\omega\alpha}d\alpha\right)\left(\int_{-\infty}^{\infty} h_j(\beta)e^{j\omega\beta}d\beta\right). \tag{12.25}$$

Using (12.14) we obtain for $S_{\eta_i\eta_j}(\omega)$

$$S_{\eta_i\eta_j}(\omega) = \frac{H_i(\omega)\{\phi_i\}^T\langle A\rangle[S_{p_ip_j}(\omega)]\langle A\rangle\{\phi_j\}H_j^*(\omega)}{\omega_i^2\omega_j^2 m_i m_j}. \tag{12.26}$$

The matrix of cross-power spectral densities of the generalised coordinates η_i and η_j is

$$[S_{\eta_i\eta_j}(\omega)] = \langle\frac{H_i(\omega)}{\omega_i^2 m_i}\rangle[\Phi]^T\langle A\rangle[S_{p_ip_j}(\omega)]\langle A\rangle[\Phi]\langle\frac{H_j^*(\omega)}{\omega_i^2 m_j}\rangle, \tag{12.27}$$

with

- $H_i^*(\omega) = \dfrac{\omega_i^2}{(\omega_i^2-\omega^2)-2j\zeta_i\omega_i\omega}$ the conjugate frequency response function and $H_i(\omega)H_i^*(\omega) = |H_i(\omega)|^2$.

Equation (12.23) can be written as

$$[S_{p_ip_j}(\omega)] = S_p(\omega)\begin{bmatrix} C_{p_1p_1}(\omega) & C_{p_1p_2}(\omega) & .. & C_{p_1p_n}(\omega) \\ C_{p_2p_1}(\omega) & C_{p_2p_2}(\omega) & .. & C_{p_2p_n}(\omega) \\ . & . & .. & . \\ . & . & .. & . \\ C_{p_np_1}(\omega) & C_{p_np_2}(\omega) & .. & C_{p_np_n}(\omega) \end{bmatrix}, \tag{12.28}$$

with

- $S_p(\omega)$ a reference power spectral density function of the applied pressures. This reference power spectral density of the pressure is, in general, related to the sound pressure levels of the sound field exposed to the surface of the structure.

- $C_{p_ip_j}(\omega)$ the correlation (coherence) function between the pressures at the nodes i and j.

Some typical pressure fields can be described:

1. If the dimension of the surface is less than a quarter of the wavelength $C_{p_ip_j}(\omega) \approx 1.0$. The wavelength $\lambda = \frac{2\pi}{k}$ with the wave-number $k = \frac{\omega}{c}$ and c is the speed of sound. At room temperature the speed of sound under 1 Bar pressure is $c \approx 340$ m/s
2. If the sound pressure field is completely random (rain on the roof) the off-diagonal terms in the correlation matrix $[C_{p_ip_j}(\omega)]$ are zero and the diagonal terms unity, hence $[C_{p_ip_j}(\omega)] = [I]$.
3. A three-dimensional wave field with uniform intensity $\left(I = \frac{p^2}{\rho c}\right)$ in all directions is commonly called a reverberent field. ρ is the density of the air, $\rho \approx 1.2$ kg/m^3. The coherence function $C_{p_ip_j}(\omega)$ is $C_{p_ip_j}(\omega, x, x') = \frac{\sin(k|x-x'|)}{(k|x-x'|)}$, [JASA 55], and $|x-x'|$ is the distance between two points.

Equation (12.26) can thus be written

$$S_{\eta_i\eta_j}(\omega) = \frac{H_i(\omega)S_p(\omega)J_{ij}^2(\omega)H_j^*(\omega)}{\omega_i^2\omega_i^2 m_i m_j}, \tag{12.29}$$

where

- $J_{ij}^2(\omega)$ the joint acceptance $J_{ij}^2(\omega) = \{\phi_i\}^T\langle A\rangle[C_{p_ip_j}(\omega)]\langle A\rangle\{\phi_j\}$

In accordance with (12.27)

$$[S_{\eta_i\eta_j}(\omega)] = S_p(\omega)\langle\frac{H_i(\omega)}{\omega_i^2 m_i}\rangle[J_{ij}^2(\omega)]\langle\frac{H_j^*(\omega)}{\omega_i^2 m_j}\rangle, \tag{12.30}$$

with the matrix of joint acceptances

$$[J_{ij}^2(\omega)] = [\Phi]^T\langle A\rangle[C_{p_ip_j}(\omega)]\langle A\rangle[\Phi]. \tag{12.31}$$

Smith [Smith 65] introduces the modal impedance $Z_i(\omega)$

$$Z_i(\omega) = \frac{m_i\omega_i^2}{H_i(\omega)\omega}. \tag{12.32}$$

and (12.30) becomes

$$[S_{\eta_i\eta_j}(\omega)] = \frac{S_p(\omega)}{\omega^2}\langle\frac{1}{Z_i(\omega)}\rangle[J_{ij}^2(\omega)]\langle\frac{1}{Z^*_i(\omega)}\rangle. \tag{12.33}$$

The matrix of cross-power spectral density of the physical displacements $S_{x_ix_j}(\omega)$ can now be calculated

$$[S_{x_ix_j}(\omega)] = [\Phi][S_{\eta_i\eta_j}(\omega)][\Phi]^T. \tag{12.34}$$

The matrix of cross power spectral densities of the velocities is

$$[S_{\dot{x}_i\dot{x}_j}(\omega)] = \omega^2[S_{x_ix_j}(\omega)], \tag{12.35}$$

and the matrix of cross power spectral densities of accelerations becomes

$$[S_{\ddot{x}_i\ddot{x}_j}(\omega)] = \omega^4[S_{x_ix_j}(\omega)]. \tag{12.36}$$

The matrix of cross-power spectral densities of the physical stresses $S_{\sigma_i\sigma_j}(\omega)$ can also be calculated

$$[S_{\sigma_i\sigma_j}(\omega)] = [\Phi_\sigma][S_{\eta_i\eta_j}(\omega)][\Phi_\sigma]^T, \tag{12.37}$$

where
- $[\Phi_\sigma]$ is the matrix of the stress modes

The power spectral density is symmetric with respect to $\omega = 0$ and, if we replace the circular frequency ω (rad/s) by the number of cycles per second f (Hz), the power spectral density function $S(\omega)$ can by replaced by

$$W(f) = 2S(\omega) . \tag{12.38}$$

In all equations the power spectral $S(\omega)$ density must be replaced by $W(f)$ and ω by $\omega = 2\pi f$.

A rigid plate with a total area $4A$ (m^2) and a total mass of 4m (kg) is vibrating supported on a spring with a spring stiffness $4k$ (N/m). The rigid plate is idealised in a finite element model with 6 nodes and 4 quadrilateral elements. The damping ratio is ζ. We will calculate the responses of the mass–spring system assuming:

1. a full correlation of the pressures between all nodes (full correlation matrix)
2. a white-noise type of correlation between all nodes (diagonal correlation matrix)

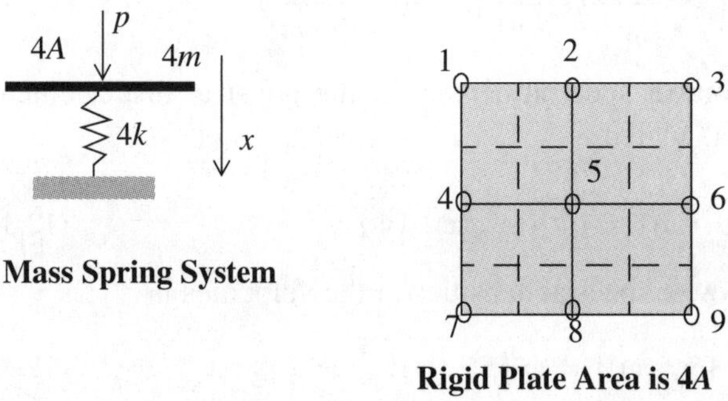

Fig. 12.3. Vibration rigid plate exposed to an acoustic pressure

If a unit static pressure is applied to the plate the equivalent nodal forces represent the associated areas of the nodes. The nodal forces depend on the shape functions describing the finite element. The natural frequency is $f = \frac{1}{2\pi}\sqrt{\frac{k}{m}}$ (Hz). For the rigid plate the matrix of the associated areas $\langle A \rangle$, the lumped mass matrix $[M]$ and the mode $\{\phi\}$ are very obvious.

The power spectral density function of the pressure $W_p(f)$ $\left(\frac{Pa^2}{Hz}\right)$ is

- $W_p(f) = W_o$ for $\frac{1}{3}f_o \leq f \leq 2f_o$
- $W_p(f) = 0$ elsewhere

$$\langle A\rangle = 4A\begin{bmatrix} \frac{1}{4} & 0 & 0 & 0 & 0 & 0 & 0 & 0 & 0 \\ 0 & \frac{1}{2} & 0 & 0 & 0 & 0 & 0 & 0 & 0 \\ 0 & 0 & \frac{1}{4} & 0 & 0 & 0 & 0 & 0 & 0 \\ 0 & 0 & 0 & \frac{1}{2} & 0 & 0 & 0 & 0 & 0 \\ 0 & 0 & 0 & 0 & 1 & 0 & 0 & 0 & 0 \\ 0 & 0 & 0 & 0 & 0 & \frac{1}{2} & 0 & 0 & 0 \\ 0 & 0 & 0 & 0 & 0 & 0 & \frac{1}{4} & 0 & 0 \\ 0 & 0 & 0 & 0 & 0 & 0 & 0 & \frac{1}{2} & 0 \\ 0 & 0 & 0 & 0 & 0 & 0 & 0 & 0 & \frac{1}{4} \end{bmatrix}, [M] = 4m\begin{bmatrix} \frac{1}{4} & 0 & 0 & 0 & 0 & 0 & 0 & 0 & 0 \\ 0 & \frac{1}{2} & 0 & 0 & 0 & 0 & 0 & 0 & 0 \\ 0 & 0 & \frac{1}{4} & 0 & 0 & 0 & 0 & 0 & 0 \\ 0 & 0 & 0 & \frac{1}{2} & 0 & 0 & 0 & 0 & 0 \\ 0 & 0 & 0 & 0 & 1 & 0 & 0 & 0 & 0 \\ 0 & 0 & 0 & 0 & 0 & \frac{1}{2} & 0 & 0 & 0 \\ 0 & 0 & 0 & 0 & 0 & 0 & \frac{1}{4} & 0 & 0 \\ 0 & 0 & 0 & 0 & 0 & 0 & 0 & \frac{1}{2} & 0 \\ 0 & 0 & 0 & 0 & 0 & 0 & 0 & 0 & \frac{1}{4} \end{bmatrix}, \{\phi\} = \begin{Bmatrix} 1 \\ 1 \\ 1 \\ 1 \\ 1 \\ 1 \\ 1 \\ 1 \\ 1 \end{Bmatrix}.$$

The generalised mass $m_1 = 4m$.

The joint acceptance $J_{11}^2(f)$ becomes

$$J_{11}^2(f) = \{\phi_1\}^T\langle A\rangle[C_{p_ip_j}(f)]\langle A\rangle\{\phi_1\}$$

1. For a full correlation matrix (plane wave) with components $C_{p_ip_j}(f) = 1$, the joint acceptance becomes $J_{11}^2(f) = 16A^2$
2. For a diagonal correlation matrix (rain on the roof correlation function) with components $C_{p_ip_j}(f) = \delta_{ij}$ the joint acceptance becomes $J_{11}^2(f) = 4A^2$

The power spectral density of the generalised coordinate η becomes, (12.33),

$$W_{\eta\eta}(f) = \frac{|H_1(f)|^2 W_p(f) J_{11}^2}{(2\pi f_o)^4 m_1^2},$$

The power spectral density of the accelerations of the generalised coordinate η is

$$W_{\ddot{\eta}\ddot{\eta}}(f) = \left(\frac{f}{f_o}\right)^4 \frac{|H_1(f)|^2 W_p(f) J_{11}^2}{m_1^2},$$

with

- $|H_1(f)|^2 = \dfrac{1}{\left[1-\left(\frac{f}{f_o}\right)^2\right]^2 + \left[2\zeta\left(\frac{f}{f_o}\right)\right]^2}$
- f_o the natural frequency (Hz)

The power spectral density function of the responses of the nodes is

$$[W_{\ddot{x}_i\ddot{x}}(f)] = [\phi]W_{\eta_i\eta_j}(f)[\phi]^T = \left(\frac{f}{f_o}\right)^4 \frac{|H_1(f)|^2 W_p(f) J_{11}^2}{m_1^2}[\phi][\phi]^T.$$

Using the formula of Miles [Miles 54] the root mean square value of the acceleration of the generalised co-ordinate η is

$$\ddot{\eta}_{rms} = \frac{J_{11}}{m_1}\sqrt{\frac{\pi}{2}f_o Q W_p(f_o)} = \frac{J_{11}}{m_1}\sqrt{\frac{\pi}{2}f_o Q W_o}$$

with the amplification factor $Q = \dfrac{1}{2\zeta}$, and ζ the damping ratio. The rms value of the 9 dofs becomes (diagonal terms)

$$\begin{Bmatrix} \ddot{x}_1 \\ \ddot{x}_2 \\ \ddot{x}_3 \\ \ddot{x}_4 \\ \ddot{x}_5 \\ \ddot{x}_6 \\ \ddot{x}_7 \\ \ddot{x}_8 \\ \ddot{x}_9 \end{Bmatrix}_{rms} = [\phi]\ddot{\eta}_{rms} = \frac{J_{11}}{m_1}\sqrt{\frac{\pi}{2}f_o Q W_o}\begin{Bmatrix} 1 \\ 1 \\ 1 \\ 1 \\ 1 \\ 1 \\ 1 \\ 1 \\ 1 \end{Bmatrix}$$

A simply supported beam with length $4L$ (m) is idealised by 5 nodes (Fig. 12.4). The beam has a unit width b (m). The total area is $4Lb$. The total mass of the beam is $4mLb$ (kg), where m is mass per unit of area (kg/m^2). The first mode shape $\phi_1(x) = \sin\dfrac{\pi x}{4L}$ is associated with the natural frequency $f_1 = \dfrac{(\pi)^2}{2\pi}\sqrt{\dfrac{EI}{mb(4L)^2}}$ (Hz) and the second mode shape $\phi_2(x) = \sin\dfrac{2\pi x}{4L}$ is associated with the second natural frequency

$f_2 = 4f_1 = \frac{(2\pi)^2}{2\pi}\sqrt{\frac{EI}{mb(4L)^2}}$ (Hz). The damping ratio is ζ. The beam is exposed by a random pressure field (raindrops on the roof), hence the coherence matrix is $[C_{p_ip_j}(f)] = [I]$.

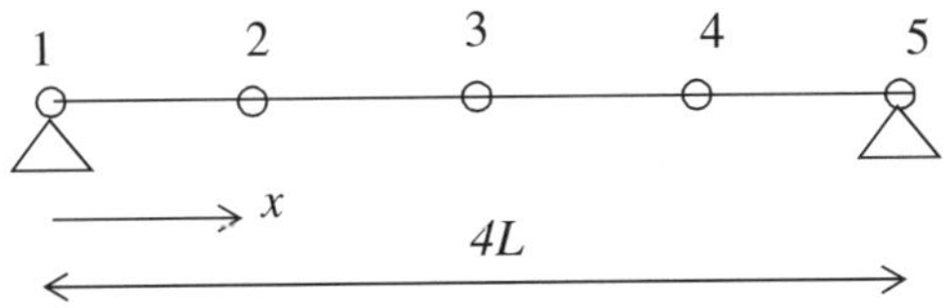

Fig. 12.4. Supported beam subjected to acoustic load

The modal matrix of both modes, the mass matrix $[M]$ and the diagonal matrix of areas $\langle A\rangle$ are

$$[\Phi] = \begin{bmatrix} 0 & 0 \\ \frac{1}{2}\sqrt{2} & 1 \\ 1 & 0 \\ \frac{1}{2}\sqrt{2} & -1 \\ 0 & 0 \end{bmatrix}, \ [M] = mLb\begin{bmatrix} \frac{1}{2} & 0 & 0 & 0 & 0 \\ 0 & 1 & 0 & 0 & 0 \\ 0 & 0 & 1 & 0 & 0 \\ 0 & 0 & 0 & 1 & 0 \\ 0 & 0 & 0 & 0 & \frac{1}{2} \end{bmatrix}, \text{ and } \langle A\rangle = Lb\begin{bmatrix} \frac{1}{2} & 0 & 0 & 0 & 0 \\ 0 & 1 & 0 & 0 & 0 \\ 0 & 0 & 1 & 0 & 0 \\ 0 & 0 & 0 & 1 & 0 \\ 0 & 0 & 0 & 0 & \frac{1}{2} \end{bmatrix}.$$

The generalised masses become

$$[\Phi]^T[M][\Phi] = 2mLb\begin{bmatrix} 1 & 0 \\ 0 & 1 \end{bmatrix},$$

and the joint acceptance matrix is

$$[J_{ij}^2(f)] = [\Phi]^T\langle A\rangle[C_{p_ip_j}(f)]\langle A\rangle[\Phi] = \begin{bmatrix} J^2_{11} & J^2_{12} \\ J^2_{12} & J^2_{22} \end{bmatrix} = 2(Lb)^2\begin{bmatrix} 1 & 0 \\ 0 & 1 \end{bmatrix}.$$

The power spectral density function of the generalised coordinates becomes

$$[W_{\eta_i\eta_j}(f)] = W_p(f)\left\langle\frac{H_i(f)}{\omega_i^2 m_i}\right\rangle[J_{ij}^2(f)]\left\langle\frac{H_j^*(f)}{\omega_i^2 m_j}\right\rangle = \frac{W_p(f)}{2m^2(2\pi)^4}\begin{bmatrix} \frac{|H_1|^2}{f_1^4} & 0 \\ 0 & \frac{|H_2|^2}{f_2^4} \end{bmatrix},$$

with

- $$|H_1(f)|^2 = \frac{1}{\left[1-\left(\frac{f}{f_1}\right)^2\right]^2 + \left[2\zeta\left(\frac{f}{f_1}\right)\right]^2}$$

  and

  $$|H_2(f)|^2 = \frac{1}{\left[1-\left(\frac{f}{f_2}\right)^2\right]^2 + \left[2\zeta\left(\frac{f}{f_2}\right)\right]^2}$$

Furthermore we find

$$[W_{x_i x_j}(f)] = [\Phi][W_{\eta_i\eta_j}(f)][\Phi]^T$$

$$= \frac{W_p(f)}{m^2(2\pi)^4}\begin{bmatrix} 0 & 0 & 0 & 0 & 0 \\ 0 & 0.25\frac{|H_1|^2}{f_1^4}+0.5\frac{|H_2|^2}{f_2^4} & 0.353\frac{|H_1|^2}{f_1^4} & 0.25\frac{|H_1|^2}{f_1^4}-0.5\frac{|H_2|^2}{f_2^4} & 0 \\ 0 & 0.353\frac{|H_1|^2}{f_1^4} & 0.5\frac{|H_1|^2}{f_1^4} & 0.353\frac{|H_1|^2}{f_1^4} & 0 \\ 0 & 0.25\frac{|H_1|^2}{f_1^4}-0.5\frac{|H_2|^2}{f_2^4} & 0.353\frac{|H_1|^2}{f_1^4} & 0.25\frac{|H_1|^2}{f_1^4}+0.5\frac{|H_2|^2}{f_2^4} & 0 \\ 0 & 0 & 0 & 0 & 0 \end{bmatrix}.$$

12.4 Problems

12.4.1 Problem 1

Recalculate the second example (simply supported beam, Fig. 12.4) with a fully populated correlation matrix $C_{p_i p_j}(f) = 1$ for every i=1,...,5 and j=1,...,5.

12.4.2 Problem 2

Recalculate the second example (simply supported beam, Fig. 12.4) assuming a reverberent sound field. The correlation matrix of the reverberent sound field is $C_{p_i p_j}(\omega, x, x') = \frac{\sin(|x_i - x'_j|)}{(|x_i - x'_j|)}$ for every i=1,...,5 and j=1,...,5. Assume for example $|x_1 - x'_2| = L$, etc.

13 Statistical Energy Analysis

13.1 Introduction

Statistical energy analysis (SEA) originated in the aerospace industry in the early 1960s. Today, SEA is applied to a large variety of products, from cars and trucks to aircraft, spacecraft, electronic equipment, buildings, consumer products and more.

SEA is based on the principle of energy conservation. All the energy inputting to a system, through mechanical or acoustic excitation, must leave the system through structural damping or acoustic radiation. The method is fast and is applicable over a wide frequency range. SEA is very good for problems that combine many different sources of excitation, whether mechanical or acoustic.

With the SEA a statistical description of the structural vibrational behaviour of elements (systems) is described. In the high-frequency band a deterministic modal description of the dynamic behaviour of structures is not very useful. The modes (oscillators) are grouped statistically and the energy transfer from one group of modes to another group of modes is statistically proportional to the difference in the subsystem total energies.

Readers who are interested in a more detailed description of the SEA method are encouraged to read the following interesting literature [Lyon 62, Ungar 66, Barnoski 69, Cremer 73, Lyon 75, Norton 89, Lyon 95, Keane 94, Eaton 96, Woodhouse 81]. A very clear discussion and explanation about the SEA can be found in [Nigam 94, Chapter 10].

SEA is attractive in high-frequency regions where a deterministic analysis of all resonant modes of vibration is not practical.

13.2 Some Basics about Averaged Quantities

The average over all time of the square of the displacement $x(t)$ is defined as [Smith 65, Keltie 01]

$$\langle x^2 \rangle = \lim_{T\to\infty}\frac{1}{T}\int_0^T x^2(t)dt = \frac{1}{2}\Re(X(\omega)X^*(\omega)), \tag{13.1}$$

with

$$x(t) = \Re[X(\omega)e^{j\omega t}] = A\cos(\omega t - \vartheta), \tag{13.2}$$

or

$$x(t) = \Im[X(\omega)e^{j\omega t}] = A\sin(\omega t - \vartheta), \tag{13.3}$$

where

- $X(\omega)$ the complex amplitude dependent upon the frequency
- A the amplitude of the oscillation
- ϑ the phase angle

We can write $x(t)$ as follows

$$x(t) = \Re[X(\omega)e^{j\omega t}] = \Re(X)\cos(\omega t) - \Im(X)\sin(\omega t), \tag{13.4}$$

and

$$x(t) = \Im[X(\omega)e^{j\omega t}] = \Re(X)\sin(\omega t) + \Im(X)\cos(\omega t). \tag{13.5}$$

From this we can conclude that

- $\Re\{X(\omega)\} = A\cos\vartheta$
- $\Im\{X(\omega)\} = A\sin\vartheta$
- $\tan\vartheta = -\dfrac{\Im\{X(\omega)\}}{\Re\{X(\omega)\}}$

The average value of $\langle x^2 \rangle$ becomes

$$2\langle x^2 \rangle = [\Re\{X(\omega)\}]^2\langle\{\cos(\omega t)\}^2\rangle - 2\Re\{X(\omega)\}\Im\{X(\omega)\}\langle\sin(\omega t)\cos(\omega t)\rangle$$
$$+([\Im\{X(\omega)\}]^2)\{\sin(\omega t)\}^2. \tag{13.6}$$

The mean value of

$$\frac{1}{\frac{2\pi}{\omega}}\int_0^{\frac{2\pi}{\omega}}(\sin\omega t)^2 dt = 1, \quad \frac{1}{\frac{2\pi}{\omega}}\int_0^{\frac{2\pi}{\omega}}(\cos\omega t)^2 dt = 1 \text{ and}$$

$$\frac{1}{\frac{2\pi}{\omega}}\int_0^{\frac{2\pi}{\omega}} \sin\omega t\cos\omega t dt = 0, \tag{13.7}$$

resulting in

$$\langle x^2\rangle = \frac{1}{2}\{[\Re\{X(\omega)\}]^2 + [\Im\{X(\omega)\}]^2\}. \tag{13.8}$$

We know that the modulus of a complex number $X(\omega)$ can be obtained from

$$|X(\omega)|^2 = \Re\{X(\omega)X(\omega)^*\} = [\Re\{X(\omega)\}]^2 + [\Im\{X(\omega)\}]^2. \tag{13.9}$$

So we can express (13.8), using (13.9), as

$$\langle x^2\rangle = \frac{1}{2}\{[\Re\{X(\omega)\}]^2 + [\Im\{X(\omega)\}]^2\} = \frac{1}{2}\Re\{X(\omega)X(\omega)^*\} \tag{13.10}$$

$$\langle x^2\rangle = \frac{1}{2}\Re(X(\omega)X^*(\omega)) = \frac{1}{2}|X(\omega)|^2. \tag{13.11}$$

Furthermore we can derive from the previous equations that

$$\langle x\dot{x}\rangle = \frac{1}{2}\Re(X(\omega)\dot{X}^*(\omega)) = \frac{1}{2}\Re(j\omega X(\omega)X^*(\omega)) = \frac{1}{2}\Re(j\omega|X(\omega)|^2) = 0, \tag{13.12}$$

and

$$\langle \dot{x}\ddot{x}\rangle = \frac{1}{2}\Re(\ddot{X}(\omega)\dot{X}^*(\omega)) = \frac{1}{2}\Re(j\omega\dot{X}(\omega)\dot{X}^*(\omega)) = \frac{1}{2}\Re(j\omega|\dot{X}(\omega)|^2) = 0. \tag{13.13}$$

With
$X_1(\omega) = \Re X_1 + j\Im X_1$ and $X_2(\omega) = \Re X_2 + j\Im X_2$
and with

$$\langle x_1\dot{x}_2\rangle = \frac{1}{2}\Re(X_1(\omega)\dot{X}_2^*(\omega)) = \frac{1}{2}\Re(j\omega X_1(\omega)X_2^*(\omega)) = \frac{\omega}{2}(\Re X_1\Im X_2 - \Im X_1\Re X_2), \tag{13.14}$$

and

$$\langle x_2\dot{x}_1\rangle = \frac{1}{2}\Re(X_2(\omega)\dot{X}_1^*(\omega)) = \frac{1}{2}\Re(j\omega X_2(\omega)X_1^*(\omega)) = \frac{-\omega}{2}(\Re X_1\Im X_2 - \Im X_1\Re X_2), \tag{13.15}$$

we can prove that

$$\langle x_1\dot{x}_2\rangle = -\langle x_2\dot{x}_1\rangle, \ \langle \dot{x}_1\ddot{x}_2\rangle = -\langle \dot{x}_2\ddot{x}_1\rangle \text{ and } \langle x_1\ddot{x}_1\rangle = -\langle \dot{x}_1^2\rangle. \tag{13.16}$$

The equation of motion of an sdof dynamic system is given by

$$m\ddot{x} + c\dot{x} + k = F(t) \tag{13.17}$$

The average input power Π_{in} is, with $x(t) = X(\omega)e^{j\omega t}$ and $F(t) = F(\omega)e^{j\omega t}$

$$\Pi_{in} = \langle F\dot{x}\rangle = \frac{1}{2}\Re\left\{F(\omega)\dot{X}^*(\omega)\right\}. \tag{13.18}$$

If the averages are applied to (13.17) we obtain

$$m\langle\ddot{x}\dot{x}\rangle + c\langle\dot{x}^2\rangle + k\langle x\dot{x}\rangle = \langle F\dot{x}\rangle = \Pi_{in}. \tag{13.19}$$

With $\langle\ddot{x}\dot{x}\rangle = 0$ and $\langle x\dot{x}\rangle = 0$ (13.19) becomes

$$\Pi_{in} = c\langle\dot{x}^2\rangle = \Pi_{diss}. \tag{13.20}$$

Equation (13.20) means that the average power Π_{in} introduced in the sdof of (13.17) is equal to the average dissipated power Π_{diss}. The dissipated power Π_{diss} can be written as

$$\Pi_{diss} = c\langle\dot{x}^2\rangle = 2\zeta\sqrt{km}\langle\dot{x}^2\rangle = 2m\zeta\omega_n\langle\dot{x}^2\rangle = m\zeta\omega_n\left|\dot{X}(\omega)\right|^2, \tag{13.21}$$

or

$$\Pi_{diss} = m\zeta\omega_n\left|\dot{X}(\omega)\right|^2 = 2m\zeta\omega_n\langle\dot{x}^2\rangle = m\eta\omega_n\langle\dot{x}^2\rangle, \tag{13.22}$$

with

- η the loss factor $\eta = 2\zeta$

The mobility function $Y(\omega)$ is defined as

$$Y(\omega) = \frac{\dot{X}(\omega)}{F(\omega)}. \tag{13.23}$$

Thus, $\dot{X}(\omega) = Y(\omega)F(\omega)$ and

$$\langle\dot{x}^2\rangle = \frac{1}{2}\left|\dot{X}(\omega)\right|^2 = \frac{1}{2}\{Y(\omega)F(\omega)(Y^*(\omega)F^*(\omega))\}$$

$$= |Y(\omega)|^2\left(\frac{1}{2}\{(F(\omega)F^*(\omega))\}\right) = |Y(\omega)|^2\langle F^2\rangle. \tag{13.24}$$

The average power $\Pi_{in} = \langle F\dot{x}\rangle$ can be written as

$$\Pi_{\text{in}} = \langle F\dot{x}\rangle = \frac{1}{2}\Re\left\{F(\omega)\dot{X}^*(\omega)\right\} = \frac{1}{2}|F|^2\Re\{Y^*(\omega)\} = \frac{1}{2}m\eta\omega_{\text{n}}|F|^2|Y|^2 \quad (13.25)$$

$$\Pi_{\text{in}} = \frac{1}{2}|F|^2\Re\{Y^*(\omega)\} = \frac{1}{2}|F|^2\Re\left\{\frac{1}{Z^*(\omega)}\right\} = \frac{1}{2}|F|^2|Y|^2\Re\{Z(\omega)\} \quad (13.26)$$

$$\Pi_{\text{in}} = \frac{1}{2}|F(\omega)|^2|Y(\omega)|^2\Re\{Z(\omega)\} = \frac{1}{2}m\eta\omega_n|F(\omega)|^2|Y(\omega)|^2 = \frac{1}{2}m\eta\omega_n\left|\dot{X}(\omega)\right|^2, \quad (13.27)$$

with

- $Y(\omega) = \dfrac{j\omega}{m\omega_{\text{n}}^2\left[\left(1-\dfrac{\omega^2}{\omega_{\text{n}}^2}\right)+j\eta\dfrac{\omega}{\omega_{\text{n}}}\right]}$
- $Z(\omega) = \dfrac{1}{Y(\omega)}$ the impedance function

A compedium of approximate mobilities for built-up structures in given in [Pinnington 86].

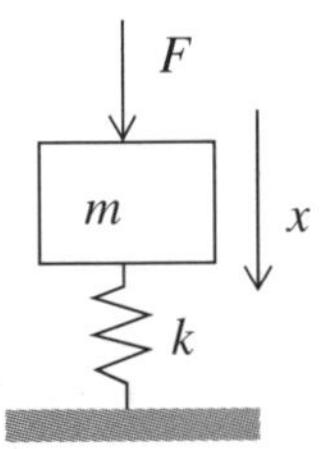

Fig. 13.1. Sdof dynamic system

The sdof dynamic system, shown in Fig. 13.1, has a natural frequency $\omega_{\text{n}} = 2\pi f_{\text{n}} = 208.2$ Rad/s. The mass is $m = 150$ kg. The steady-sate oscillating force $F = 100$ N produces a velocity $v = \dot{X} = 0.2$ m/s.

- Estimate the loss factor η.

The average input power is given by $\Pi_{\text{in}} = \frac{1}{2}F\dot{X}$.

The average input power is also given by (13.27), $\Pi_{\text{in}} = \frac{1}{2}m\eta\omega_n\dot{X}^2$.

When both previous expressions for the input power are compared with each other we can write an expression for the loss factor η

$$\eta = \frac{F}{m\omega_{\text{n}}\dot{X}} = \frac{100}{150x208.2x0.2} = 0.016\,.$$

The averaged power of the velocity $\dot{x}$ can be expressed as

$$\langle \dot{x}^2 \rangle = \int_0^\infty W_{\dot{x}\dot{x}}(f)df, \tag{13.28}$$

however, assuming an ergodic and stationary random process and the averaged power of the force $F(t)$ conform to (13.28), we can state

$$\langle F^2 \rangle = \int_0^\infty W_F(f)df, \tag{13.29}$$

with

- $W_{\dot{x}\dot{x}}(f)$ the one-sided power spectral density function of $\dot{x}(t)$
- $W_F(f)$ the one-sided power spectral density function of $F(t)$

In a certain frequency band with a bandwidth Δf and a centre frequency f we can express (13.28) and (13.29) as

$$\langle \dot{x}^2 \rangle_{\Delta f} = W_{\dot{x}\dot{x}}(f)\Delta f \tag{13.30}$$

and

$$\langle F^2 \rangle_{\Delta f} = W_F(f)\Delta f. \tag{13.31}$$

Using (13.24) we may write

$$W_{\dot{x}\dot{x}}(f) = |Y(f)|^2 W_F(f). \tag{13.32}$$

The averaged power of $\dot{x}(t)$ becomes, $W_F(f) = 2S_F(\omega)$,

$$\langle \dot{x}^2 \rangle = \int_0^\infty W_{\dot{x}\dot{x}}(f)df = \int_0^\infty |Y(f)|^2 W_F(f)df = \frac{1}{4\pi}\int_{-\infty}^\infty |Y(\omega)|^2 S_F(\omega)d\omega, \tag{13.33}$$

and with the admittance function $Y(\omega) = \dfrac{j\omega}{m[(\omega_n^2 - \omega^2) + 2j\zeta\omega\omega_n]}$.

The integral $\int_{-\infty}^{\infty} |Y(\omega)|^2 d\omega$ has a closed-form solution [Newland 94]. If

$$Y(\omega) = \frac{B_0 + j\omega B_1}{A_0 + j\omega A_1 - \omega^2 A_2} \text{ then } \int_{-\infty}^{\infty} |Y(\omega)|^2 d\omega = \frac{\pi(A_0 B_1^2 + A_2 B_0^2)}{A_0 A_1 A_2}.$$

With

- $A_0 = m\omega_n^2$
- $A_1 = m2\zeta\omega_n$
- $A_2 = m$
- $B_0 = 0$
- $B_1 = 1$

The integral $\int_{-\infty}^{\infty} |Y(\omega)|^2 d\omega$ evaluated becomes

$$\int_{-\infty}^{\infty} |Y(\omega)|^2 d\omega = \frac{\pi(A_0 B_1^2 + A_2 B_0^2)}{A_0 A_1 A_2} = \frac{\pi}{2m^2\zeta\omega_{\mathrm{n}}}. \tag{13.34}$$

Equation (13.33) can be further approximated assuming $|Y(f)|^2$ is very peaked and reasonably constant PSD function $W_F(f) = 2S_F(\omega)$ in the neighbourhood of the natural frequency f_{n}. This gives

$$\langle \dot{x}^2 \rangle = \int_0^{\infty} W_{\dot{x}\dot{x}}(f)df = \frac{1}{2\pi}\int_{-\infty}^{\infty} |Y(\omega)|^2 S_F(\omega) d\omega \approx \frac{W_F(f_{\mathrm{n}})}{4\pi}\int_{-\infty}^{\infty} |Y(\omega)|^2 d\omega. \tag{13.35}$$

Thus

$$\langle \dot{x}^2 \rangle = \frac{W_F(f_{\mathrm{n}})}{4\pi}\int_{-\infty}^{\infty} |Y(\omega)|^2 d\omega = \frac{W_F(f_{\mathrm{n}})}{16m^2\pi\zeta f_{\mathrm{n}}} = \frac{W_F(f_{\mathrm{n}})}{8m^2\pi\eta f_{\mathrm{n}}}. \tag{13.36}$$

Equation (13.36) is in accordance with [Keltie 01].

The total average energy $\langle E \rangle$ becomes

$$\langle E \rangle = \langle T + U \rangle = \langle \frac{1}{2}m\dot{x}^2 + \frac{1}{2}kx^2 \rangle = \frac{1}{2}\left(m\langle \dot{x}^2 \rangle + k\langle \frac{\dot{x}^2}{\omega_{\mathrm{n}}^2} \rangle \right) = m\langle \dot{x}^2 \rangle \tag{13.37}$$

because $\langle \dot{x}^2 \rangle = \omega_{\mathrm{n}}^2 \langle x^2 \rangle$ and $k = m\omega_{\mathrm{n}}^2$. Thus, the average energy $\langle E \rangle$ is now

$$\langle E \rangle = \frac{W_F(f_{\mathrm{n}})}{8m\pi\eta f_{\mathrm{n}}} = \frac{S_F(\omega_{\mathrm{n}})}{2m\eta\omega_{\mathrm{n}}}. \tag{13.38}$$

Referring to (13.22) the average damping energy can be expressed as follows

$$\Pi_{\mathrm{diss}} = m\eta\omega_{\mathrm{n}}\langle \dot{x}^2 \rangle = \eta\omega_{\mathrm{n}}\langle E \rangle. \tag{13.39}$$

Furthermore we can derive an equation to calculate $\langle \ddot{x}^2 \rangle$

$$\langle \ddot{x}^2 \rangle = \omega_n^2 \langle \dot{x}^2 \rangle = \frac{\pi}{2} f_n Q \frac{W_F(f_n)}{m^2}, \tag{13.40}$$

with

- Q the amplification factor with $Q = \dfrac{1}{2\zeta} = \dfrac{1}{\eta}$.

If we express the average (rms) input power Π_{in} (13.27) in the PSD function of the random force we obtain the following expression

$$\Pi_{in} = \frac{m\eta\omega_n}{2\pi} \int_{-\infty}^{\infty} S_F(\omega) |Y(\omega)|^2 d\omega \approx \frac{W_F(f_n)}{4m}. \tag{13.41}$$

The average power put into the sdof system will be dissipated in the damper, thus in accordance with (13.22) we find

$$\langle \dot{x}^2 \rangle = \frac{W_F(f_n)}{8m^2 \pi \eta f_n}. \tag{13.42}$$

This is equal to (13.36).

13.3 Two Coupled Oscillators

Consider a simple two-sdof system as shown in Fig. 13.2. The coupling element between the two sdof's is a linear spring and is nondissipative. The quantities of interest in this section are the average energies of each oscillator and the average energy flow between them.

If the coupling element is a spring with spring stiffness k_{12} the equations of motion for the coupled oscillators, as shown in Fig. 13.2, become [Elishakoff 83]

$$m_1 \ddot{x}_1 + c_1 \dot{x}_1 + k_1 x_1 + k_{12}(x_1 - x_2) = F_1 \tag{13.43}$$

$$m_2 \ddot{x}_2 + c_2 \dot{x}_2 + k_2 x_1 + k_{12}(x_2 - x_1) = F_2, \tag{13.44}$$

with

- $\omega_1 = \sqrt{\dfrac{k_1}{m_1}}$, $\omega_2 = \sqrt{\dfrac{k_2}{m_2}}$, $\omega_{12} = \sqrt{\dfrac{k_{12}}{m_1 m_2}}$
- $c_1 = 2\zeta_1 \sqrt{k_1 m_1}$, $c_2 = 2\zeta_2 \sqrt{k_2 m_2}$

- $\dfrac{c_1}{m_1} = 2\zeta_1\omega_1$, $\dfrac{c_2}{m_2} = 2\zeta_2\omega_2$

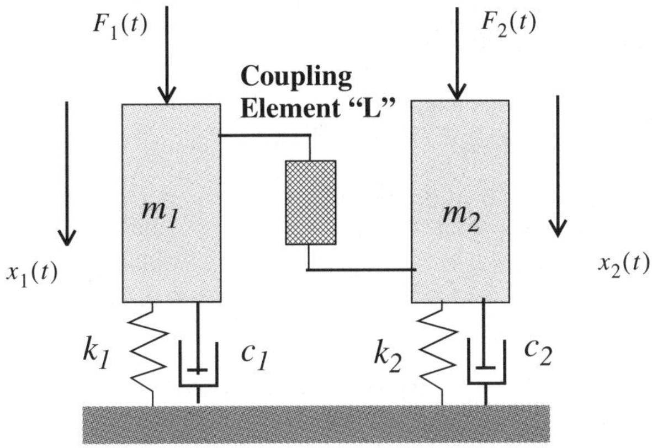

Fig. 13.2. Two coupled oscillators [Ungar 66]

We did not take into account mass coupling and gyroscopic coupling. This is discussed in [Lyon 62]. Adding the mass and gyroscopic coupling complicates the analysis unnecessarily.
Equations (13.43) and (13.44) become

$$\ddot{x}_1 + 2\zeta_1\omega_1\dot{x}_1 + \omega_1^2 x_1 + m_2\omega_{12}^2(x_1 - x_2) = \frac{F_1}{m_1} \tag{13.45}$$

$$\ddot{x}_2 + 2\zeta_2\omega_2\dot{x}_2 + \omega_2^2 x_2 + m_1\omega_{12}^2(x_2 - x_1) = \frac{F_2}{m_2}. \tag{13.46}$$

We assume that the forces F_1 and F_2 are independent of each other. The power supplied by force F_1 is $\langle F_1\dot{x}_1\rangle$ and the power supplied by F_2 is $\langle F_2\dot{x}_2\rangle$.

$$\langle F_1\dot{x}_1\rangle = m_1\langle\ddot{x}_1\dot{x}_1\rangle + 2m_1\zeta_1\omega_1\langle\dot{x}_1^2\rangle + (m_1\omega_1^2 + m_2\omega_{12}^2)\langle x_1\dot{x}_1\rangle - m_1m_2\omega_{12}^2\langle x_2\dot{x}_1\rangle \tag{13.47}$$

$$\langle F_2\dot{x}_2\rangle = m_2\langle\ddot{x}_2\dot{x}_2\rangle + 2m_2\zeta_2\omega_2\langle\dot{x}_2^2\rangle + (m_2\omega_2^2 + m_1\omega_{12}^2)\langle x_2\dot{x}_2\rangle - m_1m_2\omega_{12}^2\langle x_1\dot{x}_2\rangle \tag{13.48}$$

We can easily prove that

$$\langle F_1\dot{x}_1\rangle + \langle F_2\dot{x}_2\rangle = 2m_1\zeta_1\omega_1\langle\dot{x}_1^2\rangle + 2m_2\zeta_2\omega_2\langle\dot{x}_2^2\rangle\,, \tag{13.49}$$

which states that the total power supplied is dissipated in the two dampers. The coupling element (spring) is nondissipative. The power flow Π_{12} from oscillator 1 to oscillator 2 is defined as (negative sign introduced to be consistent with average damping energy)

$$\Pi_{12} = -m_1m_2\omega_{12}^2\langle x_2\dot{x}_1\rangle = -\frac{m_1m_2\omega_{12}^2}{2}\Re(X_2\dot{X}_1^*)\,. \tag{13.50}$$

We will proceed to evaluate these averages in term of spectral densities. Assume

$$\begin{Bmatrix} x_1 \\ x_2 \end{Bmatrix} = \begin{Bmatrix} X_1(\omega) \\ X_2(\omega) \end{Bmatrix} e^{j\omega t}.$$

Equations (13.44) and (13.45) can now be written as follows

$$\begin{bmatrix} -\omega^2 + 2j\zeta_1\omega\omega_1 + \omega_1^2 + m_2\omega_{12}^2 & -m_2\omega_{12}^2 \\ -m_1\omega_{12}^2 & -\omega^2 + 2j\zeta_2\omega\omega_2 + \omega_2^2 + m_1\omega_{12}^2 \end{bmatrix} \begin{Bmatrix} X_1(\omega) \\ X_2(\omega) \end{Bmatrix} = \begin{Bmatrix} \dfrac{F_1(\omega)}{m_1} \\ \dfrac{F_2(\omega)}{m_2} \end{Bmatrix}, \tag{13.51}$$

with the determinant $D(\omega)$

$$D(\omega) = A_4\omega^4 - jA_3\omega^3 - A_2\omega^2 + jA_1\omega^1 + A_0 \tag{13.52}$$

with

- $A_4 = 1$
- $A_3 = 2(\zeta_1\omega_1 + \zeta_2\omega_2)$
- $A_2 = \{\omega_1^2 + \omega_2^2 + \omega_{12}^2(m_1 + m_2) + 4\zeta_1\zeta_2\omega_1\omega_2\}$
- $A_1 = 2(\zeta_1\omega_1\omega_2^2 + \zeta_2\omega_2\omega_1^2 + \zeta_1 m_1\omega_1\omega_{12}^2 + \zeta_2 m_2\omega_2\omega_{12}^2)$
- $A_0 = (\omega_1^2\omega_2^2 + m_1\omega_1^2\omega_{12}^2 + m_2\omega_2^2\omega_{12}^2)$

With the aid of Cramer's rule we obtain

$$X_1(\omega) = \frac{\begin{vmatrix} \dfrac{F_1(\omega)}{m_1} & -m_2\omega_{12}^2 \\ \dfrac{F_2(\omega)}{m_2} & -\omega^2 + 2j\zeta_2\omega\omega_2 + \omega_2^2 + m_1\omega_{12}^2 \end{vmatrix}}{D(\omega)}$$

$$X_1(\omega) = \frac{1}{m_1 m_2 D(\omega)}[\{m_2 F_1(\omega)(-\omega^2 + 2j\zeta_2\omega\omega_2 + \omega_2^2 + m_1\omega_{12}^2)\} + m_1 m_2\omega_{12}^2 F_2(\omega)]\,, \tag{13.53}$$

and

$$X_2(\omega) = \frac{\begin{vmatrix} -\omega^2 + 2j\zeta_1\omega\omega_1 + \omega_1^2 + m_2\omega_{12}^2 & \dfrac{F_1(\omega)}{m_1} \\ -m_1\omega_{12}^2 & \dfrac{F_2(\omega)}{m_2} \end{vmatrix}}{D(\omega)}$$

$$X_2(\omega) = \frac{1}{m_1 m_2 D(\omega)}[\{m_1 F_2(\omega)(-\omega^2 + 2j\zeta_1\omega\omega_1 + \omega_1^2 + m_2\omega_{12}^2)\} + m_1 m_2\omega_{12}^2 F_1(\omega)]\,, \tag{13.54}$$

If we express $X_1(\omega) = X_1$, etc., (13.53) and (13.54) are written as

$$X_1 = \frac{1}{m_1 m_2 D(\omega)}[G_{11}F_1 + G_{12}F_2] \tag{13.55}$$

$$X_2 = \frac{1}{m_1 m_2 D(\omega)}[G_{21}F_1 + G_{22}F_2]\,, \tag{13.56}$$

with

- $G_{11} = m_2(-\omega^2 + 2j\zeta_2\omega\omega_2 + \omega_2^2 + m_1\omega_{12}^2)$
- $G_{12} = G_{21} = m_1 m_2\omega_{12}^2$
- $G_{22} = m_1(-\omega^2 + 2j\zeta_1\omega\omega_1 + \omega_1^2 + m_2\omega_{12}^2)$

Equations (13.55) and (13.56) can be written as, [Lyon 75]

$$X_1 = H_{11}F_1 + H_{12}F_2 \tag{13.57}$$

$$X_2 = H_{21}F_1 + H_{22}F_2\,, \tag{13.58}$$

with

- $H_{11} = \frac{1}{m_1 D}(-\omega^2 + 2j\zeta_2\omega\omega_2 + \omega_2^2 + m_1\omega_{12}^2)$
- $H_{12} = H_{21} = \frac{\omega_{12}^2}{D}$
- $H_{22} = \frac{1}{m_2 D}(-\omega^2 + 2j\zeta_1\omega\omega_1 + \omega_1^2 + m_2\omega_{12}^2)$

The velocities can be expressed as

$$\dot{X}_1 = j\omega[H_{11}F_1 + H_{12}F_2] = Y_{11}F_1 + Y_{12}F_2 \tag{13.59}$$

$$\dot{X}_2 = j\omega[H_{21}F_1 + H_{22}F_2] = Y_{21}F_1 + Y_{22}F_2 . \tag{13.60}$$

The average power flow from oscillator 1 to oscillator 2 Π_{12} is given by (13.50)

$$\Pi_{12} = -\frac{m_1 m_2 \omega_{12}^2}{2}\Re(X_2\dot{X}_1^*)$$

$$\langle x_2\dot{x}_1\rangle = \frac{1}{2}\Re(X_2\dot{X}^*{}_1) = \frac{1}{2}\Re(H_{21}Y_{11}^*|F_1|^2 + H_{12}Y_{12}^*|F_2|^2) . \tag{13.61}$$

In (13.61) we assume no correlation between $F_1(t)$ and $F_2(t)$, thus $\langle F_1F_2\rangle = 0$.

$$\langle x_2\dot{x}_1\rangle = \frac{\omega^2\omega_{12}^2}{m_2|D|^2}\{\zeta_1\omega_1\}|F_2|^2 - \frac{\omega^2\omega_{12}^2}{m_1|D|^2}\{\zeta_2\omega_2\}|F_1|^2$$

$$\langle x_2\dot{x}_1\rangle = \frac{\omega^2\omega_{12}^2}{m_1 m_2|D|^2}\{m_1\zeta_1\omega_1|F_2|^2 - m_2\zeta_2\omega_2|F_1|^2\}$$

$$\langle x_2\dot{x}_1\rangle = \frac{W_{F_1}}{4\pi}\int_{-\infty}^{\infty}\Re(H_{21}Y_{11}^*)d\omega + \frac{W_{F_2}}{4\pi}\int_{-\infty}^{\infty}\Re(H_{22}Y_{12}^*)d\omega .$$

We also assume no crosscorrelation between the two applied forces $F_1(t)$ and $F_1(t)$, hence the cross-power spectral function is

$$W_{F_1F_2}(f) = 0 , \tag{13.62}$$

with

$$H_{21}Y_{11}^* = \frac{\omega_{12}^2}{m_1|D|^2}\{j\omega^3 - 2\zeta_2\omega_2\omega^2 - j\omega(\omega_2^2 + m_1\omega_{12}^2)\} ,$$

and

$$H_{21}Y_{11}^* = \frac{\omega_{12}^2}{m_2|D|^2}\{j\omega^3 + 2\zeta_1\omega_1\omega^2 - j\omega(\omega_1^2 + m_2\omega_{12}^2)\}.$$

We obtain for $\langle x_2\dot{x}_1\rangle$

$$\langle x_2\dot{x}_1\rangle = \frac{W_{F_1}}{4\pi}\int_{-\infty}^{\infty}\frac{-\omega_{12}^2\eta_2\omega_2\omega^2}{m_1|D|^2}d\omega + \frac{W_{F_2}}{4\pi}\int_{-\infty}^{\infty}\frac{\omega_{12}^2\eta_1\omega_1\omega^2}{m_2|D|^2}d\omega. \tag{13.63}$$

Thus the power flow Π_{12} becomes

$$\Pi_{12} = \frac{-m_1m_2\omega_{12}^4}{8\pi}\left[\frac{\eta_2\omega_2W_{F_1}}{m_1} - \frac{\eta_1\omega_1W_{F_2}}{m_2}\right]\int_{-\infty}^{\infty}\frac{(j\omega)^2}{|D|^2}d\omega,$$

or

$$\Pi_{12} = \frac{-m_1m_2\omega_{12}^4\eta_1\omega_1\eta_2\omega_2}{2\pi}\int_{-\infty}^{\infty}\frac{(j\omega)^2}{|D|^2}d\omega\left[\frac{W_{F_1}}{4\eta_1\omega_1m_1} - \frac{W_{F_2}}{4\eta_2\omega_2m_2}\right]. \tag{13.64}$$

Using (13.38) we can write

$$\Pi_{12} = \beta_{12}[\langle E_1\rangle - \langle E_2\rangle]$$

and

$$\Pi_{21} = \beta_{21}[\langle E_2\rangle - \langle E_1\rangle].$$

We can easily prove that $\beta_{12} = \beta_{21}$.

With

$$H_4(\omega) = \frac{j\omega B_1}{A_4\omega^4 - jA_3\omega^3 - A_2\omega^2 + jA_1\omega^1 + A_0}$$

$$\int_{-\infty}^{\infty}\frac{(j\omega)^2}{|D|^2}d\omega = \int_{-\infty}^{\infty}|H_4(\omega)|^2 d\omega = \frac{-\pi A_3B_1^2}{A_0A_3^2 + A_1^2A_4 - A_1A_2A_3}$$

The A_i coefficients are from (13.52).

- $A_4 = 1$
- $A_3 = (\eta_1\omega_1 + \eta_2\omega_2)$
- $A_2 = \{\omega_1^2 + \omega_2^2 + \omega_{12}^2(m_1 + m_2) + \eta_1\eta_2\omega_1\omega_2\}$
- $A_1 = (\eta_1\omega_1\omega_2^2 + \eta_2\omega_2\omega_1^2 + \eta_1m_1\omega_1\omega_{12}^2 + \eta_2m_2\omega_2\omega_{12}^2)$
- $A_0 = (\omega_1^2\omega_2^2 + m_1\omega_1^2\omega_{12}^2 + m_2\omega_2^2\omega_{12}^2)$
- $B_1 = 1$

$$\beta_{12} = \frac{(m_1m_2\omega_{12}^4\eta_1\omega_1\eta_2\omega_2)}{2}\frac{A_3B_1^2}{A_0A_3^2 + A_1^2A_4 - A_1A_2A_3}. \tag{13.65}$$

If two conservatively coupled oscillators are identical and excited by independent random forces, the spectrum of power flow between them is proportional to the difference in the spectra of their energies. For oscillators that are not identical, the total energy flow will be proportional to the difference in energies, provided the excitation spectra are relatively flat near the resonance frequencies [Barnoski 69].

If sdof 2 is blocked, i.e. $\langle E_2\rangle = 0$, then $\overline{\Pi}_{12} = \beta_{12}\langle E_1\rangle$. The coupling spring k_{12} is grounded. Analogons to the average damping energy (13.39) we may express the average energy transferred from sdof 1 to sdof 2 and sdof 2 blocked as follows

$$\overline{\Pi}_{12} = \beta_{12}\langle E_1\rangle = \eta_{12}\omega_1\langle E_1\rangle . \tag{13.66}$$

If sdof 1 is blocked, i.e. $\langle E_1\rangle = 0$, then $\overline{\Pi}_{21} = \beta_{21}\langle E_2\rangle = \beta_{12}\langle E_2\rangle$, thus

$$\overline{\Pi}_{21} = \beta_{21}\langle E_2\rangle = \eta_{21}\omega_2\langle E_2\rangle . \tag{13.67}$$

The average energy Π_{12} can be expressed as follows

$$\Pi_{12} = \overline{\Pi}_{12} - \overline{\Pi}_{21} = \eta_{12}\omega_1\langle E_1\rangle - \eta_{21}\omega_2\langle E_2\rangle , \tag{13.68}$$

with

- η_{12} the coupling loss factor from sdof 1 to sdof 2 with a blocked sdof 2, $\langle E_2\rangle = 0$
- η_{21} the coupling loss factor from sdof 2 to sdof 2 with a blocked sdof 1, $\langle E_1\rangle = 0$
- ω_i the circular natural frequency of sdof i
- $\langle E_i\rangle$ average total energy of sdof i, with $\langle E_i\rangle = m_i\langle \dot{x}_i^2\rangle$, (13.38)

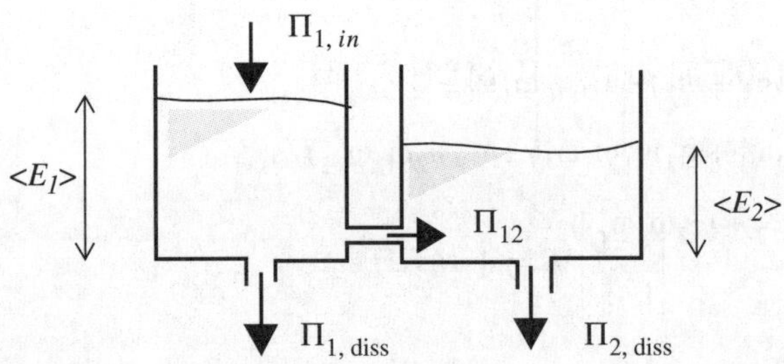

Fig. 13.3. Power flow analogy (courtesy R.G. Dejong, Calvin College, USA)

From the fact that $\beta_{12} = \beta_{21}$ we find the reciprocity relation of the coupling loss factors

$$\eta_{12}\omega_1 = \eta_{21}\omega_2 \tag{13.69}$$

We can write the power-balance equations of both system 1 and system 2. The systems are shown in Fig. . For system 1 the power-balance equation yields

$$\Pi_{1,\,\mathrm{in}} = \eta_{12}\omega_1\langle E_1\rangle - \eta_{21}\omega_2\langle E_2\rangle + \eta_1\omega_1\langle E_1\rangle\,, \tag{13.70}$$

and for system 2 the power-balance equation becomes

$$\eta_{12}\omega_1\langle E_1\rangle - \eta_{21}\omega_2\langle E_2\rangle = \eta_2\omega_2\langle E_2\rangle\,, \tag{13.71}$$

or

$$\eta_{21}\omega(\langle E_1\rangle - \langle E_2\rangle) = \eta_2\omega\langle E_2\rangle\,. \tag{13.72}$$

If (13.70) and (13.71) are written in matrix notation we get

$$\begin{bmatrix} (\eta_{12}+\eta_1)\omega_1 & -\eta_{21}\omega_2 \\ -\eta_{12}\omega_1 & (\eta_{21}+\eta_2)\omega_2 \end{bmatrix} \begin{Bmatrix} \langle E_1\rangle \\ \langle E_2\rangle \end{Bmatrix} = \begin{Bmatrix} \Pi_{1,\,\mathrm{in}} \\ 0 \end{Bmatrix}. \tag{13.73}$$

From (13.72) and (13.69) the ratio between the energy $\langle E_1\rangle$ and $\langle E_2\rangle$ can be expressed as follows

$$\frac{\langle E_2\rangle}{\langle E_1\rangle} = \frac{\eta_{21}}{(\eta_{21}+\eta_2)}\quad . \tag{13.74}$$

If we substitute the result of (13.74) into (13.70) we will find

$$\Pi_{1,\,in} = \eta_1\omega_1\langle E_1\rangle + \eta_2\omega_2\langle E_2\rangle\,. \tag{13.75}$$

The average power supplied is equal to the total dissipated energies.

13.4 Multimode Subsystems

We assume two coupled linear multimode elastic structural systems, subsystem 1 and subsystem 2. Both subsystems are coupled via their common junction. This is illustrated in Fig. 13.4. The theory described in this section is based on [Keltie 01].

In a frequency band $\Delta\omega$, each subsystem has a number of active modes $N_1(\Delta\omega)$ and $N_2(\Delta\omega)$. We introduce the term modal density $n(\omega)$ (number of modes per unit of frequency, (modes/rad/s). The number of modes in the frequency band $\Delta\omega$ can be written as

$$N_1(\Delta\omega) = n_1(\omega)\Delta\omega \text{ and } N_2(\Delta\omega) = n_2(\omega)\Delta\omega . \tag{13.76}$$

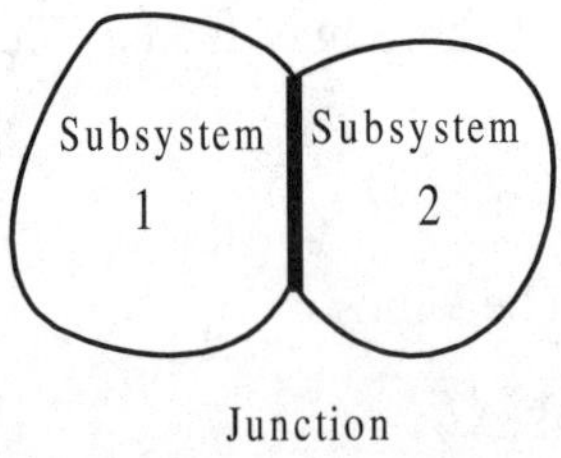

Fig. 13.4. Two coupled mdof subsystems

We count the modes in subsystem 1, denoted by α, $1 \le \alpha \le N_1$ and in the same manner for subsystem 2, denoted by σ, $1 \le \sigma \le N_2$. There are $N_1 N_2$ interacting modal pairs α and β. This is shown in Fig. 13.5. This modal pair will be considered as two coupled oscillators.

We will derive the equation for the power flow Π_{12} between subsystem 1 and subsystem 2 taking into account the following assumptions:

- Each mode of each subsystem has a (circular) natural frequency equally distributed (probability) in the frequency range $\Delta\omega$ (rad/s)
- Every mode in a subsystem has equal energy $\langle E_\alpha \rangle$ or $\langle E_\sigma \rangle$, hence $\langle E_1 \rangle = N_1 \langle E_\alpha \rangle$ and $\langle E_2 \rangle = N_2 \langle E_\sigma \rangle$
- Each mode in a subsystem has the same damping η_α or η_σ

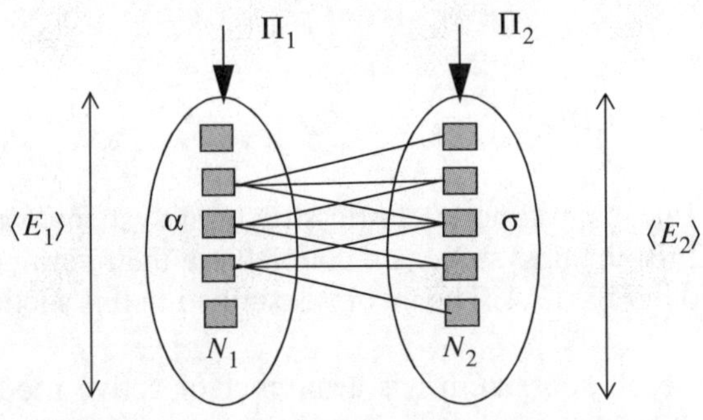

Fig. 13.5. Illustration of energy flow in multimode systems

The average intermodal power flow $\Pi_{\alpha\sigma}$ between mode α and mode σ, in the frequency band $\Delta\omega$, is

$$\Pi_{\alpha\sigma} = \langle\beta_{\alpha\sigma}\rangle(\langle E_{\alpha,\,\Delta\omega}\rangle - \langle E_{\sigma,\,\Delta\omega}\rangle)\,. \tag{13.77}$$

The total power flow $\Pi_{1\sigma}$ from all modes N_1 of subsystem 1 in the frequency band $\Delta\omega$ to mode σ is given by

$$\Pi_{1\sigma} = \langle\beta_{\alpha\sigma}\rangle N_1(\langle E_\alpha\rangle - \langle E_\sigma\rangle)\,. \tag{13.78}$$

The total power flow Π_{12} between all modes N_1 of subsystem 1 and all modes N_2 of subsystem 2 in the frequency band $\Delta\omega$ now becomes

$$\Pi_{12} = \langle\beta_{\alpha\sigma}\rangle N_1 N_2(\langle E_\alpha\rangle - \langle E_\sigma\rangle)\,. \tag{13.79}$$

In terms of the total subsystem energies

$$\langle E_1\rangle = N_1\langle E_\alpha\rangle \quad \text{and} \quad \langle E_2\rangle = N_2\langle E_\sigma\rangle \quad . \tag{13.80}$$

Equation (13.79) can be rewritten in subsystem total energies in the frequency band $\Delta\omega$ using (13.80),

$$\Pi_{12} = \langle\beta_{\alpha\sigma}\rangle N_1 N_2\left(\frac{\langle E_1\rangle}{N_1} - \frac{\langle E_2\rangle}{N_2}\right). \tag{13.81}$$

We define the coupling loss factor η_{12}, assuming a blocked subsystem 2, as

$$\omega\eta_{12} = \langle\beta_{\alpha\sigma}\rangle N_2\,, \tag{13.82}$$

and the coupling loss factor η_{21}, assuming subsystem 1 blocked, as

$$\omega\eta_{21} = \langle\beta_{\alpha\sigma}\rangle N_1\,. \tag{13.83}$$

We can derive the reciprocity relationship from (13.82) and (13.83)

$$\eta_{12}N_1 = \eta_{21}N_2\,. \tag{13.84}$$

Equation (13.81) can be rewritten as follows using the definition for the coupling loss factor

$$\Pi_{12} = \omega\eta_{12}\langle E_1\rangle - \omega\eta_{21}\langle E_1\rangle\,. \tag{13.85}$$

The dissipated energy per subsystem will be defined, similar to (13.39), as

$$\Pi_{1,\,\text{diss}} = \omega\eta_1\langle E_1\rangle \text{ and } \Pi_{2,\,\text{diss}} = \omega\eta_2\langle E_2\rangle\,. \tag{13.86}$$

The power flow between two subsystem is illustrated in Fig. 13.6.

We define the subsystem modal density (number of modes per unit of radial frequency) n_i with (13.76)

$$n_i = \frac{N_i}{\Delta\omega}. \tag{13.87}$$

With use of (13.84) we can prove the reciprocity relation

$$\eta_{12}n_1 = \eta_{21}n_2 . \tag{13.88}$$

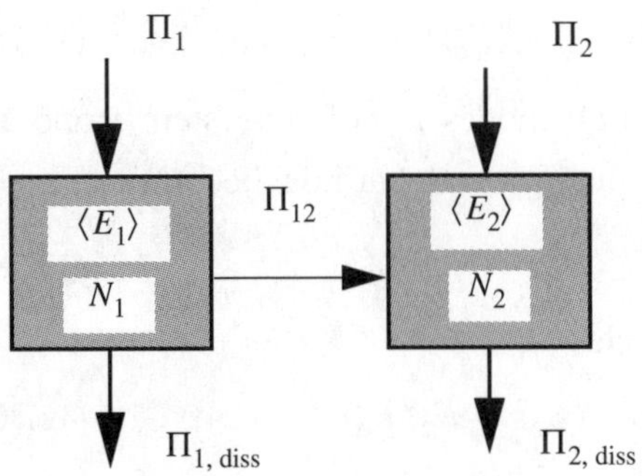

Fig. 13.6. Two interacting subsystems

If we go back to (13.85) and substitute the result of (13.88) we find

$$\Pi_{12} = \omega\eta_{12}\left[\langle E_1\rangle - \frac{n_1}{n_2}\langle E_2\rangle\right]. \tag{13.89}$$

Finally, we obtain the following equation

$$\Pi_{12} = \omega\eta_{12}n_1\left[\frac{\langle E_1\rangle}{n_1} - \frac{\langle E_2\rangle}{n_2}\right]. \tag{13.90}$$

The dissipated energy $\Pi_{i,\,\text{diss}}$ of subsystem i is expressed as

$$\Pi_{i,\,\text{diss}} = \omega\eta_i\langle E_i\rangle , \tag{13.91}$$

with

- $\Pi_{i,\,\text{diss}}$ the dissipated energy of subsystem i
- ω the centre frequency in a certain frequency band (octave or one-third octave)
- η_i the damping loss factor (DLF)
- $\langle E_i\rangle$ total average energy of subsystem i

The power-balance equation of the two subsystems, shown in Fig. 13.6 can be written as

$$\Pi_1 = \Pi_{12} + \Pi_{1,\,\mathrm{diss}}\,, \tag{13.92}$$

and

$$\Pi_2 = \Pi_{21} + \Pi_{2,\,\mathrm{diss}}\,. \tag{13.93}$$

Equations (13.92) and (13.93) can be written

$$\Pi_1 = \omega\eta_{12}n_1\left[\frac{\langle E_1\rangle}{n_1} - \frac{\langle E_2\rangle}{n_2}\right] + \omega\eta_1\langle E_1\rangle\,, \tag{13.94}$$

and

$$\Pi_2 = \omega\eta_{21}n_2\left[\frac{\langle E_2\rangle}{n_2} - \frac{\langle E_1\rangle}{n_1}\right] + \omega\eta_2\langle E_2\rangle\,. \tag{13.95}$$

In matrix notation

$$\begin{bmatrix} \eta_{12}+\eta_1 & \dfrac{-\eta_{12}n_1}{n_2} \\ \dfrac{-\eta_{21}n_2}{n_1} & \eta_{21}+\eta_2 \end{bmatrix} \begin{Bmatrix} \langle E_1\rangle \\ \langle E_2\rangle \end{Bmatrix} = \frac{1}{\omega}\begin{Bmatrix} \Pi_1 \\ \Pi_2 \end{Bmatrix}. \tag{13.96}$$

Applying (13.88) we may write (13.96) as follows

$$\begin{bmatrix} \eta_{12}+\eta_1 & -\eta_{21} \\ -\eta_{12} & \eta_{21}+\eta_2 \end{bmatrix} \begin{Bmatrix} \langle E_1\rangle \\ \langle E_2\rangle \end{Bmatrix} = \frac{1}{\omega}\begin{Bmatrix} \Pi_1 \\ \Pi_2 \end{Bmatrix}, \tag{13.97}$$

with

- Π_i the source power input of subsystem i
- ω the centre frequency is a certain frequency band (octave or one-third octave)
- η_{ij} the coupling loss factor (CLF) between subsystem i and subsystem j

Equation (13.97) is not symmetric.
The following set of equations is written in a symmetric form because of (13.88)

$$\begin{bmatrix} (\eta_{12}+\eta_1)n_1 & -\eta_{21}n_2 \\ -\eta_{12}n_1 & (\eta_{21}+\eta_2)n_2 \end{bmatrix} \begin{Bmatrix} \dfrac{\langle E_1\rangle}{n_1} \\ \dfrac{\langle E_2\rangle}{n_2} \end{Bmatrix} = \frac{1}{\omega}\begin{Bmatrix} \Pi_1 \\ \Pi_2 \end{Bmatrix}. \tag{13.98}$$

As a first approximation, a spacecraft structure can be modelled as a flat aluminium platform that is coupled to a large aluminium cylinder as illustrated in Fig. 13.7 [Norton 98]. The density of Al-alloy $\rho = 2700$ kg/m^3. The alumininium plate is 5 mm thick and is $3.5x\,3$ m^2. The cylinder is 2 m long, has a mean diameter of 1.5 m and has a 3 mm wall thickness. The following information is available about the structure in the range 500 Hz octave band: the platform is directly driven and the cylinder is only driven via the coupling joints; the internal loss factor of the platform (system 1) $\eta_1 = 4.4x10^{-3}$, the internal loss factor of the cylinder (system 2), $\eta_2 = 2.4x10^{-3}$; the platform rms vibrational velocity is $\langle v_1 \rangle = 27.2$ mm/s; and the cylinder rms velocity $\langle v_2 \rangle = 13.2$ mm/s

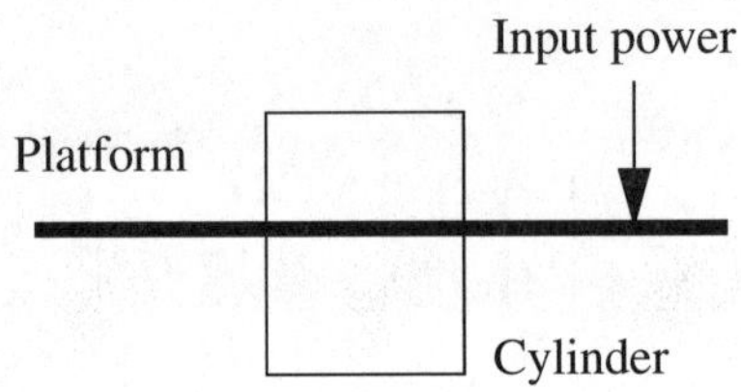

Fig. 13.7. Simple spacecraft structure

Estimate the coupling loss factors η_{12} and η_{21}, and the input power Π_1.

Table 13.1. Spacecraft properties

Structural part	Mass (kg)	Velocity (rms) (m/s)	Energy (Nm/s)
Platform	$M_1 = 117.8935$	$27.2x10^{-3}$	$\langle E_1 \rangle = 8.7222x10^{-2}$
Cylinder	$M_2 = 76.3407$	$13.2x10^{-3}$	$\langle E_2 \rangle = 1.3302x10^{-2}$

The total power put in the spacecraft will be dissipated in the platform and the cylinder, thus

$$\Pi_1 = \omega\eta_1\langle E_1 \rangle + \omega\eta_2\langle E_2 \rangle = 1.3060 \text{ Nm, W.}$$

The power-balance equations for the platform and the cylinder can be written as (13.97)

$$\begin{bmatrix} \eta_{12}+\eta_1 & -\eta_{21} \\ -\eta_{12} & \eta_{21}+\eta_2 \end{bmatrix} \begin{Bmatrix} \langle E_1 \rangle \\ \langle E_2 \rangle \end{Bmatrix} = \frac{1}{\omega} \begin{Bmatrix} \Pi_1 \\ 0 \end{Bmatrix}.$$

From the second equation it follows that

$$\frac{\langle E_2\rangle}{\langle E_1\rangle} = \frac{\eta_{12}}{\eta_{21}+\eta_2} = 0.1525,$$

and with the power-balance equation for the platform

$$(\eta_{12}+\eta_1)\langle E_1\rangle - \eta_{21}\langle E_2\rangle = \frac{\Pi_1}{\omega},$$

we obtain for the coupling loss factors $\eta_{12} = 4.26x10^{-4}$ and $\eta_{21} = 3.92x10^{-4}$.

13.5 SEA Parameters

To generate the SEA equations (e.g. (13.97) and (13.98)) a number of parameters must be calculated:

- Modal density
- Source power input
- Subsystem energy
- Damping loss factor
- Coupling loss factor

The mentioned SEA parameters are discussed in the following sections.

13.5.1 Subsystem Modal Densities

The modal density $n(\omega)$ is the number of modes per radian frequency. The modal density $n(f)$ is the number of modes per frequency (Hz). The relation between $n(\omega)$ and $n(f)$ is

$$n(\omega) = \frac{dN(\omega)}{d\omega} = \frac{dN(f)}{df}\frac{df}{d\omega} = \frac{n(f)}{2\pi}. \tag{13.99}$$

Bending Beam

The circular natural frequency of a simply supported bending beam is defined as

$$\omega_p = \left(\frac{p\pi}{L}\right)^2\sqrt{\frac{EI}{m}},\ p = 1, 2, 3,\dots, \tag{13.100}$$

assuming a mode shape $\sin\left(\frac{p\pi x}{L}\right)$, with

- EI the bending stiffness of a bending beam (Nm^2)
- m the mass per unit of length (kg/m)
- L the length of the beam (m)

We may express the number p in ω_p

$$p = \frac{L}{\pi}\sqrt{\omega_p}\sqrt[4]{\frac{m}{EI}}. \tag{13.101}$$

The modal density can be easily obtained

$$n(\omega) = \frac{dp}{d\omega} = \frac{L}{2\pi}\frac{1}{\sqrt{\omega}}\sqrt[4]{\frac{m}{EI}}. \tag{13.102}$$

The modal density $n(\omega)$ in (13.102) depends on the square root of frequency. The expression (13.102) can be taken as the general modal density expression for beams in flexure.

Bending Plate

The circular natural frequency of the rectangular plate, simply supported, along all edges, and no prestress, is given by

$$\omega_n^2 = \frac{\pi^4 D}{\tilde{m}}\left[\left(\frac{m}{a}\right)^2 + \left(\frac{n}{b}\right)^2\right]^2 = \frac{D}{\tilde{m}}k_p^4 \tag{13.103}$$

$$k_p = \sqrt{\omega_n}\sqrt[4]{\frac{\tilde{m}}{D}}, \tag{13.104}$$

with

- $D = \dfrac{Et^3}{12(1-\nu^2)}$ the flexural rigidity of the isotropic plate (Nm^2/m)
- a, b the length and width of the plate (m)
- t the thickness of the plate (m)
- $\tilde{m}$ the mass per unit of area (kg/m^2)
- m, n wave number
- $k_p^4 = \pi^4\left[\left(\frac{m}{a}\right)^2 + \left(\frac{n}{b}\right)^2\right]^2$

The number of modes $N(k)$ in a quarter of the circle with radius k, see Fig. 13.8, is given by

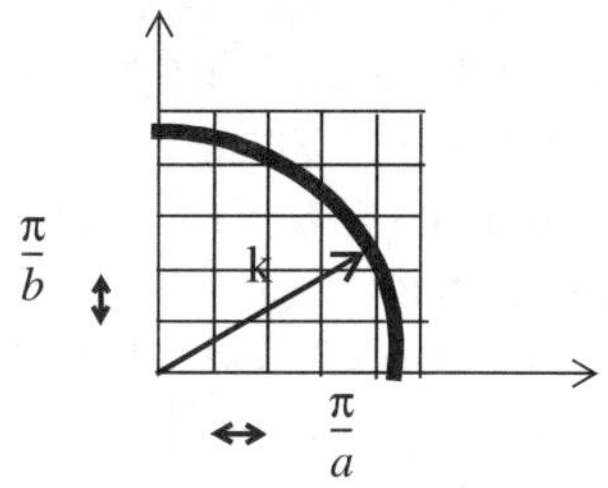

$$N(k) = \frac{\pi k^2}{4}\frac{ab}{\pi^2} = \frac{k^2 ab}{4\pi} . \qquad (13.105)$$

Fig. 13.8. Modal lattice and constant-frequency contours for supported plate [Smith 65]

The modal density $n(\omega)$ can be calculated from

$$n(\omega) = \frac{dN(k)}{dk}\frac{dk}{d\omega} . \qquad (13.106)$$

Using the result of (13.104) and (13.105) in the previous (13.106) we find the following relationship for the modal density of the bending plate

$$n(\omega) = \frac{ab\sqrt{\omega}\sqrt[4]{\frac{\tilde{m}}{D}}}{2\pi}\frac{1}{2\sqrt{\omega}}\sqrt[4]{\frac{\tilde{m}}{D}} = \frac{ab}{4\pi}\sqrt{\frac{\tilde{m}}{D}} . \qquad (13.107)$$

We can replace the area ab by $A_p = ab$, thus (13.107) becomes

$$n(\omega) = \frac{A_p}{4\pi}\sqrt{\frac{\tilde{m}}{D}} . \qquad (13.108)$$

The modal density $n(\omega)$ in (13.108) is independent of the frequency.

Consider an Al-alloy flat plate with a length a=2.5 m and a width b=2.0 m with a plate thickness of 10 mm. The Al-alloy has the following mechanical properties, Young's modulus $E = 70x\,10^9$ Pa, the density is 2700 kg/m^3 and the Poisson's ratio $\upsilon = 0.3$. Calculate, in the octave band from 32–1000 Hz the modal density and the number of modes in the octave frequency bandwidth. The modal density of the plate is constant over the frequency range. For the circular frequencies the modal density becomes

$$n(\omega) = \frac{A_p}{4\pi}\sqrt{\frac{\tilde{m}}{D}} = \frac{2.5x2.0}{4\pi}\sqrt{\frac{2700x0.01}{\frac{70x10^9x0.01^2}{12(1-0.3^2)}}} = 0.0142 \text{ modes/rad/s},$$

and in the frequency domain (Hz) the modal density is

$$n(f) = 2\pi n(\omega) = 0.0893 \text{ modes/Hz.}$$

In Table 13.2 the number of modes per one-octave frequency band is calculated and shown.

Table 13.2. Mode numbers

One-octave Frequency band f (Hz)	Bandwidth $\Delta f = \frac{1}{2}f\sqrt{2}$ (Hz)	Modal density (modes/Hz)	Number of modes in Δf
31.5	22	0.0893	2.0
63	45	0.0893	4.0
125	88	0.0893	7.9
250	177	0.0893	15.8
500	354	0.0893	31.6
1000	707	0.0893	63.1

We showed the derivation of the modal density $n(\omega)$ for the bending beam and the bending plate. In Table 13.3 other frequently used modal densities for structures and acoustic cavities will be given.

Table 13.3. Modal densities

Structural subsystem	Motion	Model density $n(\omega)$
Beam	Longitudinal	$\frac{L}{\pi}\sqrt{\frac{\rho}{E}}$
Beam	Flexure, bending	$\frac{L}{2\pi}\frac{1}{\sqrt{\omega}}\sqrt[4]{\frac{m}{EI}}$
Membrane	Lateral	$\frac{A_p\omega}{2\pi}\frac{\tilde{m}}{S}$
Plate	Bending, flexure	$n_p = \frac{A_p}{4\pi}\sqrt{\frac{\tilde{m}}{D}}$

Table 13.3. Modal densities (Continued)

Structural subsystem	Motion	Model density $n(\omega)$
Sandwich plate [Nigam 94, Chapter 10]	Bending, flexure	$\frac{A_p}{\pi g \beta}\omega\left(1 + \frac{\{\tilde{m}\omega^2 + 2g^2\beta(1-\upsilon^2)\}}{\sqrt{\tilde{m}^2\omega^4\tilde{m}\omega^2 g^2\beta(1-\tilde{m}^2)}}\right)$
Acoustic chamber		$\frac{V\omega^2}{2\pi^2 c^3} + \frac{A\omega}{8\pi^2 c^2} + \frac{P}{16\pi c}$
Cylinder	Bending, flexure	$\begin{cases} n_p & \left(\frac{\omega}{\omega_r} > 1\right) \\ n_p\left(\frac{\omega}{\omega_r}\right)^{\frac{2}{3}} & \left(\frac{\omega}{\omega_r} < 1\right) \end{cases}$

Symbols

- A_p surface area
- $d = h_c + \frac{(t_1 + t_2)}{2}$ (m)
- $g = \frac{\sqrt{G_x G_y}}{h_c}\left[\frac{1}{E_1 t_1} + \frac{1}{E_2 t_2}\right]$
- h_c core height (m)
- m mass per unit of length (kg/m)
- m mass per unit of area (kg/m^2)
- t_1, t_2 thickness of face sheets (m)
- A total surface of the acoustic cavity (m^2)
- D flexural rigidity $D = \frac{Et^3}{12(1-\nu^2)}$ (Nm^2/m)
- E, E_1, E_2 Young's modulus (Pa)
- G, G_x, G_y Shear modulus (Pa)
- R radius of ylinder (m)
- S membrane tension force per unit of length (N/m)
- I second moment of area (m^4)

- P the perimeter of the acoustic cavity (m)
- V volume of the cavity (acoustic chamber) (m^3)
- ω circular frequency (rad/s)
- $\omega_r = \frac{1}{R}\sqrt{\frac{E}{\rho(1-\upsilon^2)}}$ ring frequency is frequency whose wavelength is equal to the cylinder circumference [Norton 98]
- $\beta = d^2\frac{E_1t_1E_2t_2}{E_1t_1+E_2t_2}$
- ρ density of material (kg/m^3)
- ν Poisson's ratio

13.5.2 Source Power Input

Mechanical Random Loads

The average (rms) input power Π_{in} of an sdof, is expressed in the PSD function of the random force, (13.41)

$$\Pi_{\text{in}} = \frac{m\eta\omega_{\text{n}}}{2\pi}\int_{-\infty}^{\infty} S_F(\omega)|Y(\omega)|^2 d\omega \approx \frac{W_F(f_{\text{n}})}{4m} = \frac{S_F(\omega_{\text{n}})}{2m}.$$

If we assume the power spectral density is constant over a frequency bandwidth $\Delta\omega$, the mean-square value of the random applied load is

$$\langle F^2\rangle_{\Delta\omega} = S_F(\omega)\Delta\omega. \tag{13.109}$$

Equation (13.41) can now be written as

$$\Pi_{\text{in}} = \frac{S_F(\omega_{\text{n}})}{2m} = \frac{\langle F^2\rangle_{\Delta\omega}}{2m\Delta\omega}. \tag{13.110}$$

We have $N(\omega)$ lightly damped modes in the frequency band $\Delta\omega$, the total input power using (13.110), [Kenny 02a, 02b],

$$\Pi_{\text{in},\Delta\omega} = \frac{\langle F^2\rangle_{\Delta\omega}}{2M\Delta\omega}N(\omega) = n(\omega)\frac{\langle F^2\rangle_{\Delta\omega}}{2M}, \tag{13.111}$$

with M the total mass. The average input power is given by (13.25) and can be rewritten as the average input power in the frequency band $\Delta\omega$

$$\Pi_{in,\,\Delta\omega} = \langle F^2\rangle_{\Delta\omega}\Re\{Y^*(\omega)\}\ . \tag{13.112}$$

If we compare (13.111) with (13.112) we find

$$n(\omega) = 2M\Re\{Y^*(\omega)\}. \tag{13.113}$$

The modal density can be averaged (smoothed) over the frequency band $\Delta\omega = \omega_2 - \omega_1$ with the centre frequency ω using the following expression

$$n(\omega) = \frac{1}{\Delta\omega}\int_{\omega_1}^{\omega_2} 2M\Re\{Y^*(\omega)\}. \tag{13.114}$$

[Skudryk 68] derived an expression with respect to the frequency average input power of a bending plate excited by a force $F(\omega)$

$$\Pi_{\text{plate}} = \frac{1}{2}|F(\omega)|^2\frac{1}{8\sqrt{D\tilde{m}}} = \langle F^2\rangle\frac{1}{8\sqrt{D\tilde{m}}}, \tag{13.115}$$

with $\tilde{m}$ is the mass per unit of area.

13.5.3 Subsystem Energies

Mechanical Systems

The average energy $\langle E\rangle$ of an sdof system is given by (13.38)

$$\langle E\rangle = \frac{W_F(f_n)}{8m\pi\eta f_n} = \frac{S_F(\omega_n)}{2m\eta\omega_n}.$$

If we assume the power spectral density is constant over a frequency bandwidth $\Delta\omega$, the mean-square value of the random applied load is (13.105)

$$\langle F^2\rangle_{\Delta\omega} = S_F(\omega)\Delta\omega.$$

We have $N(\omega)$ lightly damped modes in the frequency band $\Delta\omega$, the total energy $\langle E\rangle_{\Delta\omega}$ using (13.38) is

$$\langle E\rangle_{\Delta\omega} = \frac{\langle F^2\rangle_{\Delta\omega}}{2M\eta\omega\Delta\omega}N(\omega) = n(\omega)\frac{\langle F^2\rangle_{\Delta\omega}}{2M\eta\omega}, \tag{13.116}$$

with M the mass of the subsystem. If we define the average energy $\langle E\rangle_{\Delta\omega}$ as

$$\langle E\rangle_{\Delta\omega} = M\langle v^2\rangle_{\Delta\omega}, \tag{13.117}$$

with $\langle v^2\rangle_{\Delta\omega}$ the mean square of the velocity, averaged in place and in the frequency band $\Delta\omega$. The averaged mean square $\langle v^2\rangle_{\Delta\omega}$ now becomes

$$\langle v^2\rangle_{\Delta\omega} = n(\omega)\frac{\langle F^2\rangle_{\Delta\omega}}{2M^2\eta\omega}. \tag{13.118}$$

The average acceleration $\langle a^2 \rangle_{\Delta\omega}$ can be calculated using

$$\langle a^2 \rangle_{\Delta\omega} = \frac{\langle v^2 \rangle_{\Delta\omega}}{\omega^2}, \tag{13.119}$$

with

- ω the centre frequency of the bandwidth $\Delta\omega$.

The power spectral density is $S_a(\omega)$ now becomes

$$S_a(\omega) = \frac{\langle a^2 \rangle_{\Delta\omega}}{\Delta\omega} = \frac{W_a(f)}{2}. \tag{13.120}$$

Acoustic Systems

The total energy in $\langle E_{ar} \rangle$ for a reverberant acoustic room (chamber) is, [Beranek 71]

$$\langle E_{ar} \rangle = \frac{\langle p^2 \rangle}{\rho c} V, \tag{13.121}$$

where

- $\langle p^2 \rangle$ the mean-square sound pressure (space-time average) (Pa, N/m)^2 which can be calculated with $\langle p^2 \rangle = p_{ref}^2 10^{\frac{SPL}{10}}$, with $p_{ref} = 2x10^{-5}$ (Pa, N/m^2).
- SPL sound pressure level, $SPL = 10\log\left(\frac{p^2}{p_{ref}^2}\right)$ (dB)
- V volume of acoustic room (m^3).
- ρ the ambient density of the fluid (air $\rho = 1.2$ kg/m^3 at room temperature and 1 Bar).
- c the ambient speed of sound in fluid (air $c \approx 340$ m/s at room temperature).

Sound Radiation

A vibrating panel with average velocity $\langle v \rangle_{\Delta\omega}$, space and time averaged, surrounded by a fluid, will radiate power. This power radiation Π_{rad}, in the frequency band $\Delta\omega$, is [Beranek 71]

$$\Pi_{rad} = A_p \rho c \sigma_{rad} \langle v^2 \rangle_{\Delta\omega} \tag{13.122}$$

with

• σ_{rad} the radiation efficiency

The radiation efficiency depends on the wavelength of bending in the plate structure and the wave number of the acoustic field. If both wavelengths (wave numbers) are equal we talk about the critical (coincidence) frequency. For an isotropic plate [Smith 65] the critical frequency is

$$f_{crit} = \frac{c^2}{2\pi}\sqrt{\frac{\tilde{m}}{D}} \text{ (Hz)}. \tag{13.123}$$

The wave number k (wave constant or propagation constant) (1/m) of the acoustic field is [Smith 65]

$$k = \frac{\omega}{c}. \tag{13.124}$$

The relation of the plate wave number k_p to the circular natural frequency ω_n of a simply supported isotropic plate is given by (13.104)

$$k_p = \sqrt{\omega_n}\sqrt[4]{\frac{\tilde{m}}{D}} = \frac{\omega_n}{c} = k. \tag{13.125}$$

Equation (13.125) will result in (13.123). The radiation efficiency σ_{rad}, in several frequency ranges is given in Table 13.4 [Beranek 71].

Table 13.4. Radiation efficiency

Frequency region	Single-sided[a] radiation efficiency (ratio) σ_{rad}
Well below coincidence frequency	f (Hz)
$k_p \ll k$, $a, b \ll \lambda = \frac{c}{f}$,	
$ka, kb \ll 1$, $A_p = ab$	
1. Up to f_{11}[b], $f_{11}f_{crit} = \frac{c^2\beta}{2A_p}$,	$\sigma_{rad} = \frac{4A_p}{c^2}f^2$
$\beta = \frac{1}{2}\left(\frac{a}{b} + \frac{b}{a}\right)$	
2. Just above f_{11}	σ_{rad} fluctuates

Table 13.4. Radiation efficiency (Continued)

Frequency region	Single-sided[a] radiation efficiency (ratio) σ_{rad}
Below coincidence frequency $(k < k_p)$, multimodal region $(k_p a, k_p b) > 2$, corner modes and edge modes occur when frequency is such $f \geq \frac{c}{P}$ $P = 2(a+b))$ $\lambda_{crit} = \frac{c}{f_{crit}}$ $\alpha = \frac{f}{f_{crit}}$	$\sigma_{rad} = \frac{\lambda_{crit}^2}{A_p} g_1(\alpha) + \frac{P\lambda_{crit}}{A_p} g_2(\alpha)$ $g_1(\alpha) = \begin{cases} \frac{8}{\pi^4} \frac{(1-2\alpha^2)}{\alpha\sqrt{(1-\alpha^2)}} & f < \frac{f_{crit}}{2} \\ 0 & f > \frac{f_{crit}}{2} \end{cases}$ $g_2(\alpha) = \frac{1}{4\pi^2}\left[\frac{(1-\alpha^2)\ln\left(\frac{1+\alpha}{1-\alpha}\right) + 2\alpha}{\sqrt{(1-\alpha^2)^3}}\right]$
At the critical frequency	$\sigma_{rad} = \left(\frac{a}{\lambda_{crit}}\right)^{\frac{1}{2}} + \left(\frac{b}{\lambda_{crit}}\right)^{\frac{1}{2}}$
Above the critical frequency	$\sigma_{rad} = \left(1 - \frac{f_{crit}}{f}\right)^{-\frac{1}{2}}$

a. For double-sided radiation the radiation efficiency will be multiplied by a factor 2

b. f_{11} is the first natural frequency of the panel with the highest volume displacement.

Diffuse (Reverberant) Sound Field Driving a Freely Hung Panel

We will solve the power transfer from a reverberant room to a free-hanging plate. The SEA parameters are illustrated in Fig. 13.9. System 1 is the reverberant room and system 2 is the plate. We assume the energy in the reverberant chamber will be kept constant. The power-balance equation for the plate can be written as

$$\Pi_{12} = \Pi_{2,\,\mathrm{diss}}\,, \tag{13.126}$$

or

$$\omega\eta_{12}n_1\left[\frac{\langle E_1\rangle}{n_1} - \frac{\langle E_2\rangle}{n_2}\right] = \omega\eta_2\langle E_2\rangle = 0\,. \tag{13.127}$$

We will now focus on the power transfer from the reverberant room to the vibrating plate

$$\Pi_{12}(\omega) = \omega\eta_{12}n_1\left[\frac{\langle E_1\rangle}{n_1} - \frac{\langle E_2\rangle}{n_2}\right]. \tag{13.128}$$

The radiated power Π_{rad} of the plate into the reverberant room is expressed in (13.128) by

$$\Pi_{\text{rad}} = \omega\eta_{12}n_1\frac{\langle E_2\rangle}{n_2} = \omega\eta_{21}\langle E_2\rangle = A_p\rho c\sigma_{\text{rad}}\langle v^2\rangle_{\Delta\omega}\ , \tag{13.129}$$

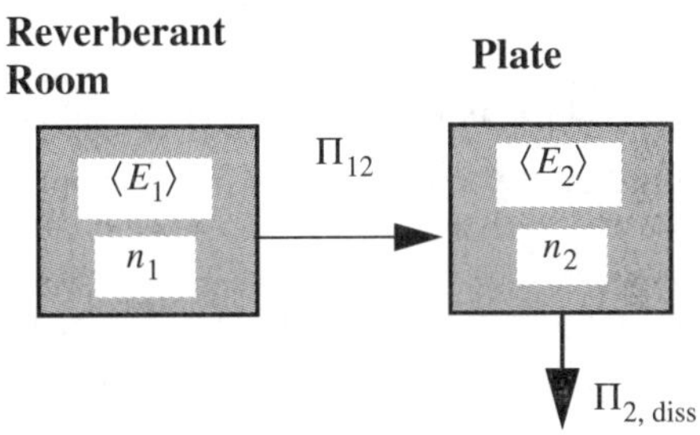

Fig. 13.9. Plate hanging in reverberant room

with $\langle E_2\rangle = M_p\langle v^2\rangle_{\Delta\omega}$ (13.117) we obtain the following expression for the coupling loss factor η_{21}

$$\eta_{21} = \frac{A_p\rho c\sigma_{\text{rad}}}{M_p\omega}. \tag{13.130}$$

The power transferred from the acoustic chamber to the plate can be expressed as $\omega\eta_{12}\langle E_1\rangle$. With

- $\langle E_{\text{ar}}\rangle = \langle E_1\rangle = \dfrac{\langle p^2\rangle}{\rho c}V$
- $n_2 = n_{ar} = \dfrac{V\omega^2}{2\pi^2c^3}$
- $n_p = \dfrac{A_p}{4\pi}\sqrt{\dfrac{\tilde{m}}{D}}$
- $\eta_{12} = \eta_{21}\dfrac{n_2}{n_1}$

The power transferred from the acoustic room to the plate now becomes

$$\Pi_{ar} = \omega\eta_{12}\langle E_1\rangle = \frac{2\pi^2 A_p \sigma_{rad}^2 \langle p^2\rangle c^2 n_p}{M_p \omega^2} \tag{13.131}$$

13.5.4 Damping Loss Factor

[Trudell 80] proposed to apply loss factors, more or less dependent upon the frequency. The damping value originated from acoustic tests on Saturn IVB/V interstage panels.

$$\eta = \begin{cases} 0.01 & f \le 250\text{Hz} \\ 4.7687 f^{-1.11640} & f > 250\text{Hz} \end{cases} . \tag{13.132}$$

For a cavity an equivalent internal loss factor is obtained from an average acoustic absorption coefficient $\bar{\alpha}$ using the expression

$$\eta_{cav} = \frac{cA_w\bar{\alpha}}{4V\omega}, \tag{13.133}$$

where:

- c speed of sound in air
- A_w total wall area of the cavity (m^2)
- V volume of the cavity (m^3)

The proportion of incident energy that is reflected into the room is $1-\bar{\alpha}$, thus

$$\frac{\Pi_{refl}}{\Pi_{inc}} = (1-\bar{\alpha}). \tag{13.134}$$

Every time a wave strikes a wall, a quantity of sound energy $\bar{\alpha}\frac{V\langle p^2\rangle}{\rho c}$ is lost from the reverberant field. Statistically, this reflection occurs $\frac{cA_w}{4V}$ per second. $\frac{4V}{A_w}$ is statistically the average length the wave will travel. The number of strikes per second against the wall is $\frac{cA_w}{4V}$. The rate of energy loss in the reverberant room is

$$\frac{cA_w}{4V}\bar{\alpha}\frac{V\langle p^2\rangle}{\rho c} = \eta_{cav}\omega\frac{V\langle p^2\rangle}{\rho c}. \tag{13.135}$$

From (13.135) we can derive (13.133).

13.5.5 Coupling Loss Factor

In this section the coupling loss factors associated with some of the more common coupling joints are summarised. For more information one may want to read [Norton 89, Section 6.6].

Structure-to-Structure Coupling Loss Factors

The most commonly encountered structure-to-structure coupling is a line junction between two plate structures. The coupling loss factor for a line junction is given in terms of the wave transmission coefficient τ for the line junction [Norton 98]. The coupling loss factor of a line junction from plate 1 to plate 2 is given by

$$\eta_{12} = \frac{2c_B L \tau_{12}}{\pi\omega A_{p,1}}, \tag{13.136}$$

where

- c_B the bending wave velocity (or phase velocity),

$$c_B = \omega^{\frac{1}{2}}\left\{\frac{Et^3}{12(1-\upsilon^2)}\right\}^{\frac{1}{4}}$$

- t thickness of the bending plate
- E Young's modulus of the plate material (Pa)
- υ Poisson's ratio
- L length of the line junction (m)
- ω radian frequency (rad/s)
- $A_{p,1}$ the area of plate 1 (m^2)
- τ_{12} wave transmission coefficient

The normal incidence transmission coefficient for two coupled flat plates at right angles to each other is given by [Norton 98]

$$\tau_{12}(0) = 2\left(\psi^{\frac{1}{2}} + \psi^{-\frac{1}{2}}\right)^{-2}, \tag{13.137}$$

where

- $\psi = \dfrac{\rho_1 c_{L,1}^{\frac{3}{2}} t_1^{\frac{5}{2}}}{\rho_2 c_{L,2}^{\frac{3}{2}} t_2^{\frac{5}{2}}}$
- ρ density of the plate material (kg/m^3)
- c_L longitudinal wave velocity, $c_L = \left(\frac{E}{\rho}\right)^{\frac{1}{2}}$ (m/s)

The random incidence transmission coefficient τ_{12} is approximated by [Norton 98]

$$\tau_{12} = \tau_{12}(0)\frac{2.754\frac{t_1}{t_2}}{1 + 3.24\frac{t_1}{t_2}}. \tag{13.138}$$

The coupling loss factor for two homogeneous plates coupled by point connections (e.g. bolts) is approximated by, [Norton 98]

$$\eta_{12} = \frac{4Nt_1c_{L,1}(\rho_1 t_1 c_{L,1})^2(\rho_2 t_2 c_{L,2})^2}{\sqrt{3}\omega A_{p,1}\{(\rho_1 t_1 c_{L,1})^2 + (\rho_2 t_2 c_{L,2})^2\}} \tag{13.139}$$

with

- N the number of bolts

Consider two flat Al-alloy plates that are coupled at right angles to each other. The first plate is 3 mm thick and is $2.5x\,1.2$ m^2, and the second plate is 5.5mm thick and $2.0x\,1.2$ m^2. Evaluate the coupling loss factors η_{12} and η_{21} in all the octave band frequencies from 125 to 2000 Hz for

1. A welded joint along 1.2 m edge
2. A bolted joint with twelve bolts along the 1.2 m edge

The results of the coupling loss factor calculations are shown in Table 13.5.

Table 13.5. Coupling loss factors two for plates connected with an angle of 90°

	Welded joint		Bolted joint	
Octave band (Hz)	$\eta_{12}\, x\, 10^3$	$\eta_{21}\, x\, 10^3$	$\eta_{12}\, x\, 10^3$	$\eta_{21}\, x\, 10^3$
125	3.15	5.77	14.4	26.4
250	2.23	4.09	7.19	13.2
500	1.58	2.90	3.60	6.60
1000	1.12	2.05	1.80	3.30
2000	0.789	1.45	0.899	1.65

For a cross-sectional change the transmission coefficient τ_{12} becomes [Francesconi 96]

$$\tau_{12} = \left(\frac{\psi^{-\frac{5}{4}} + \psi^{-\frac{3}{4}} + \psi^{\frac{3}{4}} + \psi^{\frac{5}{4}}}{\frac{1}{2}\psi^{-2} + \psi^{-\frac{1}{2}} + 1 + \psi^{\frac{1}{2}} + \frac{1}{2}\psi^{2}} \right)^2, \tag{13.140}$$

with

- $\psi = \frac{t_1}{t_2}$

Acoustic Radiation

The coupling loss factor for a structure–acoustic volume coupling (see (13.130)) is given by

$$\eta_{\mathrm{sa}} = \frac{A_p \rho c \sigma_{\mathrm{rad}}}{M_p \omega}. \tag{13.141}$$

Using the reciprocity relation of coupling loss factors, (13.88), we obtain the following expression for the coupling factor from the acoustic space to the structure

$$\eta_{\mathrm{as}} = \frac{A_p \rho c \sigma_{\mathrm{rad}}}{M_p \omega} \frac{n_s}{n_{as}}, \tag{13.142}$$

where

- n_s the modal density of the structure

- n_{as} the modal density of the acoustic space

13.6 Stresses and Strains

The strain energy U^* per unit of volume in a elastic body in the principal stress directions is given by

$$U^* = \frac{1}{2}[\sigma_1\varepsilon_1 + \sigma_2\varepsilon_2 + \sigma_3\varepsilon_3] . \tag{13.143}$$

For plate and shell structures we have 2-D stress states (plane stress) and we assume $\sigma_3 = 0$, hence the strain energy per unit. volume becomes

$$U^* = \frac{1}{2}[\sigma_1\varepsilon_1 + \sigma_2\varepsilon_2] . \tag{13.144}$$

In the worst case situation the total strain energy is a line strain state, so

$$U^* = \frac{1}{2}[\sigma\varepsilon] . \tag{13.145}$$

For isotropic materials, with Young's modulus E (Pa) the strain energy per unit of volume is expressed in stresses

$$U^* = \frac{1}{2E}[\sigma^2] . \tag{13.146}$$

In pure bending the stress can be expressed as (see Fig. 13.10)

$$\sigma(z) = \frac{2z}{t}\sigma_{max} . \tag{13.147}$$

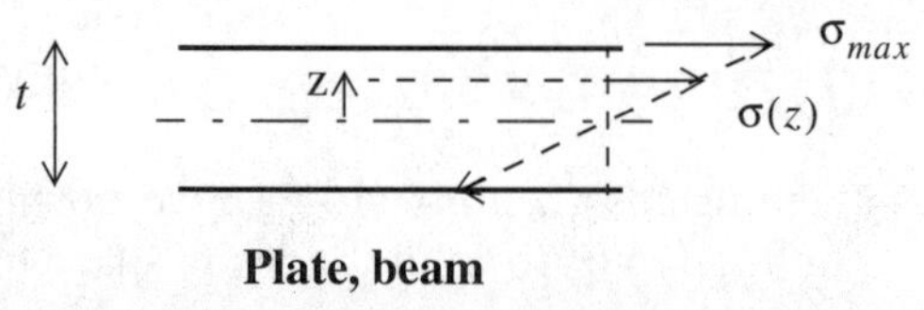

Fig. 13.10. Bending stress in plate and beam, in one direction

The strain energy per unit of area now becomes

$$\int_{-\frac{1}{2}t}^{\frac{1}{2}t} U^* dz = \frac{1}{2E}\int_{-\frac{1}{2}t}^{\frac{1}{2}t} \sigma^2(z)dz = \frac{2}{Et^2}\sigma_{max}^2\int_{-\frac{1}{2}t}^{\frac{1}{2}t} z^2 dz = \frac{t\sigma_{max}^2}{6E} . \tag{13.148}$$

The average strain energy per unit of area is equal to the average kinetic energy per unit of area of a plate, thus

$$\frac{t\langle\sigma_{max}^2\rangle}{6E} = \frac{1}{2}\frac{M_p}{A_p}\langle v^2\rangle . \quad (13.149)$$

For $\langle\sigma_{max}^2\rangle$ we find the following expression

$$\langle\sigma_{max}^2\rangle = \frac{3\mathrm{E}M_p}{tA_p}\langle v^2\rangle . \quad (13.150)$$

For a sandwich plate with a face sheet thickness t_f (13.148) becomes

$$\int_{-\frac{1}{2}t}^{\frac{1}{2}t}\langle U^*\rangle dz = \frac{t_f\langle\sigma_{max}^2\rangle}{\mathrm{E}} = \frac{1}{2}\frac{M_p}{A_p}\langle v^2\rangle . \quad (13.151)$$

For $\langle\sigma_{max}^2\rangle$ we now obtain the following relation

$$\langle\sigma_{max}^2\rangle = \frac{\mathrm{E}M_p}{2t_fA_p}\langle v^2\rangle . \quad (13.152)$$

13.7 Problems

13.7.1 Problem 1

A spring-mounted rigid body with a 100 kg mass can be modelled as an oscillator with a spring stiffness k=6.25x 10^{6}N/m. A steady-state applied force of 75 N produces a velocity of 0.15 (m/s)]. Estimate the damping ratio ζ, the loss factor η and the amplification factor Q (quality factor).
Answers: $\zeta = 0.0125$, $\eta = 0.025$ and $Q = 40$.

13.7.2 Problem 2

Consider two coupled groups of oscillators with similar modal densities, in which only the first group is directly driven in the steady state. Using the steady-state power-balance equations, show that

$$\frac{\langle E_2\rangle}{\langle E_1\rangle} = \frac{\eta_{21}}{\eta_2 + \eta_{21}} .$$

Now, assuming that the oscillators are strongly coupled, the first group is lightly damped, the second is heavily damped, and that one wishes to minimise the vibrational levels transmitted to the second group, what should one do [Norton 89]? Hint: use (13.75).

13.7.3 Problem 3

Show that

$$\frac{\Pi_{1,in}}{\omega\langle E_1\rangle} = \eta_1 + \frac{\frac{n_2}{n_1}\eta_2\eta_{21}}{n_2 + \eta_{21}}$$

for two coupled groups of oscillators in steady-state vibration.

13.7.4 Problem 4

Consider two coupled oscillators (sdofs) where only one is directly driven by external forces and the other is driven only through coupling. Derive expressions for the total vibrational energies $\langle E_1\rangle$ and $\langle E_2\rangle$ of each of the oscillators in terms of the input power $\Pi_{1,in}$, the loss factors η_1 and η_2, the coupling loss factors η_{12}, η_{21} and the natural frequencies ω_1 and ω_2 of the oscillators.

Discuss the following situations:

1. $\eta_{21} \gg \eta_1$ and $\eta_{21} \gg \eta_2$
2. $\eta_{21} \ll \eta_1$ and $\eta_{21} \ll \eta_1$
3. $\eta_2 \ll \eta_{21} \ll \eta_1$
4. $\eta_1 \ll \eta_{21} \ll \eta_2$

Answers.

$$\frac{\langle E_1\rangle}{\Pi_{1,in}} = \frac{\left(1+\frac{\eta_{21}}{\eta_2}\right)}{\omega_1\eta_1\left(1+\frac{\eta_{21}}{\eta_2}\right)+\omega_2\eta_2}, \quad \frac{\langle E_2\rangle}{\Pi_{1,in}} = \frac{\frac{\eta_{21}}{\eta_2}}{\omega_1\eta_1\left(1+\frac{\eta_{21}}{\eta_2}\right)+\omega_2\eta_2}.$$

13.7.5 Problem 5

Evaluate (1) the modal density and (2) the number of modes in each of the octave bands from 125 to 4000Hz for a 5-m long Al-alloy I of-shaped bar with flanges 75 mm width and a web plate with a height of 200mm and a constant thickness of 3 mm. (E-Al-alloy=70x 10^9 Pa, $\rho_{\text{Al-alloy}} = 2700$ kg/m^3) for (1) longitudinal, and (2) flexural (bending) vibrations.

13.7.6 Problem 6

The Large European Acoustic Facility (LEAF) has a chamber volume $V = 1624$ m^3. The chamber dimensions are 9 x 11 x 16.4 (width x length x height) m^3. Calculate the modal density of the LEAF in the one-octave frequency band from 31.5–8000Hz.

13.7.7 Problem 7

The Large European Acoustic Facility (LEAF) has a chamber volume $V = 1624$ m^3. Calculate the average energy $\langle E_{\text{LEAF}} \rangle$ of the LEAF in the one-octave band and SPL values as shown in Table 13.6.

Table 13.6. SPL LEAF

One-octave frequency band (Hz)	SPL (dB), $p_{\text{ref}} = 2x10^{-5}$ Pa
31.5	136
63	141
125	147
250	150
500	147
1000	144
2000	137
4000	131
8000	125

The speed of sound in air c=340m/s and the density of air ρ =1.2kg/m^3.

13.7.8 Problem 8

A panel is placed in a reverberant room with an average pressure $\langle p^2 \rangle$. Show that the average acceleration $\langle a^2 \rangle$ of that panel can be expressed as

$$\frac{\langle a^2 \rangle}{\langle p^2 \rangle} = \frac{2\pi^2 n_p c}{M_p \rho} \frac{\eta_{\text{rad}}}{\eta_{\text{rad}} + \eta_p},$$

where

- n_p modal density of a panel
- M_p total mass of the panel
- c speed of sound in air
- ρ density of air
- η_{rad} coupling loss factor for radiation
- η_p the loss factor in the panel

The modal density of the room is only based on the volume participation [Lyon 64].

14 Free-free Dynamic Systems, Inertia Relief

14.1 Introduction

Free-free systems can move as a rigid body through space, the structure is so-called unconstrained. The stiffness matrix $[K]$ is singular and therefore the flexibility matrix $[G] = [K]^{-1}$ does not exist. Launch vehicles, aircraft and spacecraft are examples of free-free moving dynamic systems. In this chapter, a method, the inertia relief, will be derived to analyse free-free systems. The motion as a rigid body will be eliminated and a new set of applied loads (relative forces) will be used to analyse the elastic behaviour of the free-free system. In the following sections the relative motion, relative forces will be introduced and a definition of the inertia-relief flexibility matrix $[G_f]$ will be given.

14.2 Relative Motion

For a free-free moving dynamic system the total displacement vector $\{x\}$ may be expressed in a pure rigid motion displacement vector $\{x_r\}$ and a relative elastic displacement vector $\{x_e\}$, thus

$$\{x\} = \{x_r\} + \{x_e\}. \tag{14.1}$$

The modes of the free-free dynamic system will be divided in the 6 rigid-body modes $[\Phi_r]$ and the elastic modes $[\Phi_e]$. The displacement vector $\{x\}$ will be projected on an independent set of modes

$$\{x\} = [\Phi]\{\eta\}, \tag{14.2}$$

with
- $[\Phi]$ the modal matrix
- $\{\eta\}$ the generalised coordinates

or

$$\{x\} = \{x_r\} + \{x_e\} = [\Phi_r]\{\eta_r\} + [\Phi_e]\{\eta_e\}, \tag{14.3}$$

with
- $\{\eta_r\}$ the generalised coordinates with respect to the motions as a rigid body
- $\{\eta_e\}$ the generalised coordinates with respect to the elastic behaviour of the dynamic system

The undamped equation of motion of a dynamic system loaded with the dynamic force $\{F\}$ is

$$[M]\{\ddot{x}\} + [K]\{x\} = \{F\}. \tag{14.4}$$

With the introduction of (14.3) into (14.4) and knowing that $[K][\Phi_r] = [0]$ we obtain

$$[M][\Phi_r]\{\ddot{\eta}_r\} + [M][\Phi_e]\{\ddot{\eta}_e\} + [K][\Phi_e]\{\eta_e\} = \{F\}. \tag{14.5}$$

14.3 Relative Forces

The uncoupled equations of motion of the dynamic system, exposed to the force vector $\{F\}$, and expressed in the generalised coordinates $\{\eta\}$ can be written

1. For the generalised coordinates associated with the motions as a rigid body with in general 6 zero natural frequencies ($\omega_i^2 = 0,\ i = 1, 2, ..., 6$)

$$\ddot{\eta}_{r,i} = \{\phi_{r,i}\}^T\{F\}. \tag{14.6}$$

2. For the elastic generalised coordinates associated with the elastic deformations

$$\ddot{\eta}_{e,i} + \omega_{e,i}^2\eta_{e,i} = \{\phi_{e,i}\}^T\{F\}, i = 7, 8, ..., n. \tag{14.7}$$

The rigid-body modes $[\Phi_r]$ are orthonormal with respect to the mass matrix $[M]$

$$[\Phi_r]^T[M][\Phi_r] = [I]. \tag{14.8}$$

Similarly, the elastic modes $[\Phi_e]$ are orthonormal with respect to the mass matrix $[M]$

$$[\Phi_e]^T[M][\Phi_e] = [I]. \tag{14.9}$$

The external forces are in balance with the inertia forces of the rigid body and the internal elastic forces. If we subtract the inertia forces from the external forces the elastic forces are in equilibrium and will not excite the rigid-body motions. We will define the virtual rigid-body forces $\{F_r\}$ that are in equilibrium with the inertia loads $[M]\{\ddot{x}_r\}$ in such a way that [Craig 77]

$$\{F_r\} + [M]\{\ddot{x}_r\} = \{0\}. \tag{14.10}$$

The acceleration as a rigid body $\{\ddot{x}_r\}$ will be projected onto the rigid-body modes $\{\ddot{x}_r\} = [\Phi_r]\{\ddot{\eta}_r\}$. The virtual forces $\{F_r\}$ can be expressed in the generalised coordinates $[\ddot{\eta}_r]$

$$\{F_r\} = -[M]\{\ddot{x}_r\} = -[M][\Phi_r]\{\ddot{\eta}_r\}. \tag{14.11}$$

The generalised accelerations $[\ddot{\eta}_r]$ are related to the external forces by substituting (14.6) into (14.11)

$$\{F_r\} = -[M][\Phi_r][\Phi_r]^T\{F\}. \tag{14.12}$$

From (14.5) and (14.11) we can derive that

$$[M][\Phi_e]\{\ddot{\eta}_e\} + [K][\Phi_e]\{\eta_e\} = \{F\} - [M][\Phi_r]\{\ddot{\eta}_r\} = \{F\} + \{F_r\} = \{F_e\}. \tag{14.13}$$

The relative elastic forces $\{F_e\}$ are defined as

$$\{F_e\} = \{F\} + \{F_r\} = [A]\{F\}, \tag{14.14}$$

with

$$[A] = [I] - [M][\Phi_r][\Phi_r]^T. \tag{14.15}$$

The matrix $[A]$ is called the inertia-relief projection matrix [Craig 00] or the filtering operator [Thonon 98]. With (14.8) it can be easily proved [Preumont 97] that

$$[\Phi_r]^T[A] = 0, \tag{14.16}$$

and with $[\Phi_r]^T[M][\Phi_e]$ we get

$$[\Phi_e]^T[A] = [\Phi_e]^T. \tag{14.17}$$

From (14.16) we learn that the load vector $[A]\{F\}$ is in equilibrium with respect to the point on which the rigid-body vectors are defined. The elastic motion $\{x_e\}$ can be solved with (14.13)

$$[\Phi_e]^T[M][\Phi_e]\{\ddot{\eta}_e\} + [\Phi_e]^T[K][\Phi_e]\{\eta_e\} = [\Phi_e]^T[A]\{F\}, \tag{14.18}$$

thus

$$\langle I\rangle\{\ddot{\eta}_e\} + \langle\omega_e^2\rangle\{\eta_e\} = [\Phi_e]^T\{F\} \tag{14.19}$$

The static solution of the relative or elastic motions $\{x_e\} = [\Phi_e]\{\eta_e\}$ becomes (see (14.17) and (14.18))

$$\{x_e\} = [\Phi_e]([\Phi_e]^T[K][\Phi_e])^{-1}([\Phi_e]^T\{F\}) = [\Phi_e]\langle\omega_e^{-2}\rangle[\Phi_e]^T\{F\}. \tag{14.20}$$

An unconstrained dynamic system consists of three mass–spring dynamic systems with the following mass matrix $[M]$ and stiffness matrix $[K]$. The system is loaded with an external force vector $\{F\}$

$$[M] = \begin{bmatrix} 1 & 0 & 0 \\ 0 & 1 & 0 \\ 0 & 0 & 1 \end{bmatrix}, [K] = \begin{bmatrix} 1 & -1 & 0 \\ -1 & 2 & -1 \\ 0 & -1 & 1 \end{bmatrix}, \{F\} = \begin{Bmatrix} 1 \\ 0 \\ 0 \end{Bmatrix}.$$

Suppose the first dof (x_1) has a prescribed unit displacement $x_1 = 1$. The displacement vector

$$\begin{Bmatrix} x_2 \\ x_3 \end{Bmatrix} = -\begin{bmatrix} 2 & -1 \\ -1 & 1 \end{bmatrix}^{-1}\begin{Bmatrix} -1 \\ 0 \end{Bmatrix}x_1 = \begin{Bmatrix} 1 \\ 1 \end{Bmatrix}x_1 = \begin{Bmatrix} 1 \\ 1 \end{Bmatrix}.$$

The vector

$$\begin{Bmatrix} x_1 \\ x_2 \\ x_3 \end{Bmatrix} = \begin{Bmatrix} 1 \\ 1 \\ 1 \end{Bmatrix}$$

is the displacement of the dynamic system as a rigid body, because

$$[K]\begin{Bmatrix} 1 \\ 1 \\ 1 \end{Bmatrix} = \begin{Bmatrix} 0 \\ 0 \\ 0 \end{Bmatrix}.$$

The rigid-body mode $\{\phi_r\}$ is defined as $\{\phi_r\}^T[M]\{\phi_r\} = 1$, thus

$$\{\phi_r\} = \frac{1}{3}\begin{Bmatrix} 1 \\ 1 \\ 1 \end{Bmatrix},$$

because

$$\begin{bmatrix} 1 & 1 & 1 \end{bmatrix}[M]\begin{Bmatrix} 1 \\ 1 \\ 1 \end{Bmatrix} = 3.$$

The matrix $[A]$ can now be calculated

$$[A] = [I] - [M][\Phi_r][\Phi_r]^T = \frac{1}{3}\begin{bmatrix} 2 & -1 & -1 \\ -1 & 2 & -1 \\ -1 & -1 & 2 \end{bmatrix}.$$

The components of the vector of relative elastic forces become

$$\{F_e\} = [A]\{F\} = \frac{1}{3}\begin{Bmatrix} 2 \\ -1 \\ -1 \end{Bmatrix}.$$

We denote that the components of the relative elastic force $\{F_e\}$ are in equilibrium, because $\{\phi_r\}[A]\{F\} = \{0\}$.

14.4 Flexibility Matrix

The relative elastic forces $\{F_e\}$ are in equilibrium because the inertia forces $\{F_r\}$ were subtracted from the external forces $\{F\}$. Hence we may constrain the free-free structure, to eliminate the motions as a rigid body, in any arbitrarily selected point A (maximum 6 dofs) as illustrated in Fig. 14.1.

We want to calculate the elastic displacement $\{x_{e,A}\}$ of the constrained structure, due to the relative elastic force $\{F_e\}$, with respect to point A. The displacements $\{x_{e,A}\}$ are

$$\{x_{e,A}\} = [G_e]\{F_e\}, \tag{14.21}$$

- $[G_e]$ the flexibility matrix with respect to the degrees of freedom of point A, the columns and rows corresponding to the dofs of point

A are filled with zeros, $[G_e] = \begin{bmatrix} G_{ee} & 0 \\ 0 & 0 \end{bmatrix}$. The displacements and rotations of point A are zero $\{x_A\} = 0$.

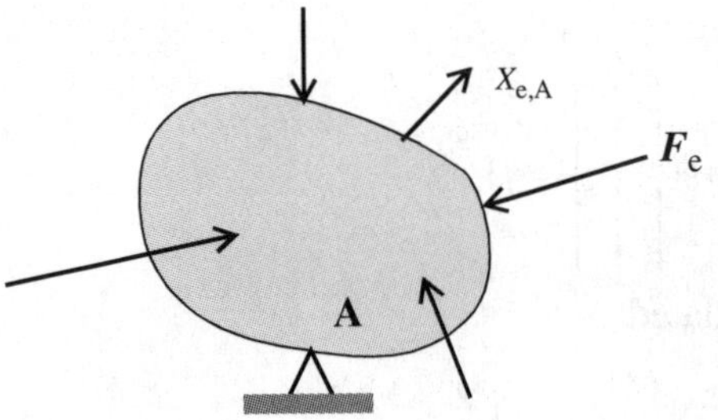

Fig. 14.1. Constrained free-free structure loaded with relative forces $\{F_e\}$

If the stiffness matrix $[K]$ is partitioned in an e-set and a c-set (maximum 6 dofs in point A) we get

$$[K] = \begin{bmatrix} K_{ee} & K_{ec} \\ K_{ce} & K_{cc} \end{bmatrix}. \tag{14.22}$$

The stiffness matrix $[K_{ee}]$ is, in general, regular and $[G_{ee}] = [K_{ee}]^{-1}$.

The flexibility matrix $[G_e]$ is

$$[G_e] = \begin{bmatrix} [K_{ee}]^{-1} & 0 \\ 0 & 0 \end{bmatrix}. \tag{14.23}$$

The real relative elastic deformation $\{x_e\}$ is a summation of the relative elastic deformation and fractions of rigid-body motions $[\Phi_r]\{\vartheta_r\}$. Thus the relative elastic deformation $\{x_r\}$ is

$$\{x_e\} = \{x_{e,A}\} + [\Phi_r]\{\vartheta_r\}, \tag{14.24}$$

- $\{\vartheta_r\}$ generalised coordinates

We will remove the fraction of rigid-body motion and force the displacement vector $\{x_e\}$ to be mass orthogonal with the rigid-body modes $[\Phi_r]$, thus

$$\{x_e\}^T[M][\Phi_r] = [0]. \tag{14.25}$$

This means that the vector of generalised coordinates $\{\vartheta_r\}$ is equal to

$$\{\vartheta_r\} = -[\Phi_r]^T[M]\{x_{e,A}\}. \tag{14.26}$$

The relative elastic deformation $\{x_e\}$ finally becomes

$$\{x_e\} = ([I]-[\Phi_r][\Phi_r]^T[M])\{x_{e,A}\} = [A]^T\{x_{e,A}\}. \tag{14.27}$$

The relative displacement $\{x_{e,A}\}$ can be expressed in relative forces by substituting (14.21) into (14.27), thus

$$\{x_e\} = [A]^T\{x_{e,A}\} = [A]^T[G_e]\{F_e\}. \tag{14.28}$$

Finally using (14.14) we can express the relative displacement vector $\{x_e\}$ in the external forces

$$\{x_e\} = [A]^T[G_e]\{F_e\} = [A]^T[G_e][A]\{F\} = [G_f]\{F\}. \tag{14.29}$$

The matrix $[G_f]$ is called the inertia-relief flexibility matrix and will be used later in the component mode synthesis (CMS) method.

An unconstrained system consists of three mass–spring dynamic systems with the following mass matrix $[M]$ and stiffness matrix $[K]$. The system is loaded by an external force vector $\{F\}$

$$[M] = \begin{bmatrix} 1 & 0 & 0 \\ 0 & 1 & 0 \\ 0 & 0 & 1 \end{bmatrix}, [K] = \begin{bmatrix} 1 & -1 & 0 \\ -1 & 2 & -1 \\ 0 & -1 & 1 \end{bmatrix}, \{F\} = \begin{Bmatrix} 1 \\ 0 \\ 0 \end{Bmatrix}.$$

If we constrain x_1 the stiffness matrix $[K_{ee}]$ is

$$[K_{ee}] = \begin{bmatrix} 2 & -1 \\ -1 & 1 \end{bmatrix}.$$

The flexibility matrix $[G_{ee}]$ is

$$[G_{ee}] = [K_{ee}]^{-1} = \begin{bmatrix} 1 & 1 \\ 1 & 2 \end{bmatrix},$$

and the total flexibility matrix $[G_e]$ is

$$[G_e] = \begin{bmatrix} 0 & 0 \\ 0 & G_{ee} \end{bmatrix} = \begin{bmatrix} 0 & 0 & 0 \\ 0 & 1 & 1 \\ 0 & 1 & 2 \end{bmatrix}.$$

The rigid-body mode $[\Phi_r]$ and $[A]$ are

$$\{\phi_r\} = \frac{1}{3}\begin{Bmatrix} 1 \\ 1 \\ 1 \end{Bmatrix}, [A] = [I] - [M][\Phi_r][\Phi_r]^T = \frac{1}{3}\begin{bmatrix} 2 & -1 & -1 \\ -1 & 2 & -1 \\ -1 & -1 & 2 \end{bmatrix}.$$

The final flexibility matrix $[G_f]$ becomes

$$[G_f] = [A]^T[G_e][A] = \begin{bmatrix} 0.5556 & -0.1111 & -0.4444 \\ -0.1111 & 0.2222 & -0.1111 \\ -0.4444 & -0.1111 & 0.5556 \end{bmatrix}.$$

The relative elastic deformation $\{x_e\}$ will be

$$\{x_e\} = [G_f]\{F\} = \begin{bmatrix} 0.5556 & -0.1111 & -0.4444 \\ -0.1111 & 0.2222 & -0.1111 \\ -0.4444 & -0.1111 & 0.5556 \end{bmatrix}\begin{Bmatrix} 1 \\ 0 \\ 0 \end{Bmatrix} = \begin{bmatrix} 0.5556 \\ -0.1111 \\ -0.4444 \end{bmatrix}.$$

14.5 Problems

14.5.1 Problem 1

A dynamic system consists of 5 dofs; $x_1 \ldots x_5$. The mass and the stiffness matrix and the external force vector $\{F\}$ are given as

$$[M] = \begin{bmatrix} 1 & 0 & 0 & 0 & 0 \\ 0 & 1 & 0 & 0 & 0 \\ 0 & 0 & 1 & 0 & 0 \\ 0 & 0 & 0 & 1 & 0 \\ 0 & 0 & 0 & 0 & 1 \end{bmatrix}, [K] = \begin{bmatrix} 1 & -1 & 0 & 0 & 0 \\ -1 & 2 & -1 & 0 & 0 \\ 0 & -1 & 2 & -1 & 0 \\ 0 & 0 & -1 & 2 & -1 \\ 0 & 0 & 0 & -1 & 1 \end{bmatrix}, \{F\} = \begin{Bmatrix} 1 \\ 0 \\ 0 \\ 0 \\ 0 \end{Bmatrix}.$$

Questions:

1. Calculate the rigid-body motion $\{\phi_r\}$ with respect to the fifth dof x_5
2. Normalise $\{\phi_r\}$ such that $\{\phi_r\}^T[M]\{\phi_r\} = 1$

3. Calculate the matrix $[A]$
4. Calculate the relative forces $\{F_e\}$
5. Check the equilibrium of the components of $\{F_e\}$
6. Calculate the flexibility matrix $[G_{ee}]$ with respect to the fifth dof x_5
7. Calculate $[G_e]$
8. Calculate the relative displacement vector $\{x_e\}$

15 Mode Acceleration Method

15.1 Introduction

The mode acceleration method (MAM) will improve the accuracy of the responses; displacements and derivatives thereof such as element forces, stresses, etc., with respect to the mode displacement method (MDM) when a reduced set of mode shapes is used [Thomson 98, McGowan 93, Craig 68,77]. The MDM is often called the mode superposition method. The MDM may only be used for linear dynamics systems. The MAM takes the truncated modes "statically" into account.
Using the MAM, less modes may be taken into account compared to the MDM.

15.2 Decomposition of Flexibility and Mass Matrix

15.2.1 Decomposition of the Flexibility Matrix

The undamped equations of motion of a multi-degrees of freedom linear dynamic system is written as

$$[M]\{\ddot{x}\} + [K]\{x\} = \{F\}, \tag{15.1}$$

with

- $[M]$ the mass matrix
- $[K]$ the stiffness matrix
- $\{x\}$ the physical degrees of freedom
- $\{F\}$ the load vector of external forces

The eigenvalue problem becomes

$$([K]-\lambda_j[M])\{\phi_j\} = \{0\}, \tag{15.2}$$

with

- λ_j the j-th eigenvalue of the eigenvalue problem
- $\{\phi_j\}$ the j-th eigenvector (mode shape) of the eigenvalue problem

In general the eigenvectors are normalised with respect to the mass matrix $[M]$ in such a way that (orthogonality relations)

$$\left.\begin{matrix}\{\phi_i\}^T[M]\{\phi_j\}\\ \{\phi_i\}^T[K]\{\phi_j\}\end{matrix}\right\} = \left\{\begin{matrix}\delta_{ij}\\ \lambda_i\delta_{ij}\end{matrix}\right., \tag{15.3}$$

with δ_{ij} the Kronecker delta function.

With the general modal matrix $[\Phi] = [\phi_1, \phi_2, \phi_3, ..., \phi_n]$ the orthogonality relations of the modal matrix are conform (15.3):

$$\left.\begin{matrix}[\Phi]^T[M][\Phi]\\ [\Phi]^T[K][\Phi]\end{matrix}\right\} = \left\{\begin{matrix}[I]\\ \langle\lambda\rangle\end{matrix}\right., \tag{15.4}$$

with $[I]$ is the square unit matrix.

Assuming the stiffness matrix $[K]$ is not singular, we can easily derive from (15.2) that:

$$(\langle\lambda\rangle^{-1}-[K]^{-1}[M])[\Phi] = \{0\}. \tag{15.5}$$

Multiplying (15.5) from the left side by $[\Phi]^T[K]$ and from the right side by $[M][\Phi]^T$ (15.5) can be written as

$$[\Phi]^T[K]\langle\lambda\rangle^{-1}[\Phi][M][\Phi]^T-[\Phi]^T[M][\Phi][M][\Phi]^T = \{0\}. \tag{15.6}$$

After some manipulations we get

$$\langle\lambda\rangle^{-1}[\Phi]^T[K]-[M][\Phi]^T = \{0\}. \tag{15.7}$$

Premultiplying (15.7) by the modal matrix $[\Phi]$ we get

$$[\Phi]\langle\lambda\rangle^{-1}[\Phi]^T[K]-[\Phi][M][\Phi]^T = \{0\}. \tag{15.8}$$

This is equal to

$$[\Phi]\langle\lambda\rangle^{-1}[\Phi]^T[K] = [I]. \tag{15.9}$$

The inverse of the nonsingular stiffness matrix is called the flexibility matrix $[G] = [K]^{-1}$ and therefore

$$[G] = [\Phi]\langle\lambda\rangle^{-1}[\Phi]^T. \tag{15.10}$$

A linear undamped dynamic system is described by the following equation of motion

$$\begin{bmatrix} 1 & 0 & 0 \\ 0 & 2 & 0 \\ 0 & 0 & 3 \end{bmatrix} \begin{Bmatrix} \ddot{x}_1 \\ \ddot{x}_2 \\ \ddot{x}_3 \end{Bmatrix} + \begin{bmatrix} 15 & -5 & -10 \\ -5 & 10 & -5 \\ -10 & -5 & 25 \end{bmatrix} \begin{Bmatrix} x_1 \\ x_2 \\ x_3 \end{Bmatrix} = \begin{Bmatrix} 2 \\ 0 \\ 0 \end{Bmatrix}.$$

The stiffness matrix $[K]$ and the flexibility matrix $[G]$ are

$$[K] = \begin{bmatrix} 15 & -5 & -10 \\ -5 & 10 & -5 \\ -10 & -5 & 25 \end{bmatrix} \text{ and } [K]^{-1} = [G] = \begin{bmatrix} 0.1800 & 0.1000 & 0.1000 \\ 0.1400 & 0.2200 & 0.1000 \\ 0.1000 & 0.1000 & 0.1000 \end{bmatrix}.$$

The diagonal matrix of the system eigenvalues $\langle\lambda\rangle$ and the unit normalised associated eigenvectors or mode shapes $[\Phi]$, with $[\Phi]^T[M][\Phi] = [I]$, are calculated as

$$\langle\lambda\rangle = \begin{bmatrix} 1.3397 & 0 & 0 \\ 0 & 8.3333 & 0 \\ 0 & 0 & 18.6603 \end{bmatrix}, [\Phi] = \begin{bmatrix} 0.4206 & 0.2402 & 0.8749 \\ 0.5067 & -0.4804 & -0.1117 \\ 0.3212 & 0.4003 & -0.2644 \end{bmatrix}.$$

Finally the flexibility matrix $[G] = [\Phi]\langle\lambda\rangle^{-1}[\Phi]^T$ becomes:

$$[G] = \begin{bmatrix} 0.1800 & 0.1000 & 0.1000 \\ 0.1400 & 0.2200 & 0.1000 \\ 0.1000 & 0.1000 & 0.1000 \end{bmatrix}.$$

15.2.2 Decomposition of the Mass Matrix

The modal matrix $[\Phi]$ is normalised with respect to the mass matrix such that (15.4):

$$[\Phi]^T[M][\Phi] = [I].$$

Premultiplying the previous equation with the modal matrix $[\Phi]$ and postmultiplying the previous equation with $[M][\Phi]^T$ the result will be:

$$[\Phi][\Phi]^T[M][\Phi][M][\Phi]^T = [\Phi][M][\Phi]^T. \tag{15.11}$$

Using the orthogonality relations of the modal matrix with respect to the mass matrix we obtain:

$$[\Phi][\Phi]^T[M] = [I]. \tag{15.12}$$

Thus finally the inverse of the mass matrix can be calculated as:

$$[M]^{-1} = [\Phi][\Phi]^T. \tag{15.13}$$

The mass matrix is:

$$[M] = \begin{bmatrix} 1 & 0 & 0 \\ 0 & 2 & 0 \\ 0 & 0 & 3 \end{bmatrix},$$

The unit normalised modal matrix is: $[\Phi] = \begin{bmatrix} 0.4206 & 0.2402 & 0.8749 \\ 0.5067 & -0.4804 & -0.1117 \\ 0.3212 & 0.4003 & -0.2644 \end{bmatrix}$.

The reconstructed mass matrix becomes: $[M] = ([\Phi][\Phi]^T)^{-1} = \begin{bmatrix} 1 & 0 & 0 \\ 0 & 2 & 0 \\ 0 & 0 & 3 \end{bmatrix}$.

15.2.3 Convergence Properties of Reconstructed Matrices

The number of eigenvalues m of a multi-degrees of freedom linear dynamic system, consisting of n degrees of freedom is, in general, much less than the number of degrees of freedom, hence $m \ll n$.

The modal matrix $[\Phi]$ can be partitioned in the kept modes and the deleted modes:

$$[\Phi] = [\Phi_k, \Phi_d]. \tag{15.14}$$

Reconstructing the flexibility matrix $[G]$ gives

$$[G] = [\Phi_k]\langle\lambda_k\rangle^{-1}[\Phi_k]^T + [\Phi_d]\langle\lambda_d\rangle^{-1}[\Phi_d]^T = [G_k] + [G_r], \tag{15.15}$$

with

- $[G_r]$ the residual flexibility matrix

and reconstructing the inverse of the mass matrix using only the 'kept" modes will give

$$[M_k]^{-1} = [\Phi_k][\Phi_k]^T. \tag{15.16}$$

In reconstructing the flexibility matrix the eigenvalues can be found in the denominator. These eigenvalues will increase in ascending order and the influence on the participation in the flexibility matrix will decrease rapidly.

The diagonal matrix of the system eigenvalues $\langle\lambda\rangle$ and the unit normalised associated eigenvectors or mode shapes $[\Phi]$ are calculated as

$$\langle\lambda\rangle = \begin{bmatrix} 1.3397 & 0 & 0 \\ 0 & 8.3333 & 0 \\ 0 & 0 & 18.6603 \end{bmatrix}, \ [\Phi] = \begin{bmatrix} 0.4206 & 0.2402 & 0.8749 \\ 0.5067 & -0.4804 & -0.1117 \\ 0.3212 & 0.4003 & -0.2644 \end{bmatrix}.$$

The convergence of the inverse of the mass and stiffness matrices will be shown when one and two eigenvectors are taken into account. The measure of convergence is 100% if the measure of error

$$\varepsilon_M = \frac{\|\delta M^{-1}\|}{\|M^{-1}\|}, \ \varepsilon_G = \frac{\|\delta G\|}{\|G\|} = \frac{\|G_r\|}{\|G\|}, \tag{15.17}$$

with:

$\|A\|$ the norm of the matrix $[A]$. The norm of the matrix is defined as the largest singular value of the matrix $[A]$, [Strang 88]. The singular value (SVD) is closely associated with the eigenvalue-eigenvector factorisation of a symmetric matrix. The delta matrices are defined by

- $[\delta M^{-1}]$ The difference between $[M]^{-1} - [\Phi_k][\Phi_k]^T$
- $[\delta G]$ The difference between $[G] - [\Phi_k]\langle\lambda_k\rangle^{-1}[\Phi_k]^T$

Table 15.1. Convergence properties

Number of ascending modes	$\varepsilon_M = \frac{\|\delta M^{-1}\|}{\|M^{-1}\|}$ %	$\varepsilon_G = \frac{\|\delta G\|}{\|G\|}$ %
k=1	90.3	15.5
k=1&2	84.8	11.2
k=1&2&3	0.0	0.0

The convergence errors are illustrated in Table 15.1. One sees the rapid convergence of the flexibility matrix.

15.3 Mode Acceleration Method

The basis of the MAM are the damped matrix equations of motion:

$$[M]\{\ddot{x}\} + [C]\{\dot{x}\} + [K]\{x\} = \{F(t)\}, \tag{15.18}$$

with

- $[C]$ the damping matrix
- $\{x\}$ the physical degrees of freedom; displacements, velocities and acceleration
- $\{F(t)\}$ the applied dynamic loads

The MAM is, in fact, rearranging the matrix equations of motion of (15.18) in the following manner

$$\{x\} = [K]^{-1}(\{F(t)\} - [M]\{\ddot{x}\} - [C]\{\dot{x}\}). \tag{15.19}$$

Applying the MDM in linear dynamic systems the physical degrees of freedom $\{x\}$ are depicted on the modal matrix $[\Phi]$

$$\{x\} = [\Phi]\{\eta\}, \tag{15.20}$$

with

- $\{\eta\}$ the vector of generalised coordinates

In real life the displacement vector $\{x\}$ will be depicted on a reduced set of 'kept" modes $[\Phi_k]$ and the "deleted" modes $[\Phi_d]$ are not considered. The displacement vector $\{x\}$ will be approximated by

$$\{x\} = [\Phi_k \Phi_d]\begin{Bmatrix} \eta_k \\ \eta_d \end{Bmatrix} \approx [\Phi_k]\{\eta_k\}. \tag{15.21}$$

With the aid of the orthogonality relations (15.4) of the normal modes the mass and stiffness matrices can be made diagonal. Assuming the same for the damping matrix $[C]$, we get

$$[\Phi_k]^T[C][\Phi_k] = \langle 2\xi_j\omega_j \rangle, \tag{15.22}$$

and with $\lambda_j = \omega_j^2$ the coupled equation of motions can be expressed in uncoupled equations of motion for the generalised coordinates

$$\ddot{\eta}_j + 2\xi_j\omega_j\dot{\eta} + \omega_j^2\eta = \{\phi_j\}^T\{F(t)\}. \tag{15.23}$$

We can now write

$$\ddot{\eta}_j = \{\phi_j\}^T\{F(t)\} - 2\xi_j\omega_j\dot{\eta} - \omega_j^2\eta\,. \tag{15.24}$$

Equation (15.19) can be written

$$\{x\} = [K]^{-1}\{F(t)\} - [K]^{-1}[M][\Phi_k]\{\ddot{\eta}\} - [K]^{-1}[C][\Phi_k]\{\dot{\eta}\}\,. \tag{15.25}$$

From (15.5) it follows that

$$[K]^{-1}[M][\Phi_k] = \langle\lambda_k\rangle^{-1}[\Phi_k]\,, \tag{15.26}$$

and (15.10) and (15.22) yield

$$[K]^{-1}[C][\Phi_k] = [\Phi_k]\langle\lambda_k\rangle^{-1}\langle 2\xi_k\omega_k\rangle\,. \tag{15.27}$$

after substituting, (15.26) and (15.27) into (15.25) it becomes

$$\{x\} = ([G] - [\Phi_k]\langle\lambda_k\rangle^{-1}[\Phi_k]^T)\{F(t)\} + [\Phi_k]\{\eta_k\}. \tag{15.28}$$

Finally, we obtain the MAM equation

$$\{x\} = [G_r]\{F(t)\} + [\Phi_k]\{\eta_k\}. \tag{15.29}$$

The displacement vector $\{x\}$ consists of the MDM plus a static contribution of all "deleted "modes $[\Phi_d]$. Quite often (15.29) is written as

$$\{x\} = [\Psi_{MAM}] + [\Phi_k]\{\eta_k\}\,, \tag{15.30}$$

with

- $[G_r]$ the residual flexibility matrix
- $[\psi_{MAM}] = [G_r]\{F(t)\}$ the "residual flexibility attachment modes", in fact, the MAM correction on the MDM

In [Chung 98] an efficient manner to create the residual attachment modes is suggested. Using [Chung 98] the displacement vector $\{x\}$ becomes

$$\{x\} = [\hat{\psi}_{MAM}]\{F(t)\} + [\Phi_k]\{\eta_k\}\,, \tag{15.31}$$

with

$$[\hat{\Psi}_{MAM}] = [G_r][T]\,, \tag{15.32}$$

and with

- $[\hat{\Psi}_{MAM}]$ the "Chung" residual flexibility attachment modes.
- $[T]$ the square load-distribution matrix, with unit loads in the columns at locations where the loads $\{F(t)\}$ are applied.

To generate the residual attachment modes the following procedure can be followed:

1. Solve the linear system $[G] = [K]^{-1}[I]$
2. Generate the flexibility matrix $[G_k] = [\Phi_k]\langle\lambda_k\rangle^{-1}[\Phi_k]^T$ with $[\Phi_k]^T[M][\Phi_k] = [I]$
3. Generate the residual flexibility matrix $[G_r] = [G]-[\Phi_k]\langle\lambda_k\rangle^{-1}[\Phi_k]^T$
4. Generate the load distribution matrix $[T]$
5. Finally generate the "Chung" attachment modes $[\hat{\Psi}_{MAM}] = [G_r][T]$

A linear undamped dynamic system is described by the following matrix equation of motion

$$\begin{bmatrix} 1 & 0 & 0 \\ 0 & 2 & 0 \\ 0 & 0 & 3 \end{bmatrix} \begin{Bmatrix} \ddot{x}_1 \\ \ddot{x}_2 \\ \ddot{x}_3 \end{Bmatrix} + \begin{bmatrix} 15 & -5 & -10 \\ -5 & 10 & -5 \\ -10 & -5 & 25 \end{bmatrix} \begin{Bmatrix} x_1 \\ x_2 \\ x_3 \end{Bmatrix} = \begin{Bmatrix} 2 \\ 0 \\ 0 \end{Bmatrix}.$$

The load vector and the load-distribution matrix are

$$\{F\} = \begin{Bmatrix} 2 \\ 0 \\ 0 \end{Bmatrix} \text{ and } [T] = \begin{bmatrix} 1 & 0 & 0 \\ 0 & 0 & 0 \\ 0 & 0 & 0 \end{bmatrix}.$$

The first eigenvalue λ_1 and the unit normalised associated first eigenvector or mode shapes $\{\phi_1\}$ are calculated as

$$\lambda_1 = 1.3397\,,\ \{\phi_1\} = \begin{Bmatrix} 0.4206 \\ 0.5067 \\ 0.3212 \end{Bmatrix}.$$

The residual-flexibility matrix

$$[G_r] = [G] - \frac{\{\phi_1\}\{\phi_1\}^T}{\lambda_1} = \begin{bmatrix} 0.0479 & -0.0191 & -0.0009 \\ -0.0191 & 0.0284 & -0.0215 \\ -0.0009 & -0.0215 & 0.0230 \end{bmatrix}.$$

The residual-flexibility attachment modes; classical and according to [Chung 98] become

$$\{\Psi_{MAM}\} = \begin{Bmatrix} 0.0959 \\ -0.0382 \\ -0.0017 \end{Bmatrix} \text{ and } \{\hat{\Psi}_{MAM}\} = \begin{bmatrix} 0.0479 & 0 & 0 \\ -0.0191 & 0 & 0 \\ -0.0009 & 0 & 0 \end{bmatrix}.$$

The following example illustrates the influence of the MAM. The transient responses of the dynamic system, consisting of three degrees of freedom, when the first degree of freedom is exposed to a constant force, will be calculated:

1. The complete system, i.e. all three modes included
2. The first mode only
3. The first mode only with MAM correction

The transient responses are calculated numerically with the aid of the Wilson-θ method [D'Souza 84], with $\theta = 1.4$. The initial displacements and velocities at $t = 0$ are zero.

The undamped equations of motion of the 3 mass–spring system are

$$\begin{bmatrix} 1 & 0 & 0 \\ 0 & 1 & 0 \\ 0 & 0 & 1 \end{bmatrix} \begin{Bmatrix} \ddot{x}_1 \\ \ddot{x}_2 \\ \ddot{x}_3 \end{Bmatrix} + \begin{bmatrix} 1 & -1 & 0 \\ -1 & 2 & -1 \\ 0 & -1 & 2 \end{bmatrix} \begin{Bmatrix} x_1 \\ x_2 \\ x_3 \end{Bmatrix} = \begin{Bmatrix} 1 \\ 0 \\ 0 \end{Bmatrix},$$

or

$$[M]\{\ddot{x}\} + [K]\{x\} = \{F(t)\}.$$

The force vector $\{F(t)\} = 0$ when $t < 0$ and $\{F(t)\} = \begin{Bmatrix} 1 \\ 0 \\ 0 \end{Bmatrix}$ when $t \geq 0$.

The damping matrix $[C]$ will be introduced as modal damping after the equations of motion are uncoupled using the orthogonality relations of the modes, $[\Phi_k]^T[C][\Phi_k] = \langle 2\xi_j\omega_j \rangle$.

The diagonal matrix of the system eigenvalues $\langle \lambda \rangle$ and the unit normalised associated eigenvectors or mode shapes $[\Phi]$ are calculated as

$$\langle \lambda \rangle = \begin{bmatrix} 0.1981 & 0 & 0 \\ 0 & 1.5550 & 0 \\ 0 & 0 & 3.2470 \end{bmatrix}, \ [\Phi] = \begin{bmatrix} 0.7370 & -0.5910 & 0.3280 \\ 0.5910 & 0.3280 & -0.7370 \\ 0.3280 & 0.4003 & 0.5910 \end{bmatrix}.$$

The Wilson-θ procedure is used with $\{x\} = [\Phi]\{\eta\}$, the generalised mass matrix $[mg] = [\Phi]^T[M][\Phi] = [I]$, the generalised stiffness matrix $[kg] = [\Phi]^T[K][\Phi] = \langle\lambda_k\rangle = \langle\omega^2{}_k\rangle$, $\{fg\} = [\Phi]^T\{F(t)\}$ and the diagonal modal damping matrix $[cg] = 2\xi\langle\omega_k\rangle$. The following activities will be done:

- Solve the acceleration $\{\ddot{x}(0)\} = [\Phi][mg]^{-1}[\Phi]^T\{F(0)\}$. The initial values for the generalised coordinates can be calculated with
  $\{\eta(0)\} = ([\Phi]^T[\Phi])^{-1}[\Phi]^T\{x(0)\}$.
- Set up the effective stiffness with a time step Δt. With a constant Δt the effective stiffness is constant, $[\bar{k}] = \dfrac{6}{\theta^2\Delta t^2}[mg] + \dfrac{3}{\theta\Delta t}[cg] + [kg]$.
- Set up the effective force vector

$$\{\bar{f}(t+\theta\Delta t)\} = \{fg(t+\theta\Delta t)\} + \left(\frac{6}{\theta^2\Delta t^2}[mg] + \frac{3}{\theta\Delta t}[cg]\right)\{\eta(t)\}$$

$$+\left(\frac{6}{\theta\Delta t}[mg] + 2[cg]\right)\{\dot{\eta}(t)\} + \left(2[mg] + \frac{\theta\Delta t}{2}[cg]\right)\{\ddot{\eta}(t)\}.$$

- Solve $[\bar{k}]\{\eta(t+\theta\Delta t)\} = \{\bar{f}(t+\theta\Delta t)\}$.
- Calculate the acceleration, velocities and displacements for the generalised coordinates

$$\{\ddot{\eta}(t+\Delta t)\} = \frac{6}{\theta^3\Delta t^2}(\{\eta(t+\theta\Delta t)\} - \{\eta(t)\}) - \frac{6}{\theta^2\Delta t}\{\dot{\eta}(t)\} + \left(1-\frac{3}{\theta}\right)\{\ddot{\eta}(t)\}$$

$$\{\dot{\eta}(t+\Delta t)\} = \{\dot{\eta}(t)\} + \frac{\Delta t}{2}(\{\ddot{\eta}(t+\Delta t)\} + \{\ddot{\eta}(t)\})$$

$$\{\eta(t+\Delta t)\} = \{\eta(t)\} + \Delta t\{\dot{\eta}(t)\} + \frac{\Delta t^2}{6}(\{\ddot{\eta}(t+\Delta t)\} + 2\{\ddot{\eta}(t)\}).$$

- Calculate the physical displacement, etc.

$\{x\} = [\Psi_{MAM}] + [\Phi_k]\{\eta_k\}$ or $\{x\} = [\hat{\psi}_{MAM}]\{F(t)\} + [\Phi_k]\{\eta_k\}$,
$\{\dot{x}\} = [\Phi_k]\{\dot{\eta}_k\}$ and $\{\ddot{x}\} = [\Phi_k]\{\ddot{\eta}_k\}$.

The simulation time is 200 s with a time increment $\Delta t = 0.05$ s and the modal damping ratio for all modes is $\xi = 0.05$.

Solution question 1; all 3 modes taken into account.
When the vibrations are damped out the displacements will converge to the static displacements

$$\begin{Bmatrix} x_1 \\ x_2 \\ x_3 \end{Bmatrix} = [G]\{F\} = \frac{1}{3}\begin{Bmatrix} 3 \\ 2 \\ 1 \end{Bmatrix}$$

The plots of the displacements $\{x\}$ are given in Fig. 15.1.

Solution question 2; one mode taken into account (no MAM).
When the vibrations are damped out the displacements do not converge to the static displacements

$$\begin{Bmatrix} x_1 \\ x_2 \\ x_3 \end{Bmatrix} = [G]\{F\} \neq \frac{1}{3}\begin{Bmatrix} 3 \\ 2 \\ 1 \end{Bmatrix}$$

The plots of the displacements $\{x\}$ are given in Fig. 15.2.

.

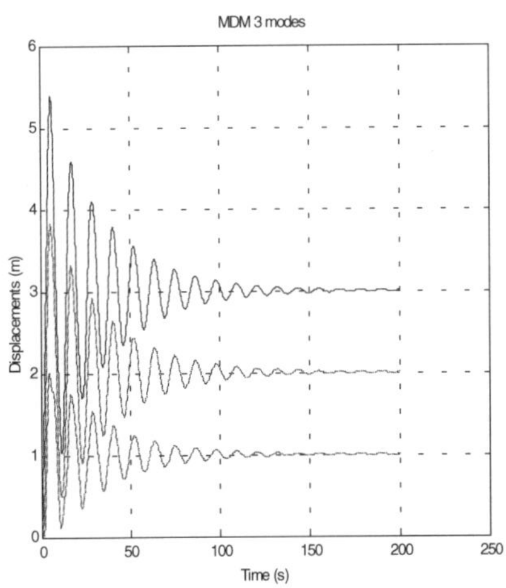

Fig. 15.1. Displacements, MDM, 3 modes

Solution question 3; one mode taken into account and MAM.
When the vibration is damped out the displacements converge to the static displacements

$$\begin{Bmatrix} x_1 \\ x_2 \\ x_3 \end{Bmatrix} = [G]\{F\} = \frac{1}{3}\begin{Bmatrix} 3 \\ 2 \\ 1 \end{Bmatrix}.$$

The plots of the displacements $\{x\}$ are given in Fig. 15.3.

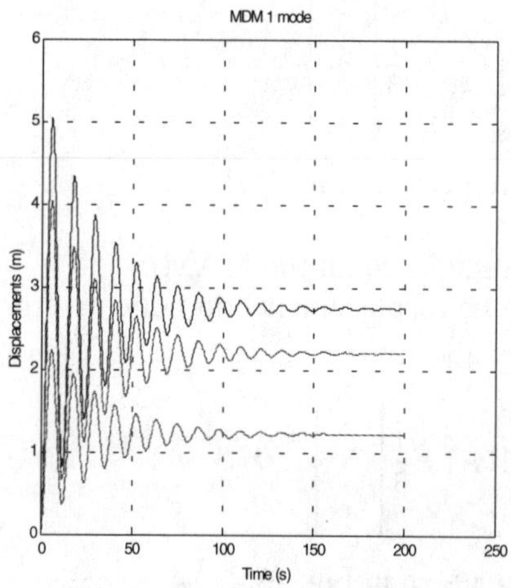

Fig. 15.2. Displacements, MDM, 1 mode

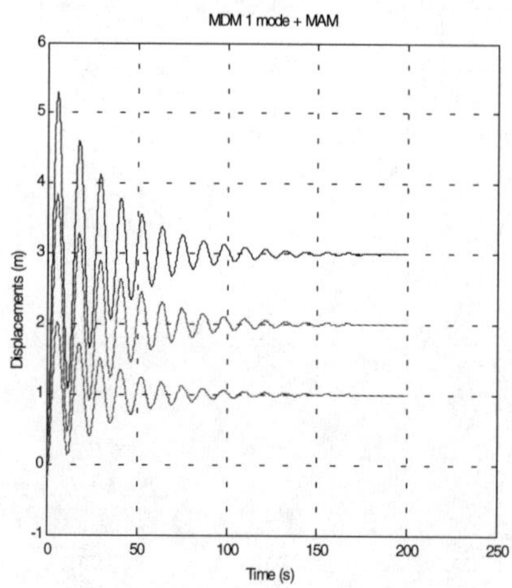

Fig. 15.3. Displacements, MDM, 1 mode + MAM

15.4 Problems

15.4.1 Problem 1

The complete set of modes is given by the modal matrix $[\Phi]$. The modes or vibration modes are orthogonal with respect to the mass matrix $[M]$, such that $[\Phi]^T[M][\Phi] = [I]$. Prove that $[M][\Phi][\Phi]^T = [I]$.

The 6 motions as a rigid-body of a linear elastic body with respect to a point A, redundant constrained, are denoted by $[\Phi_r]$. The elastic modes with respect to point A are denoted with $[\Phi_e]$. The matrix of the modal participation factor is given by $\{L_{ij}\}^T = \{\phi_{r,i}\}^T[M]\{\phi_{e,j}\}$, i=1,..,6 and j=1,2,...,m. The modal effective mass is given by $[M_{eff,j}] = \{L_{ij}\}^T\{L_{ij}\}$. Prove that $\sum_j [M_{eff,j}] = [\Phi_r]^T[M][\Phi_r] = [M_r]$, the mass matrix as a rigid body with respect to point A, if all modes are taken into account.

15.4.2 Problem 2

A linear dynamic system consists of three degrees of freedom and has the following characteristics [Lutes 96]:

- The mass matrix is given by $[M] = m\begin{bmatrix} 1 & 0 & 0 \\ 0 & 1 & 0 \\ 0 & 0 & 1 \end{bmatrix}$.
- The modal matrix is given by: $[\Phi] = [\phi_1,\phi_2,\phi_3] = \begin{bmatrix} 3 & 2 & 1 \\ 5 & 0 & -3 \\ 6 & -1 & 2 \end{bmatrix}$, with $[\Phi]^T[M][\Phi] \neq [I]$.
- The natural frequencies of the dynamic system are $[\omega_1,\omega_2,\omega_3] = \left[\sqrt{\frac{k}{m}}, 2\sqrt{\frac{k}{m}}, 3\sqrt{\frac{k}{m}}\right]$ with m (mass) and k (stiffness) as scalar values.
- The modal viscous damping is $\lfloor \xi_1,\xi_2,\xi_3 \rfloor = \lfloor 0.05, 0.05, 0.05 \rfloor$.

Reconstruct the following matrices:

- The stiffness matrix $[K]$
- The damping matrix $[C]$
- Sketch the three mass–damper–spring system

15.4.3 Problem 3

A linear dynamic system consists of three degrees of freedom and has the following characteristics [Lutes 96]:

- The mass matrix is given by $[M] = m\begin{bmatrix}1 & 0 & 0\\0 & 1 & 0\\0 & 0 & 1\end{bmatrix}$.
- The modal matrix is given by: $[\Phi] = [\phi_1,\phi_2,\phi_3] = \begin{bmatrix}2 & 9 & 2\\5 & 0 & -17\\9 & -2 & 9\end{bmatrix}$, with $[\Phi]^T[M][\Phi] \neq [I]$.
- The natural frequencies of the dynamic system are $[\omega_1,\omega_2,\omega_3] = \left[\sqrt{\frac{k}{m}},\sqrt{\frac{3k}{m}},\sqrt{\frac{5k}{m}}\right]$ with m (mass) and k (stiffness) as scalar values.
- The modal viscous damping is $\lfloor\xi_1,\xi_2,\xi_3\rfloor = \lfloor 0.01,0.01,0.02\rfloor$.

Reconstruct the following matrices:

- The stiffness matrix $[K]$
- The damping matrix $[C]$
- Sketch the three mass–damper–spring system

15.4.4 Problem 4

A rigid bar has a length L and a second moment of intertia I_o. The bar is hinged at A and in the middle supported by a spring with spring stiffness k_1. At B a mass spring system is connected to the bar with mass M and a spring stiffness k_2. The system is illustrated in Fig. 15.4.

- Set up the equations of motion using the degrees of freedom φ and u.
- Calculate the natural frequencies, generalised masses and stiffness and the undamped modes with: I_o=3 kgm^2, L=8 m, k_1=300 N/m and k_2=200

N/m and M=2 kg. Normalise the modal matrix $[\Phi]$ such that $[\Phi]^T[M][\Phi] = [I]$.

- Calculate ε_M and ε_G (see Table 15.1) with
  - the first mode $\{\phi_1\}$
  - the complete modal matrix $[\Phi]$.
- Reconstruct $[M]$ and $[K]$ with a complete modal base $[\Phi]$.

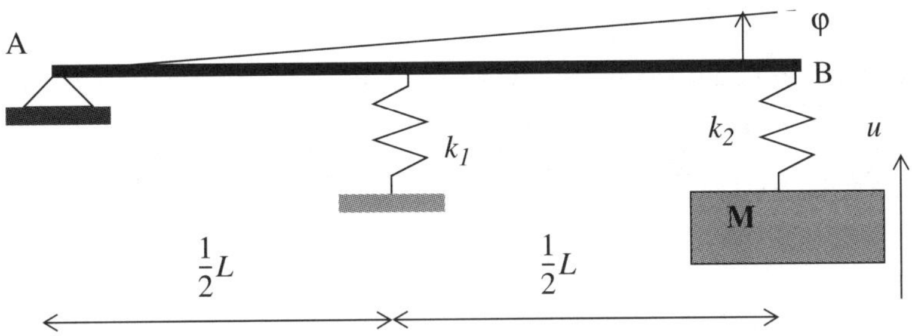

Fig. 15.4. Dynamic system

15.4.5 Problem 5

Calculate the transient responses of the dynamic system, consisting of three degrees of freedom, when the first degree of freedom is exposed to a constant force

1. The complete system, i.e. all three modes included
2. The first mode only
3. The first mode only with MAM correction

The MDM must be applied.

The transient responses must be calculated numerically with the aid of the 'Park stiff stable" method [D'Souza 84, Park 75]. The initial displacements and velocities at $t = 0$ are zero.

The undamped equations of motion of the 3 mass–spring system are

$$\begin{bmatrix} 1 & 0 & 0 \\ 0 & 1 & 0 \\ 0 & 0 & 1 \end{bmatrix} \begin{Bmatrix} \ddot{x}_1 \\ \ddot{x}_2 \\ \ddot{x}_3 \end{Bmatrix} + \begin{bmatrix} 10 & -10 & 0 \\ -10 & 20 & -10 \\ 0 & -10 & 20 \end{bmatrix} \begin{Bmatrix} x_1 \\ x_2 \\ x_3 \end{Bmatrix} = \begin{Bmatrix} 10 \\ 0 \\ 0 \end{Bmatrix}$$

or

$$[M]\{\ddot{x}\} + [K]\{x\} = \{F(t)\}.$$

The force vector $\{F(t)\} = 0$ when $t < 0$ and $\{F(t)\} = \begin{Bmatrix} 1 \\ 0 \\ 0 \end{Bmatrix}$ when $t \geq 0$.

The damping matrix $[C]$ will be introduced as modal damping after the equations of motion are uncoupled using the orthogonality relations of the modes, $[\Phi_k]^T[C][\Phi_k] = \langle 2\xi_j\omega_j \rangle$.

Use the Park procedure below with $\{x\} = [\Phi]\{\eta\}$, the generalised mass matrix $[mg] = [\Phi]^T[M][\Phi] = [I]$, the generalised stiffness matrix $[kg] = [\Phi]^T[K][\Phi] = \langle \lambda_k \rangle = \langle \omega^2{}_k \rangle$, $\{fg\} = [\Phi]^T\{F(t)\}$ and the diagonal modal damping matrix $[cg] = 2\xi\langle \omega_k \rangle$. The follwing steps will be followed:

- Solve the acceleration $\{\ddot{x}(0)\} = [\Phi][mg]^{-1}[\Phi]^T\{F(0)\}$. The initial values for the generalised coordinates can be calculated with $\{\eta(0)\} = ([\Phi]^T[\Phi])^{-1}[\Phi]^T\{x(0)\}$, and $\{\dot{\eta}(0)\} = ([\Phi]^T[\Phi])^{-1}[\Phi]^T\{\dot{x}(0)\}$.
- Set up the effective stiffness matrix with a time step Δt. With a constant Δt the effective stiffness matrix is constant.
$$[\bar{k}] = \frac{100}{36\Delta t^2}[mg] + \frac{10}{6\Delta t}[cg] + [kg].$$
- Set up the effective force vector

$$\{\bar{f}(t+\theta\Delta t)\} = \{fg(t+\Delta t)\} + \frac{15}{6\Delta t}[mg]\dot{\eta}(t) - \frac{1}{\Delta t}[mg]\{\dot{\eta}(t-\Delta t)\} + \frac{1}{6\Delta t}[mg]\{\dot{\eta}(t-2\Delta t)\} + \left(\frac{150}{36\Delta t^2}[mg] + \frac{15}{6\Delta t}[cg]\right)\{\eta(t)\} - \left(\frac{10}{6\Delta t^2}[mg] + \frac{1}{\Delta t}[cg]\right)\{\eta(t-\Delta t)\} + \left(\frac{1}{36\Delta t^2}[mg] + \frac{1}{6\Delta t}[cg]\right)\{\eta(t-2\Delta t)\}.$$

- Solve $[\bar{k}]\{\eta(t+\Delta t)\} = \{\bar{f}(t+\Delta t)\}$.
- Calculate the acceleration and velocities for the generalised coordinates

$$\{\ddot{\eta}(t+\Delta t)\} = \frac{1}{6\Delta t}[10\{\dot{\eta}(t+\Delta t)\} - 15\{\dot{\eta}(t)\} + 6\{\dot{\eta}(t-\Delta t)\} - \{\dot{\eta}(t-2\Delta t)\}]$$

$$\{\dot{\eta}(t+\Delta t)\} = \frac{1}{6\Delta t}[10\{\eta(t+\Delta t)\} - 15\{\eta(t)\} + 6\{\eta(t-\Delta t)\} - \{\eta(t-2\Delta t)\}].$$

- It is advised to start with the Newmark-beta method to obtain $\{\eta(\Delta t)\}$, $\{\eta(2\Delta t)\}$, $\{\dot{\eta}(\Delta t)\}$, and $\{\dot{\eta}(2\Delta t)\}$ to use Park's method to calculate $\{\eta(3\Delta t)\}$, etc.
- Calculate physical displacements, etc.

$\{x\} = [\Psi_{MAM}] + [\Phi_k]\{\eta_k\}$ or $\{x\} = [\hat{\psi}_{MAM}]\{F(t)\} + [\Phi_k]\{\eta_k\}$,

$\{\dot{x}\} = [\Phi_k]\{\dot{\eta}_k\}$ and $\{\ddot{x}\} = [\Phi_k]\{\ddot{\eta}_k\}$.

The simulation time is 200 s with a time increment $\Delta t = 0.05\text{ s}$ and the modal damping ratio for all modes is $\xi = 0.02$.

16 Residual Vectors

16.1 Introduction

Residual vectors have been discussed by John Dickens and Ted Rose in [Dickens 00, Rose 91]. The modal base, when the modal displacement method [MDM] is applied, will be extended by residual vectors to account for the deleted modes. This method is quite similar to the mode acceleration method (MAM). Dickens proposed to construct a static mode (displacement) with respect to the boundaries based on the residual loads. Rose constructed a static mode, again with respect to the posed boundary conditions, however, based on the static part of the dynamic loads.

Since the residual vectors are treated as modes, they will have associated modal mass, modal stiffness and damping. With the aid of artificial damping the responses due to the residual vectors will be minimised.

In this chapter two methods of residual vectors will be discussed:

- The method proposed by John Dickens, [Dickens 00].
- The method discussed by Ted Rose, [Rose 91].

16.2 Residual Vectors

16.2.1 Dickens Method

The eigenvalue problem of an undamped mdof system is written

$$([K]-\omega_i^2[M])\{\phi_i\} = \{0\} \tag{16.1}$$

$$[K]\{\phi_i\} = \omega_i^2[M]\{\phi_i\}\,. \tag{16.2}$$

The modal static force can be expressed as

$$\{F_i\} = [K]\{\phi_i\}\eta_i, \tag{16.3}$$

or using (16.1) we may write

$$\{F_i\} = \omega_i^2[M]\{\phi_i\}\eta_i. \tag{16.4}$$

The total static force is

$$\{F\} = [K]\{x\} = [K][\Phi]\{\eta\}. \tag{16.5}$$

If (16.5) is premultiplied by $[\Phi]^T$ we obtain

$$[\Phi]^T\{F\} = [\Phi]^T[K][\Phi]\{\eta\}. \tag{16.6}$$

Using the orthogonality relations of the mode shapes with respect to the mass matrix $[M]$ and the stiffness matrix $[K]$;

$$[\Phi]^T[K][\Phi] = \langle\omega_i^2 m\rangle, \quad [\Phi]^T[M][\Phi] = \langle m\rangle$$

Equation (16.6) can be written as

$$\{\eta\} = \langle\omega_i^2 m\rangle^{-1}[\Phi]^T\{F\} \tag{16.7}$$

or for one generalised coordinate η_i we achieve

$$\eta_i = \frac{1}{\omega_i^2 m}\{\phi_i\}^T\{F\}. \tag{16.8}$$

By substituting (16.8) into (16.4) the static modal force $\{F_i\}$ of (16.4), becomes

$$\{F_i\} = \omega_i^2[M]\{\phi_i\}\eta_i = \omega_i^2[M]\{\phi_i\}\frac{1}{\omega_i^2 m}\{\phi_i\}^T\{F\} = \frac{1}{m}[M]\{\phi_i\}(\{\phi_i\}^T\{F\}). \tag{16.9}$$

The summation of all modal forces $\sum_i\{F_i\}$ now becomes

$$\sum_i\{F_i\} = [M][\Phi]\langle m\rangle^{-1}[\Phi]^T\{F\}. \tag{16.10}$$

If not all modes are taken into account in the modal displacement method (MDM) the residual load $\{F_{\text{res}}\}$ is

$$\{F_{\text{res}}\} = \{F\} - \sum_i\{F_i\} = ([I] - [M][\Phi]\langle m\rangle^{-1}[\Phi]^T)\{F\}. \tag{16.11}$$

If the modal base $[\Phi]$ is complete, $[\Phi]^{-1}$ and $[M]^{-1}$ do exist, and with

$$([\Phi]^T[M][\Phi])^{-1} = \langle m\rangle^{-1} = [\Phi]^{-1}[M]^{-1}[\Phi]^{-T} \tag{16.12}$$

equation (16.11) becomes

$$\{F_{\text{res}}\} = ([I]-[M][\Phi][\Phi]^{-1}[M]^{-1}[\Phi]^{-T}[\Phi]^T)\{F\} = 0\,. \tag{16.13}$$

We will now construct the displacement vectors $[x_{res}]$ that are based on the residual load vectors $[F_{\text{res}}]$. We now assume multiple-load cases $[F_{\text{res}}]$.

$$[K][x_{\text{res}}] = [F_{\text{res}}]\,, \tag{16.14}$$

with

$$[F_{\text{res}}] = ([I]-[M][\Phi][m]^{-1}[\Phi]^T)[F]\,. \tag{16.15}$$

The stiffness matrix is regular because the boundary conditions, to prevent rigid body modes, are implemented. The vectors $[x_{\text{res}}]$ will be made orthogonal with respect to the modal base $[\Phi]$ and indicated by $[\phi_{\text{res}}]$. This can be done as follows [Dickens 00]

$$[\tilde{K}] = [x_{\text{res}}]^T[K][x_{\text{res}}] \text{ and } [\tilde{M}] = [x_{\text{res}}]^T[M][x_{\text{res}}]\,,$$

and solving the eigenvalue problem

$$[\tilde{K}]\{\psi\} = \langle\lambda\rangle[\tilde{M}]\{\psi\} \tag{16.16}$$

and, furthermore

$$[\phi_{\text{res}}] = [x_{\text{res}}]\{\psi\}\ . \tag{16.17}$$

The modal base $[\Phi]$ will be augmented by the vectors $[\phi_{res}]$, thus we get a new modal base $[\Psi]$

$$[\Psi] = ([\Phi],[\phi_{\text{res}}])\,. \tag{16.18}$$

This modal base $[\Psi]$ is used to apply the modal displacement method. The linear mdof dynamic systems is represented by the following matrix equations of motion

$$[M]\{\ddot{x}(t)\}+[C]\{\dot{x}(t)\}+[K]\{x(t)\} = \{F(t)\}\,. \tag{16.19}$$

The coupled linear equations can be decoupled using the mode displacement method (MDM) or mode superposition method. The physical displacement vector $x(t)$ is expressed as follows

$$x(t) = [\Psi]\{\eta(t)\}\,,$$

with

- $\{\eta(t)\}$ The vector of generalised coordinates.

16.2.2 Rose Method

The static residual vectors, as proposed by [Rose 91] are based on the static part of the dynamic loads $[F] = [\{F_1\}, \{F_2\}, ...]$. (16.14) is now written as

$$[K][x_{res}] = [F] \tag{16.20}$$

The stiffness matrix is regular because the boundary conditions, to prevent rigid- body modes, are implemented. The vectors $[x_{res}]$ will be made orthogonal with respect to the modal base $[\Phi]$. The new modal base $[\Psi]$ consists of

$$[\Psi] = ([\Phi], [x_{res}]). \tag{16.21}$$

The orthogalisation procedure of the modal base $[\Psi]$ will now be described. The following eigenvalue problem will be solved

$$[\tilde{K}]\{\Upsilon\} = \langle\lambda\rangle[\tilde{M}]\{\Upsilon\}, \tag{16.22}$$

with

- $[\tilde{K}] = [\Psi]^T[K][\Psi]$ and
- $[\tilde{M}] = [\Psi]^T[M][\Psi]$

The solution of the eigenvalue problem of (16.22) will result in the original modes, plus new (high-frequency) pseudomodes. The new modal base becomes

$$[\chi] = [\Psi]\{\Upsilon\}. \tag{16.23}$$

The physical displacement vector $\{x(t)\}$ is expressed as follows

$$\{x(t)\} = [\chi]\{\eta(t)\}. \tag{16.24}$$

The following examples illustrate the influence of the residual vectors. The transient responses of the dynamic system, consisting of three degrees of freedom, when the first degree of freedom is exposed to a constant force, will be calculated:

1. The complete system, i.e. all three modes included
2. The first mode only
3. The first mode only with residual vector correction [Dickens 00]

The transient responses are calculated numerically with the aid of the 'Newmark-beta" method [Wood 90], with $\gamma = 0.5$ and $\beta = 0.25$. The initial displacements and velocities at $t = 0$ are zero.

The undamped equations of motion of the 3 mass-spring system are

$$\begin{bmatrix} 1 & 0 & 0 \\ 0 & 1 & 0 \\ 0 & 0 & 1 \end{bmatrix} \begin{Bmatrix} \ddot{x}_1 \\ \ddot{x}_2 \\ \ddot{x}_3 \end{Bmatrix} + \begin{bmatrix} 1 & -1 & 0 \\ -1 & 2 & -1 \\ 0 & -1 & 2 \end{bmatrix} \begin{Bmatrix} x_1 \\ x_2 \\ x_3 \end{Bmatrix} = \begin{Bmatrix} 1 \\ 0 \\ 0 \end{Bmatrix},$$

or

$$[M]\{\ddot{x}\} + [K]\{x\} = \{F(t)\}.$$

The force vector $\{F(t)\} = 0$ when $t < 0$ and $\{F(t)\} = \begin{Bmatrix} 1 \\ 0 \\ 0 \end{Bmatrix}$ when $t \ge 0$.

The damping matrix $[C]$ will be introduced as modal damping after the equations of motion are decoupled using the orthogonality relations of the modes, $[\Phi_k]^T[C][\Phi_k] = \langle 2\xi_j\omega_j \rangle$

The diagonal matrix of the system eigenvalues $\langle \lambda \rangle$ and the unit normalised associated eigenvectors or mode shapes $[\Phi]$ are calculated as

$$\langle \lambda \rangle = \begin{bmatrix} 0.1981 & 0 & 0 \\ 0 & 1.5550 & 0 \\ 0 & 0 & 3.2470 \end{bmatrix}, \quad [\Phi] = \begin{bmatrix} 0.7370 & -0.5910 & 0.3280 \\ 0.5910 & 0.3280 & -0.7370 \\ 0.3280 & 0.4003 & 0.5910 \end{bmatrix}.$$

The Newmark-beta procedure is as with $\{x\} = [\Phi]\{\eta\}$, the generalised mass matrix $[mg] = [\Phi]^T[M][\Phi] = [I]$, the generalised stiffness matrix $[kg] = [\Phi]^T[K][\Phi] = \langle \lambda_k \rangle = \langle \omega^2{}_k \rangle$, $[fg] = [\Phi]^T[F(t)]$ and the diagonal modal damping matrix is $[cg] = 2\xi\langle \omega_k \rangle$. The following activities shall be carried out:

- Solve the acceleration $\{\ddot{x}(0)\} = [\Phi][mg]^{-1}[\Phi]^T\{F(0)\}$. The initial values for the generalised coordinates can be calculated with
  $\{\eta(0)\} = ([\Phi]^T[\Phi])^{-1}[\Phi]^T\{x(0)\}$ and
  $\{\dot{\eta}(0)\} = ([\Phi]^T[\Phi])^{-1}[\Phi]^T\{\dot{x}(0)\}$.
- Set up the effective stiffness matrix with a time step Δt. With a constant Δt the effective stiffness matrix is constant.

$$[\bar{k}] = \frac{1}{\beta\Delta t^2}[mg] + \frac{\gamma}{\beta\Delta t}[cg] + [kg].$$

- Set up the effective force vector

$$\{\bar{f}(t+\Delta t)\} = \{fg(t+\Delta t)\} + \left(\left(\frac{1}{2\beta} - 1\right)[mg] + \Delta t\left(\frac{\gamma}{2\beta} - 1\right)[cg]\right)\{\ddot{\eta}(t)\}$$

$$+\left(\frac{1}{2\Delta t}[mg] + \left(\frac{\gamma}{\beta} - 1\right)[cg]\right)\{\dot{\eta}(t)\} + \left(\frac{1}{\beta\Delta t^2}[mg] + \frac{\gamma}{\beta\Delta t}[cg]\right)\{\eta(t)\}.$$

- Solve $[\bar{k}]\{\eta(t+\Delta t)\} = \{\bar{f}(t+\Delta t)\}$.
- Calculate the accelerations and the velocities for the generalised coordinates with

$$\{\ddot{\eta}(t+\Delta t)\} = \frac{1}{\beta\Delta t^2}(\{\eta(t+\Delta t)\} - \{\eta(t)\}) - \frac{1}{\beta\Delta t}\{\dot{\eta}(t)\} - \left(\frac{1}{2\beta} - 1\right)\{\ddot{\eta}(t)\}$$

$$\{\dot{\eta}(t+\Delta t)\} = \frac{\gamma}{\beta\Delta t}(\{\eta(t+\Delta t)\} - \{\eta(t)\}) - \left(\frac{1}{2\beta} - 1\right)\{\dot{\eta}(t)\} - \Delta t\left(\frac{\gamma}{2\beta} - 1\right)\{\ddot{\eta}(t)\}.$$

- Calculate physical displacements, etc.

$\{x\} = ([\Phi_k][\phi_{res}])\{\eta_k\}$, $\{\dot{x}\} = ([\Phi_k][\phi_{res}])\{\dot{\eta}_k\}$ and $\{\ddot{x}\} = ([\Phi_k][\phi_{res}])\{\ddot{\eta}_k\}$.

The simulation time is 200 s with a time increment $\Delta t = 0.05$ s and the modal damping ratio for the elastic modes is $\xi = 0.05$ and for the residual vector $\xi = 0.05$

Solution question 1; all 3 modes taken into account.
When the vibrations are damped out the displacements will converge to the static displacements

$$\begin{Bmatrix} x_1 \\ x_2 \\ x_3 \end{Bmatrix} = [G]\{F\} = \frac{1}{3}\begin{Bmatrix} 3 \\ 2 \\ 1 \end{Bmatrix}.$$

The plots of the displacements $\{x\}$ are given in Fig. 16.1

Solution question 2; one mode taken into account (no residual vector).
When the vibrations are damped out the displacements do not converge to the static displacements

$$\begin{Bmatrix} x_1 \\ x_2 \\ x_3 \end{Bmatrix} = [G]\{F\} \neq \frac{1}{3}\begin{Bmatrix} 3 \\ 2 \\ 1 \end{Bmatrix}.$$

The plots of the displacements $\{x\}$ are given in Fig. 16.2

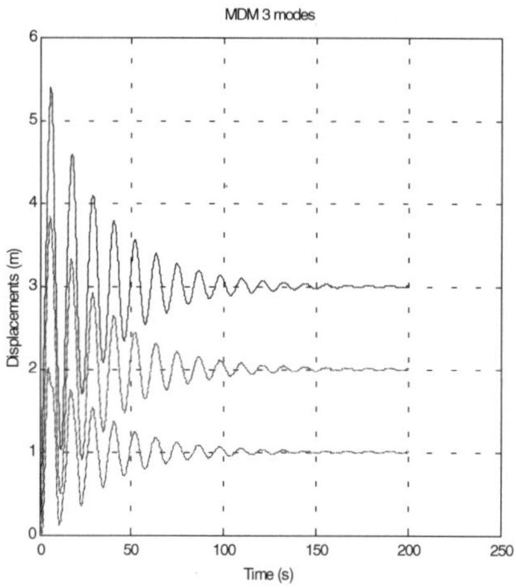

Fig. 16.1. Displacements, MDM, 3 modes

Solution question 3; one mode taken into account and one residual vector [Dickens 00].
When the vibrations are damped out the displacements converge to the static displacements

$$\begin{Bmatrix} x_1 \\ x_2 \\ x_3 \end{Bmatrix} = [G]\{F\} = \frac{1}{3}\begin{Bmatrix} 3 \\ 2 \\ 1 \end{Bmatrix}.$$

The modal base $[\Psi]$ consists of the first mode and 1 residual vector.

$$[\Psi] = \begin{bmatrix} 0.7370 & 0.2578 \\ 0.5910 & -0.1991 \\ 0.3280 & -0.2204 \end{bmatrix}.$$

The plots of the displacements $\{x\}$ are given in Fig. 16.3.
This example is the same dynamic system and has the same applied loads as in the previous example in this chapter.

This following example illustrates the influence of Rose's method of residual vectors.

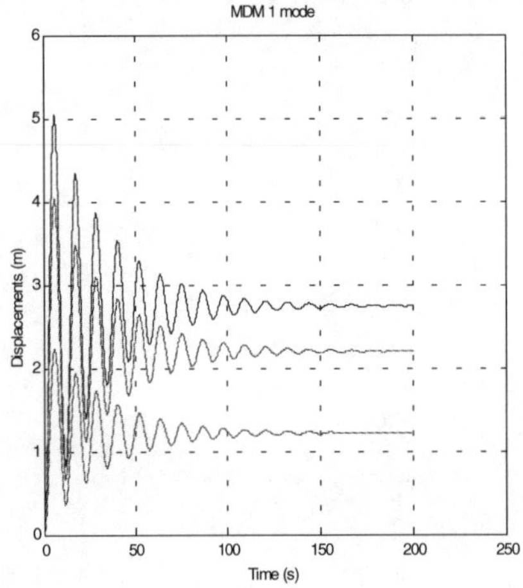

Fig. 16.2. Displacements, MDM, 1 mode

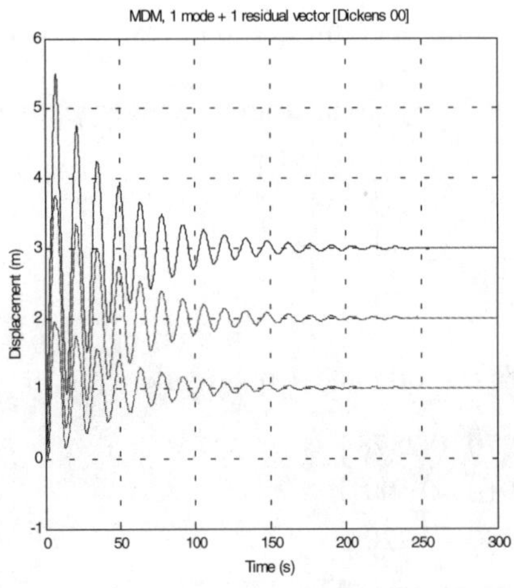

Fig. 16.3. Displacements, MDM, 1 mode + 1 residual vector [Dickens 00]

The transient responses of the dynamic system, consisting of three degrees of freedom, when the first degree of freedom is exposed to a constant force, will be calculated:

1. The complete system, i.e. all three modes included
2. The first mode only
3. The first mode only with residual vector correction [Rose 91]

The same numerical method is used as in the previous example in this chapter.

Calculate physical displacements. etc. with
$\{x\} = [\chi]\{\eta_k\}$, $\{\dot{x}\} = [\chi]\{\dot{\eta}_k\}$ and $\{\ddot{x}\} = [\chi]\{\ddot{\eta}_k\}$.
The simulation time is 300 s with a time increment $\Delta t = 0.05\text{ s}$ and the modal damping ratio for the elastic modes is $\xi = 0.05$ and for the residual vector the damping ratio has been taken as $\xi = 0.05$

Solution question 1; all 3 modes taken into account.
When the vibrations are damped out the displacements will converge to the static displacements

$$\begin{Bmatrix} x_1 \\ x_2 \\ x_3 \end{Bmatrix} = [G]\{F\} = \frac{1}{3}\begin{Bmatrix} 3 \\ 2 \\ 1 \end{Bmatrix}.$$

The results of the displacements are shown in Fig. 16.1.

Solution question 2; one mode taken into account (no residual vector)
When the vibrations are damped out the displacements do not converge to the static displacements

$$\begin{Bmatrix} x_1 \\ x_2 \\ x_3 \end{Bmatrix} = [G]\{F\} \neq \frac{1}{3}\begin{Bmatrix} 3 \\ 2 \\ 1 \end{Bmatrix}.$$

The results of the displacements are shown in Fig. 16.2.

Solution question 3; one mode taken into account and one residual vector [Rose 91].
When the vibrations are damped out the displacements converge to the static displacements

$$\begin{Bmatrix} x_1 \\ x_2 \\ x_3 \end{Bmatrix} = [G]\{F\} = \frac{1}{3}\begin{Bmatrix} 3 \\ 2 \\ 1 \end{Bmatrix}.$$

The modal base $[\Psi]$ consists of the first mode and 1 residual vector.

$$[\chi] = \begin{bmatrix} 0.7370 & 0.0669 \\ 0.5910 & -0.0517 \\ 0.3280 & -0.0572 \end{bmatrix}.$$

The plots of the displacements $\{x\}$ are given in Fig. 16.4.

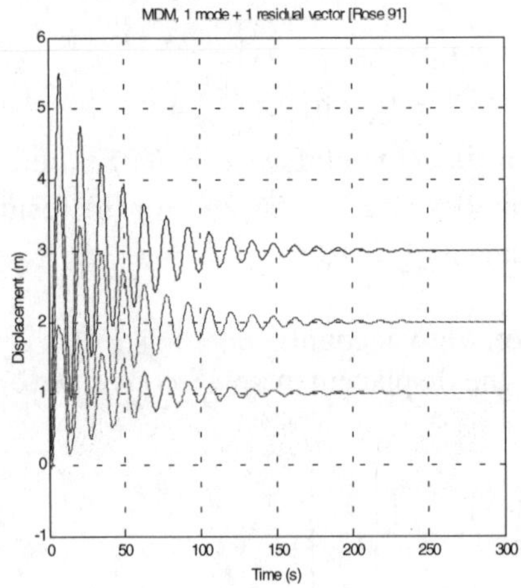

Fig. 16.4. Displacements, MDM, 1 mode + 1 residual vector [Rose 91]

16.3 Problems

16.3.1 Problem 1

The transient responses of the dynamic system, consisting of three degrees of freedom, when the first degree of freedom is exposed to a constant force, must be calculated:

1. The complete system, i.e. all three modes included
2. The first mode only
3. The first mode only with Dickens residual vector
4. The first mode only with Rose residual vector

The MDM will be applied.

The transient responses must be calculated numerically with the aid of the 'Park stiff stable" method [D'Souza 84, Park 75]. The initial displacements and velocities at t=0 are zero.

The undamped equations of motion of the 3 mass-spring system are

$$\begin{bmatrix} 1 & 0 & 0 \\ 0 & 1 & 0 \\ 0 & 0 & 1 \end{bmatrix} \begin{Bmatrix} \ddot{x}_1 \\ \ddot{x}_2 \\ \ddot{x}_3 \end{Bmatrix} + \begin{bmatrix} 10 & -10 & 0 \\ -10 & 20 & -10 \\ 0 & -10 & 20 \end{bmatrix} \begin{Bmatrix} x_1 \\ x_2 \\ x_3 \end{Bmatrix} = \begin{Bmatrix} 2 \\ 0 \\ 0 \end{Bmatrix},$$

or

$$[M]\{\ddot{x}\} + [K]\{x\} = \{F(t)\}.$$

The force vector $\{F(t)\} = 0$ when $t < 0$ and $\{F(t)\} = \begin{Bmatrix} 2 \\ 0 \\ 0 \end{Bmatrix}$ when $t \ge 0$.

The damping matrix $[C]$ will be introduced as modal damping after the equations of motion are uncoupled using the orthogonality relations of the modes, $[\Phi_k]^T[C][\Phi_k] = \langle 2\xi_j\omega_j\rangle$.
Use the Park procedure below with $\{x\} = [\Phi]\{\eta\}$, the generalised mass matrix $[mg] = [\Phi]^T[M][\Phi] = [I]$, the generalised stiffness matrix $[kg] = [\Phi]^T[K][\Phi] = \langle\lambda_k\rangle = \langle\omega^2{}_k\rangle$, $\{fg\} = [\Phi]^T\{F(t)\}$ and the diagonal modal damping matrix $[cg] = 2\xi\langle\omega_k\rangle$.

- Solve the acceleration $\ddot{\mathrm{r}}(0)\} = [\Phi][mg]^{-1}[\Phi]^T\{F(0)$. The initial values for the generalised coordinates can be calculated with
  $\{\eta(0)\} = ([\Phi]^T[\Phi])^{-1}[\Phi]^T\{x(0)\}$, and
  $\{\dot{\eta}(0)\} = ([\Phi]^T[\Phi])^{-1}[\Phi]^T\{\dot{x}(0)\}$.
- Set up the effective stiffness matrix with a time step Δt. With a constant Δt the effective stiffness matrix is constant,

$$[\bar{k}] = \frac{100}{36\Delta t^2}[mg] + \frac{10}{6\Delta t}[cg] + [kg].$$

- Set up the effective force vector

$$\{\bar{f}(t+\theta\Delta t)\} = \{fg(t+\Delta t)\} + \frac{15}{6\Delta t}[mg]\dot{\eta}(t) - \frac{1}{\Delta t}[mg]\{\dot{\eta}(t-\Delta t)\}$$
$$+\frac{1}{6\Delta t}[mg]\{\dot{\eta}(t-2\Delta t)\} + \left(\frac{150}{36\Delta t^2}[mg] + \frac{15}{6\Delta t}[cg]\right)\{\eta(t)\}$$
$$-\left(\frac{10}{6\Delta t^2}[mg] + \frac{1}{\Delta t}[cg]\right)\{\eta(t-\Delta t)\} + \left(\frac{1}{36\Delta t^2}[mg] + \frac{1}{6\Delta t}[cg]\right)\{\eta(t-2\Delta t)\}.$$

- Solve $[\bar{k}]\{\eta(t+\Delta t)\} = \{\bar{f}(t+\Delta t)\}$.
- Calculate the acceleration and velocities for the generalised coordinates

$$\{\ddot{\eta}(t+\Delta t)\} = \frac{1}{6\Delta t}[10\{\dot{\eta}(t+\Delta t)\} - 15\{\dot{\eta}(t)\} + 6\{\dot{\eta}(t-\Delta t)\} - \{\dot{\eta}(t-2\Delta t)\}]$$

$$\{\dot{\eta}(t+\Delta t)\} = \frac{1}{6\Delta t}[10\{\eta(t+\Delta t)\} - 15\{\eta(t)\} + 6\{\eta(t-\Delta t)\} - \{\eta(t-2\Delta t)\}]$$

- It is advised to start with the Newmark-beta method to obtain $\{\eta(\Delta t)\}$, $\{\eta(2\Delta t)\}$, $\{\dot{\eta}(\Delta t)\}$, and $\{\dot{\eta}(2\Delta t)\}$ to use Park's method the calculate $\{\eta(3\Delta t)\}$, etc.
- Calculate physical displacements. etc.
  - Modal base $[\Phi]$, MDM
  - Modal base $[\Psi]$, [Dickens 00]
  - Modal base $[\chi]$, [Rose 91]

The simulation time is 200 s with a time increment $\Delta t = 0.05$ s and the modal damping ratio for all modes is $\xi = 0.01$.

17 Dynamic Model Reduction Methods

17.1 Introduction

The combining of unreduced finite element models (FEMs) of subsytems to a dynamic FEM of the complete system (satellite or launcher) will in general result in a finite element model with many degrees of freedom (dofs) and therefore difficult to handle. The responsible analyst, to manipulate the total dynamic model, will ask for a reduced dynamic FEM description of the subsystem and will prescribe the allowed number of 'left' dynamic dofs of the reduced dynamic model. The reduced dynamic model is, in general, a modal description of the system involved.

The customer will prescribe the required accuracy of the reduced dynamic model, more specifically the natural frequencies, mode shapes in comparison with the complete finite element model or reference model. For example the following requirements are prescribed:

- The natural frequencies of the reduced dynamic model shall than deviate less $\pm 3\,\%$ from the natural frequencies calculated with the reference model.
- The effective masses of the reduced dynamic model shall be within $\pm 10\,\%$ of the effective masses calculated with the reference model.
- The diagonal terms at the crossorthogonality check [Ricks 91] shall be greater than or equal to 0.95 and the off-diagonal terms shall be less than or equal to 0.05. The crossorthogonality check is based upon the mass matrix.
- The diagonal terms at the modal assurance criteria (MAC) shall be greater than or equal to 0.95 and the off-diagonal terms less than or equal to 0.10.

Sometimes the requirements concern the correlation of the response curves obtained with the reduced dynamic model and the reference model.

Reduced models are also used to support the modal survey, the experimental modal analysis. The reduced dynamic model will be used to calculate the orthogonality relations between measured and analysed modes. This reduced model is called the test-analysis model (TAM) [Kammer 87].

In the following sections 7 reduction method will be discussed:

- The static condensation method [Guyan 68]
- The dynamic reduction method [Miller 80]
- The improved reduced system (IRS) [O'Callahan 89]
- Craig–Bampton (CB) reduction method [Craig 68]
- Generalised dynamic reduction (GDR) method [Gockel 83]
- System equivalent reduction expansion process (SEREP) [Kammer 87]
- Ritz Vectors [Escobedo 93]

All reduction procedures mentioned are based upon the Ritz method [Michlin 62].

17.2 Static Condensation Method

In general it is required to reduce the number of dynamic dofs of a finite element model (dynamic model) applying the static condensation method, often called the Guyan reduction [Guyan 68], to a specified number of dynamic dofs. A reduced dynamic model with 100 dynamic dofs is quite suitable. The kept dofs will be denoted by $\{x_a\}$ and the eliminated dofs by $\{x_e\}$. Furthermore we assume there are no applied external loads $\{F_e\}$.

The undamped equations of motion are

$$[M]\{\ddot{x}\} + [K]\{x\} = \{F\}. \tag{17.1}$$

In (17.1) the mass matrix [*M*], the stiffness matrix [*K*] and the displacement vector can be partitioned as follows

$$\begin{bmatrix} M_{aa} & M_{ae} \\ M_{ea} & M_{ee} \end{bmatrix} \begin{Bmatrix} \ddot{x}_a \\ \ddot{x}_e \end{Bmatrix} + \begin{bmatrix} K_{aa} & K_{ae} \\ K_{ea} & K_{ee} \end{bmatrix} \begin{Bmatrix} x_a \\ x_e \end{Bmatrix} = \begin{Bmatrix} F_a \\ F_e \end{Bmatrix} = \begin{Bmatrix} F_a \\ 0 \end{Bmatrix}. \tag{17.2}$$

The $\{x_a\}$ dofs will represent large inertia forces with respect to the inertia forces related to the $\{x_e\}$ dofs. The large masses and second mass moments of inertia are collected in the mass matrix $[M_{aa}]$. The inertia loads $[M_{aa}]\{\ddot{x}_a\}$ are significantly larger than the other inertia loads

$$[M_{ee}]\{\ddot{x}_e\}, [M_{ae}]\{\ddot{x}_e\}, [M_{ea}]\{\ddot{x}_a\} \ll [M_{aa}]\{\ddot{x}_a\}. \tag{17.3}$$

Only the inertia forces $[M_{aa}]\{\ddot{x}_a\}$ are maintained in (17.2), so

$$\begin{bmatrix} M_{aa} & 0 \\ 0 & 0 \end{bmatrix} \begin{Bmatrix} \ddot{x}_a \\ \ddot{x}_e \end{Bmatrix} + \begin{bmatrix} K_{aa} & K_{ae} \\ K_{ea} & K_{ee} \end{bmatrix} \begin{Bmatrix} x_a \\ x_e \end{Bmatrix} = \begin{Bmatrix} F_a \\ F_e \end{Bmatrix} = \begin{Bmatrix} F_a \\ 0 \end{Bmatrix}. \tag{17.4}$$

Using the equation related by the $\{x_e\}$ dofs we are able to express the $\{x_e\}$ dofs into the $\{x_a\}$ dofs.

$$[K_{ea}]\{x_a\} + [K_{ee}]\{x_e\} = \{0\}. \tag{17.5}$$

The inertia loads in (17.5) are neglected. We can express $\{x_e\}$ into $\{x_a\}$

$$\{x_e\} = -[K_{ee}]^{-1}[K_{ea}]\{x_a\} = [G_{ea}]\{x_a\}. \tag{17.6}$$

Only the stiffness is involved in (17.6) and therefore we talk about static condensation.

The total displacement vector $\{x\}$ will be projected on the kept dofs $\{x_a\}$

$$\{x\} = \begin{Bmatrix} x_a \\ x_e \end{Bmatrix} = \begin{bmatrix} I \\ G_{ea} \end{bmatrix}\{x_a\} = [T_{ea}]\{x_a\}. \tag{17.7}$$

The total kinetic energy in the dynamic system is

$$T = \frac{1}{2}\{\dot{x}\}^T[M]\{\dot{x}\} = \frac{1}{2}\{\dot{x}_a\}^T[T_{ea}]^T[M][T_{ea}]\{\dot{x}_a\}. \tag{17.8}$$

The reduced-mass matrix $[\overline{M}_{aa}]$ becomes

$$[\overline{M}_{aa}] = [T_{ea}]^T[M][T_{ea}]. \tag{17.9}$$

The total potential energy in the dynamic system is

$$U = \frac{1}{2}\{x\}^T[K]\{x\} = \frac{1}{2}\{x_a\}^T[T_{ea}]^T[K][T_{ea}]\{x_a\} \tag{17.10}$$

Analogous to the reduced-mass matrix $[\overline{M}_{aa}]$ the reduced-stiffness matrix $[\overline{K}_{aa}]$ becomes

$$[\overline{K}_{aa}] = [T_{ea}]^T[K][T_{ea}]. \tag{17.11}$$

The selection of the kept dofs $\{x_a\}$ is not always trivial. The kept dofs shall be selected in such a way that the mode shapes can be described as good as possible. The dofs associated with large masses shall be selected. As a guideline the following mathematical selection may be used to select the $\{x_a\}$ dofs. At least the dofs shall be selected for which applies

$$\frac{1}{2\pi}\sqrt{\frac{k_{ii}}{m_{ii}}} \leq 1.5 f_{\max} \tag{17.12}$$

where

- k_{ii} the diagonal term of the stiffness matrix $[K]$, translational and rotational
- m_{ii} the diagonal term of the mass matrix $[M]$, translational and rotational
- $f_{\max}$ maximum frequency of interest

Allen in [Allen 93a] described a more or less automatic way of selecting the analysis dofs $\{x_a\}$, however, it is still based upon (17.12).

The reduced eigenvalue problem can now be written as

$$\{[\overline{M}_{aa}] - \lambda_a[\overline{K}_{aa}]\}\{\phi_a\} = 0, \tag{17.13}$$

with

- $\{\phi_a\}$ the eigenvector of the reduced eigenvalue problem
- λ_a the eigenvalue associated with the eigenvector (mode shape) $\{\phi_a\}$

The eigenvectors that belong to the complete set of dofs, using (17.7), is

$$[\Phi_{GR}] = \begin{bmatrix} \Phi_a \\ \Phi_e \end{bmatrix} = \begin{bmatrix} I \\ G_{ea} \end{bmatrix}[\Phi_a], \tag{17.14}$$

where

- $[\Phi_e]$ the eigenvectors associated with the eliminated dofs
- $[G_{ea}]$ the transformation matrix as defined in (17.6)

A 10 dofs dynamic system (Fig. 17.1) will be used to illustrate the static condensation method. The constants are $m = 1$ and $k = 100000$. The dynamic system will be fixed in x_{10}. First, the natural frequencies and modes of the full system will be calculated.

The systems matrices are as follows

$$[M] = m\begin{bmatrix} 1&0&0&0&0&0&0&0&0&0\\ 0&1&0&0&0&0&0&0&0&0\\ 0&0&1&0&0&0&0&0&0&0\\ 0&0&0&1&0&0&0&0&0&0\\ 0&0&0&0&1&0&0&0&0&0\\ 0&0&0&0&0&2&0&0&0&0\\ 0&0&0&0&0&0&4&0&0&0\\ 0&0&0&0&0&0&0&3&0&0\\ 0&0&0&0&0&0&0&0&3&0\\ 0&0&0&0&0&0&0&0&0&3 \end{bmatrix},\ [K] = k\begin{bmatrix} 1&-1&0&0&0&0&0&0&0&0\\ -1&3&0&0&0&0&-2&0&0&0\\ 0&0&3&-1&0&0&-2&0&0&0\\ 0&0&-1&1&0&0&0&0&0&0\\ 0&0&0&0&1&-1&0&0&0&0\\ 0&0&0&0&-1&3&-2&0&0&0\\ 0&-2&-2&0&0&-2&9&-3&0&0\\ 0&0&0&0&0&0&-3&6&-3&0\\ 0&0&0&0&0&0&0&-3&7&-4\\ 0&0&0&0&0&0&0&0&-4&4 \end{bmatrix}.$$

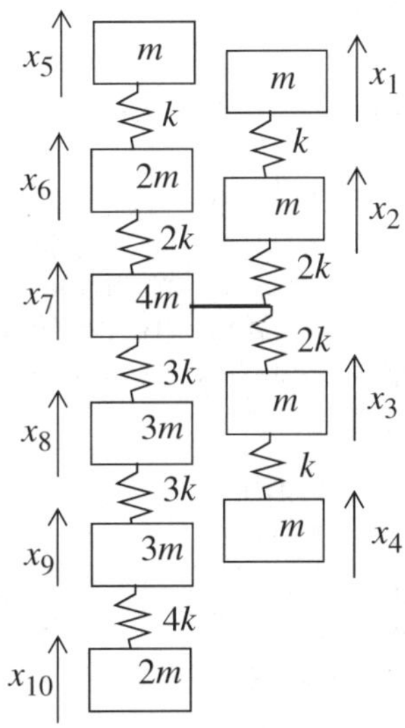

Fig. 17.1. 10 mass–spring dynamic system

The natural frequencies $\{f\}$ (Hz) calculated are

$$\{f\} = \lfloor 14.25, 36.69, 38.52, 47.43, 63.69, 75.91, 90.25, 93.00, 106.51 \rfloor^T.$$

The first four modes $[\Phi]$ are scaled such that

$$[\Phi]^T[M][\Phi] = [I] \text{ and } [\Phi]^T[K][\Phi] = \langle\omega^2\rangle = \langle(2\pi f)^2\rangle$$

$$[\Phi] = \begin{bmatrix} 0.3064 & -0.3878 & -0.6533 & -0.3762 \\ 0.2819 & -0.1817 & -0.2706 & -0.0421 \\ 0.2819 & -0.1817 & 0.2706 & -0.0421 \\ 0.3064 & -0.3878 & 0.6533 & -0.3762 \\ 0.3204 & 0.6591 & 0 & -0.3327 \\ 0.2948 & 0.3088 & 0 & -0.0373 \\ 0.2583 & -0.0304 & 0 & 0.1435 \\ 0.1750 & -0.0333 & 0 & 0.3417 \\ 0.0777 & -0.0185 & 0 & 0.2364 \\ 0 & 0 & 0 & 0 \end{bmatrix}.$$

We select the following kept dofs; $\{x_a\} = \lfloor x_1, x_4, x_5, x_7 \rfloor$.

The natural frequencies $\{f_a\}$ of the reduced eigenvalue problem, (17.13), are

$$\{f_a\} = \lfloor 14.39, 37.86, 38.98, 52.98 \rfloor^T.$$

The modes $[\Phi_{GR}]$ are scaled such that $[\Phi_{GR}]^T[M][\Phi_{GR}] = [I]$.

To compare the dynamic properties of the condensed dynamic system with the complete or reference dynamic system we can compare the natural frequencies with each other and of course the mode shapes. To do so we have to lengthen the mode shape of the reduced system (4 dofs) to a total number of 10 dofs.

$$[\Phi_{GR}] = \begin{bmatrix} 0.3134 & -0.4059 & -0.6708 & -0.4390 \\ 0.2802 & -0.1483 & -0.2236 & 0.0347 \\ 0.2802 & -0.1483 & 0.2236 & 0.0347 \\ 0.3134 & -0.4059 & 0.6708 & -0.4390 \\ 0.3269 & 0.7217 & 0 & -0.4577 \\ 0.2946 & 0.2276 & 0 & 0.0285 \\ 0.2635 & -0.0195 & 0 & 0.2716 \\ 0.1677 & -0.0124 & 0 & 0.1728 \\ 0.0719 & -0.0053 & 0 & 0.1728 \\ 0 & 0 & 0 & 0 \end{bmatrix}.$$

The modes of the reduced dynamic model will be compared to the complete dynamic model with three methods; the modal assurance criteria and the normalised crossorthogonality matrices [Friswell 95, Maia 97] and the crossorthogonality check [Ricks 91]:

- The modal assurance criteria (MAC). The absolute value of the MAC is between 0 and 1. A value of 1 means that one mode shape is a multiple of the other. The MAC matrix is defined as $MAC = \dfrac{([\Phi]^T[\Phi_{GR}])^2}{([\Phi]^T[\Phi])([\Phi_{GR}]^T[\Phi_{GR}])}$.

- The normalised crossorthogonality (NCO). The absolutes values of the NCO are between 0 and 1. A value of 1 means that one mode shape is a multiple of the other. The modified MAC is defined as NCO= $\dfrac{([\Phi]^T[M][\Phi_{GR}])^2}{([\Phi]^T[M][\Phi])([\Phi_{GR}]^T[M][\Phi_{GR}])}$.
- Crossorthogonality check. $[C]= ([\Phi]^T[M][\Phi_{GR}])$, $[\Phi]^T[M][\Phi] = \langle I \rangle$ and the terms on the main diagonal of the $([\Phi_{GR}]^T[M][\Phi_{GR}])$ are one.

The MAC becomes

$$\text{MAC}= \frac{([\Phi]^T[\Phi_{GR}])^2}{([\Phi]^T[\Phi])([\Phi_{GR}]^T[\Phi_{GR}])} = \begin{bmatrix} 0.9980 & -0.0052 & 0.0000 & 0.0251 \\ -0.0017 & 0.9897 & 0.0000 & -0.0669 \\ 0.0000 & -0.0006 & 0.9950 & 0.0000 \\ -0.0540 & 0.0459 & 0.0000 & 0.8054 \end{bmatrix},$$

and the NCO

$$\text{NCO}= \frac{([\Phi]^T[M][\Phi_{GR}])^2}{[\Phi]^T[M][\Phi]([\Phi_{GR}]^T[M][\Phi_{GR}])} = \begin{bmatrix} 0.9995 & 0.0002 & 0.0000 & -0.0013 \\ -0.0004 & 0.9775 & 0.0000 & -0.0194 \\ 0.0000 & 0.0000 & 0.9950 & 0.0000 \\ -0.0019 & 0.0612 & 0.0000 & 0.7444 \end{bmatrix},$$

and finally the crossorthogonality

$$[C]= [\Phi]^T[M][\Phi_{GR}] = \begin{bmatrix} -0.9996 & 0.0052 & 0.0000 & -0.0092 \\ -0.0004 & 0.9888 & 0.0000 & -0.0105 \\ 0.0003 & 0.0000 & -0.9975 & 0.0000 \\ -0.00151 & 0.0330 & 0.0000 & 0.8629 \end{bmatrix}.$$

The diagonal terms of the MAC, NCO and crosscorrelation $[C]$ show us the correlation of the mode shapes of the reduced dynamic model with the reference model. The off-diagonal terms shows us the coupling between the correlated modes. The first 3 modes, of both the reduced and reference dynamic modes, do correlate very well. The fourth mode of the reduced model is less correlated with the fourth mode of the complete model.

17.2.1 Improved Calculation of Eliminated Dofs

Allen in [Allen 93b] proposed a method to improve the part of the eigenvector $\{\phi_e\}$ with respect to (17.14). The eigenvalue problem of the complete dynamic system can be written as

$$\left(-\lambda\begin{bmatrix} M_{aa} & M_{ae} \\ M_{ea} & M_{ee}\end{bmatrix} + \begin{bmatrix} K_{aa} & K_{ae} \\ K_{ea} & K_{ee}\end{bmatrix}\right)\begin{Bmatrix} \phi_a \\ \phi_e \end{Bmatrix} = \begin{Bmatrix} 0 \\ 0 \end{Bmatrix}. \tag{17.15}$$

Using the second equation from (17.15) we have

$$[K_{ea}]\{\phi_a\} + [K_{ee}]\{\phi_e\} - \lambda[M_{ea}]\{\phi_a\} - \lambda[M_{ee}]\{\phi_e\} = \{0\}. \tag{17.16}$$

The vector $\{\phi_e\}$ can be related to $\{\phi_a\}$ using (17.16)

$$\{\phi_{e,j}\} = -[K_{ee} - \lambda_j M_{ee}]^{-1}[K_{ea} - \lambda_j M_{ea}]\{\phi_{a,j}\}. \tag{17.17}$$

The matrix $[K_{ee} - \lambda_j M_{ee}]$ must be inverted for every eigenvalue λ_j and that is a very inefficient method. Instead of using one mode $\{\phi_j\}$ we proceed with the complete modal matrix $[\Phi]$, hence in accordance with (17.16)

$$[K_{ea}][\Phi_a] + [K_{ee}][\Phi_e] - \langle\lambda\rangle[M_{ea}][\Phi_a] - \langle\lambda\rangle[M_{ee}][\Phi_e] = \{0\}. \tag{17.18}$$

$[\Phi_e]$ can be expressed in $[\Phi_a]$

$$[\Phi_e] = [G_{ea}][\Phi_a] + \langle\lambda_a\rangle[K_{ee}]^{-1}[M_{ea}][\Phi_a] + \langle\lambda_a\rangle[K_{ee}]^{-1}[M_{ee}][\Phi_e], \tag{17.19}$$

with

- $[G_{ea}]$ the transformation matrix as defined in (17.6)
- $\langle\lambda_a\rangle$ the diagonal matrix if eigenvalues came from (17.13)

An iterative scheme can be set up

$$[\Phi_e]^{(2)} = [G_{ea}][\Phi_a] + \langle\lambda_a\rangle[K_{ee}]^{-1}[M_{ea}][\Phi_a] + \langle\lambda_a\rangle[K_{ee}]^{-1}[M_{ee}][\Phi_e]^{(1)} \tag{17.20}$$

with

- $[\Phi_e]^{(1)}$ the first result from (17.6), $[\Phi_e] = [G_{ea}][\Phi_a]$
- $[\Phi_e]^{(2)}$ the second updated result of the part $[\Phi_e]$ of the eigenvector $[\Phi]$

The complete eigenvector becomes

$$[\Phi]^{(2)} = \begin{bmatrix} \Phi_a \\ \Phi_e^{(2)} \end{bmatrix}. \tag{17.21}$$

The iteration scheme may be repeated.

17.3 Dynamic Reduction

The undamped equations of motion are with zero external forces, (17.1)

$$[M]\{\ddot{x}\} + [K]\{x\} = \{0\}.$$

In (17.1) the mass matrix [*M*], the stiffness matrix [*K*] and the displacement vector can be partitioned as follows, (17.2)

$$\begin{bmatrix} M_{aa} & M_{ae} \\ M_{ea} & M_{ee} \end{bmatrix} \begin{Bmatrix} \ddot{x}_a \\ \ddot{x}_e \end{Bmatrix} + \begin{bmatrix} K_{aa} & K_{ae} \\ K_{ea} & K_{ee} \end{bmatrix} \begin{Bmatrix} x_a \\ x_e \end{Bmatrix} = \begin{Bmatrix} 0 \\ 0 \end{Bmatrix},$$

with

- $\{x_a\}$ are the kept degrees of freedom
- $\{x_e\}$ are the eliminated degrees of freedom

The eigenvalue problem becomes

$$\left(\begin{bmatrix} K_{aa} & K_{ae} \\ K_{ea} & K_{ee} \end{bmatrix} - \lambda \begin{bmatrix} M_{aa} & M_{ae} \\ M_{ea} & M_{ee} \end{bmatrix}\right) \begin{Bmatrix} \phi_a \\ \phi_e \end{Bmatrix} = \begin{Bmatrix} 0 \\ 0 \end{Bmatrix}. \tag{17.22}$$

The first equation of the matrix (17.22) is, [Miller 80],

$$([K_{ea}] - \lambda[M_{ea}])\{\phi_a\} + ([K_{ee}] - \lambda[M_{ee}])\{\phi_e\} = \{0\}. \tag{17.23}$$

From (17.23) we can express $\{x_e\}$ in $\{x_a\}$ and the relation is given by

$$\{\phi_e\} = -([K_{ee}] - \lambda[M_{ee}])^{-1}([K_{ea}] - \lambda[M_{ea}])\{\phi_a\} = [T_{ea}]\{\phi_a\}. \tag{17.24}$$

Unfortunately the transformation matrix $[T_{ea}]$ involves the eigenvalue λ which is not known. As a starting point we may substitute the eigenvalues obtained by (17.13) one by one to calculate the transformation matrix $[T_{ea}]$.

The transformation $[T_{ea}]$, matrix (17.24), can be developed in a series of λ, i.e. until $O(\lambda^2)$, [Miller 80]

$$[T_{ea}] = [K_{ee}]^{-1}[K_{ea}] + \lambda(-[K_{ee}]^{-1}[M_{ea}] + [K_{ee}]^{-1}[M_{ee}][K_{ee}]^{-1}[K_{ea}]) + O(\lambda^2)$$

(17.25)

The reduced-mass matrix $[\bar{M}_{aa}]$ becomes

$$[\bar{M}_{aa}] = [T_{ea}]^T[M][T_{ea}]. \tag{17.26}$$

Analogous to the reduced-mass matrix $[\bar{M}_{aa}]$ the reduced-stiffness matrix $[\bar{K}_{aa}]$ becomes

$$[\bar{K}_{aa}] = [T_{ea}]^T[K][T_{ea}]. \tag{17.27}$$

17.4 Improved Reduced System (IRS)

O'Callahan [O'Callahan 89] proposed a procedure for an improved reduced system model. The IRS method is an extension of the Static Condensation technique [Guyan 68]. The undamped homogeneous equations of motion are (17.1)

$$[M]\{\ddot{x}\} + [K]\{x\} = \{0\}.$$

Assuming harmonic displacements, etc. the displacement vector can be written

$$x(t) = X(\omega)e^{j\omega t}, \tag{17.28}$$

and (17.28) becomes

$$[K]\{X(\omega)\} = \omega^2[M]\{X(\omega)\}. \tag{17.29}$$

This means that in the absence of external and damping forces the harmonic inertia forces are in equilibrium with the harmonic elastic forces. The displacement vector $\{X(\omega)\}$ will be projected on the kept displacements $\{X_a(\omega)\}$ using (17.7), [Guyan 68], hence

$$\{X(\omega)\} = \begin{Bmatrix} X_a(\omega) \\ X_e(\omega) \end{Bmatrix} = \begin{bmatrix} I \\ G_{ea} \end{bmatrix}\{X_a(\omega)\} = [T_{ea}]\{X_a(\omega)\}. \tag{17.30}$$

After substitution of (17.30) into (17.29) becomes

$$[K]\{X(\omega)\} = \omega^2[M][T_{ea}]\{X_a(\omega)\}. \tag{17.31}$$

With the aid of (17.13) we are able to define the vector $\omega^2\{X_a(\omega)\}$

$$\omega^2\{X_a(\omega)\} = [\bar{M}_{aa}]^{-1}[\bar{K}_{aa}]\{X_a(\omega)\}. \tag{17.32}$$

Thus, (17.31) now becomes

$$[K]\{X(\omega)\} = [M][T_{ea}][\bar{M}_{aa}]^{-1}[\bar{K}_{aa}]\{X_a(\omega)\} \tag{17.33}$$

Further expanded, this gives

$$[K]\{X(\omega)\} = \begin{bmatrix} K_{aa} & K_{ae} \\ K_{ea} & K_{ee} \end{bmatrix} \begin{Bmatrix} X_a(\omega) \\ X_e(\omega) \end{Bmatrix} = [M][T_{ea}][\bar{M}_{aa}]^{-1}[\bar{K}_{aa}]\{X_a(\omega)\}. \tag{17.34}$$

The vector of eliminated dofs $\{X_e(\omega)\}$ can be recalculated

$$\{X_e(\omega)\} = [G_{ea}]\{X_a(\omega)\} + \begin{bmatrix} 0 & 0 \\ 0 & K_{ee}^{-1} \end{bmatrix} [M][T_{ea}][\bar{M}_{aa}]^{-1}[\bar{K}_{aa}]\{X_a(\omega)\}. \tag{17.35}$$

The complete displacement vector $\{X(\omega)\}$ can be related to $\{X_a(\omega)\}$

$$\{X(\omega)\} = \begin{bmatrix} \{X_a(\omega)\} \\ \{X_e(\omega)\} \end{bmatrix} = \left[[T_{ea}] + \begin{bmatrix} 0 & 0 \\ 0 & K_{ee}^{-1} \end{bmatrix} [M][T_{ea}][\bar{M}_{aa}]^{-1}[\bar{K}_{aa}] \right] \{X_a(\omega)\} \tag{17.36}$$

Equation can be written

$$\{X(\omega)\} = \begin{bmatrix} \{X_a(\omega)\} \\ \{X_e(\omega)\} \end{bmatrix} = [T_{IRS}]\{X_a(\omega)\}. \tag{17.37}$$

The reduced-mass matrix and the reduced-stiffness matrix can be achieved analogously to (17.9) and (17.11).

The reduced-mass matrix $[\bar{M}_{IRS}]$ becomes

$$[\bar{M}_{\mathrm{IRS}}] = [T_{IRS}]^T[M][T_{IRS}], \tag{17.38}$$

and analogous to the reduced-mass matrix $[\bar{M}_{\mathrm{IRS}}]$ the reduced-stiffness matrix $[\bar{K}_{\mathrm{IRS}}]$ becomes

$$[\bar{K}_{\mathrm{IRS}}] = [T_{IRS}]^T[K][T_{IRS}]. \tag{17.39}$$

The IRS reduced-mass matrix and reduced-stiffness matrix are an improvement to the reduced matrices obtained with the static condensation technique [Guyan 68]. All kept dofs have a physical meaning, no mathematical dofs are involved. The IRS reduced-mass matrix may be used to check the orthogonality relations of the mode shapes measured during the modal survey test (modal analysis) at a limited number of measurement locations (accelerometers). The number of measured dofs is, in general, equal to the kept dofs $\{x_a\}$.

We select the following kept dofs; $\{x_a\} = \lfloor x_1, x_4, x_5, x_7 \rfloor^T$ and take into account the first and second modes to reduce the model with the IRS method.

The natural frequencies $\{f_{\mathrm{IRS}}\}$ of the reduced eigenvalue problem are

$$\{f_{\mathrm{IRS}}\} = \lfloor 14.25, 36.77, 38.53, 49.20 \rfloor^T.$$

The modes $[\Phi_{IRS}]$ are scaled such that $[\Phi_{IRS}]^T[M][\Phi_{IRS}] = [I]$.

To compare the dynamic properties of the condensed dynamic system to the complete or reference dynamic system we can compare the natural frequencies with each other and, of course, the mode shapes. To do so we have lengthen the mode shape of the reduced system (4 dofs) to a total number of 10 dofs.

$$[\Phi_{\mathrm{IRS}}] = \begin{bmatrix} -0.3064 & 0.3936 & -0.6565 & 0.4280 \\ -0.2820 & 0.1770 & -0.2626 & -0.0085 \\ -0.2820 & 0.1770 & 0.2626 & -0.0085 \\ -0.3064 & 0.3936 & 0.6565 & 0.4280 \\ -0.3206 & -0.6761 & 0 & 0.4345 \\ -0.2945 & -0.2872 & 0 & -0.0139 \\ -0.2585 & 0.0285 & 0 & -0.2113 \\ -0.1748 & 0.242 & 0 & -0.2611 \\ -0.0774 & 0.0118 & 0 & 0.1428 \\ 0 & 0 & 0 & 0 \end{bmatrix}.$$

The MAC becomes

$$\mathrm{MAC}= \frac{([\Phi]^T[\Phi_{IRS}])^2}{([\Phi]^T[\Phi])([\Phi_{IRS}]^T[\Phi_{IRS}])} = \begin{bmatrix} 0.9310 & -0.0002 & 0.0000 & -0.0933 \\ -0.0006 & 1.0000 & 0.0000 & -0.0986 \\ 0.0000 & 0.0000 & 0.9999 & 0.0000 \\ 0.0623 & 0.0384 & 0.0000 & 0.9850 \end{bmatrix},$$

and the NCO is

$$\mathrm{NCO}= \frac{([\Phi]^T[M][\Phi_{IRS}])^2}{[\Phi]^T[M][\Phi]([\Phi_{IRS}]^T[M][\Phi_{IRS}])} = \begin{bmatrix} 1.0000 & 0.0001 & 0.0000 & 0.0003 \\ -0.0001 & 0.9981 & 0.0000 & -0.0168 \\ 0.0000 & 0.0000 & 0.9999 & 0.0000 \\ 0.0006 & 0.0294 & 0.0000 & 0.9117 \end{bmatrix}$$

and

$$[C]= [\Phi]^T[M][\Phi_{\text{IRS}}] = \begin{bmatrix} -1.0000 & 0.0000 & 0.0000 & -0.0002 \\ 0.0000 & -0.9991 & 0.0000 & -0.0086 \\ 0.0000 & 0.0000 & -0.9999 & 0.0000 \\ 0.0003 & -0.0150 & 0.0000 & -0.9549 \end{bmatrix}.$$

The diagonal terms of the MAC, the NCO and the modified $[C]$ show us the correlation of the mode shapes of the reduced dynamic model with the reference model. The offdiagonal terms show us the coupling between the correlated modes. The first 3 modes, of both the reduced and reference dynamic modes, do correlate very well. The third mode does not show any coupling with the other modes. The fourth mode of the reduced model is now better correlated with the fourth mode of the complete model compared with the Guyan reduction method.

17.5 Craig–Bampton Reduced Models

The Craig–Bampton method is discussed in several publications [Craig 68, Craig 77, Craig 81, Craig 00, Gordon 99] and is one of the most favourite methods for reducing the size, number of degrees of freedom, of a dynamic model (finite element model). The undamped equations of motion are (17.1)

$$[M]\{\ddot{x}\} + [K]\{x\} = \{F(t)\}.$$

We denote the external or boundary degrees of freedom with the index j and the internal degrees of freedom with the index i. The matrix equations (17.1) may be partitioned as follows

$$\begin{bmatrix} M_{ii} & M_{ij} \\ M_{ji} & M_{jj} \end{bmatrix} \begin{Bmatrix} \ddot{x}_i \\ \ddot{x}_j \end{Bmatrix} + \begin{bmatrix} K_{ii} & K_{ij} \\ K_{ji} & K_{jj} \end{bmatrix} \begin{Bmatrix} x_i \\ x_j \end{Bmatrix} = \begin{Bmatrix} F_i \\ F_j \end{Bmatrix}. \tag{17.40}$$

In [Craig 68] it is proposed to depict the displacement vector $\{x(t)\}$ on a basis of static or constraint modes $[\Phi_s]$ with $\{x_j\} = [I]$ and elastic mode shapes $[\Phi_p]$ with fixed external degrees of freedom $\{x_j\} = \{0\}$ and the eigenvalue problem $([K_{ii}] - \langle\lambda_p\rangle[M_{ii}])[\Phi_{ii}] = [0]$. We can express $\{x\}$ as

$$\{x\} = [\Phi_s]\{x_j\} + [\Phi_p]\{\eta_p\} = [\Phi_s, \Phi_p]\begin{Bmatrix} x_j \\ \eta_p \end{Bmatrix} = [\Psi]\{X\}. \tag{17.41}$$

The static modes can be obtained, assuming zero inertia effects, $\{F_i\} = \{0\}$, and successively prescribe a unit displacement for the boundary degrees of freedom, thus $\{x_j\} = [I]$. So we may write (17.40) as follows

$$\begin{bmatrix} K_{ii} & K_{ij} \\ K_{ji} & K_{jj} \end{bmatrix} \begin{Bmatrix} x_i \\ x_j \end{Bmatrix} = \begin{Bmatrix} 0 \\ R_j \end{Bmatrix}. \tag{17.42}$$

From the first equation of (17.42) we find for $\{x_i\}$

$$[K_{ii}]\{x_i\} + [K_{ij}][x_j] = 0\,, \tag{17.43}$$

hence

$$\{x_i\} = -[K_{ii}]^{-1}[K_{ij}]\{x_j\}\,, \tag{17.44}$$

and therefore

$$[\Phi_{ij}] = -[K_{ii}]^{-1}[K_{ij}][I] = -[K_{ii}]^{-1}[K_{ij}]. \tag{17.45}$$

The static transformation now becomes

$$\{x\} = \begin{Bmatrix} x_i \\ x_j \end{Bmatrix} = \begin{bmatrix} \Phi_{ij} \\ I \end{bmatrix}\{x_j\} = [\Phi_s]\{x_j\}. \tag{17.46}$$

Assuming fixed external degrees of freedom $\{x_j\} = \{0\}$ and also assuming harmonic motions $x(t) = X(\omega)e^{j\omega t}$ the eigenvalue problem can be stated as

$$([K_{ii}] - \lambda_{i,p}[M_{ii}])\{X(\lambda_{i,p})\} = \{0\}\,, \tag{17.47}$$

or, more generally, as

$$([K_{ii}] - \langle\lambda_i\rangle[M_{ii}])[\Phi_{ip}] = \{0\}. \tag{17.48}$$

The internal degrees of freedom $\{x_i\}$ will be projected on the set of orthogonal mode shapes (modal matrix) $[\Phi_{ip}]$, thus

$$\{x_i\} = [\Phi_{ip}]\{\eta_p\}. \tag{17.49}$$

The modal transformation becomes

$$\{x\} = \begin{Bmatrix} x_i \\ x_j \end{Bmatrix} = \begin{bmatrix} \Phi_{ip} \\ 0 \end{bmatrix}\{\eta_p\} = [\Phi_p]\{\eta_p\}. \tag{17.50}$$

The Craig–Bampton (CB) transformation matrix is (17.41)

$$\{x\} = [\Phi_s, \Phi_p]\begin{Bmatrix} x_j \\ \eta_p \end{Bmatrix} = [\Psi]\{X\}$$

with

- $[\Phi_s]$ the static or constraint modes
- $[\Phi_p]$ the modal matrix
- $\{x_j\}$ the external or boundary degrees of freedom
- $\{\eta_p\}$ the generalised coordinates. In general, the number of generalised coordinates p is much less than the total number of degrees of freedom $n = i + j$, $p \ll n$.

The constraint modes will introduce displacements due to adjacent structures in a static way, while the elastic modes will introduce dynamic effects generated internally in the structure.

The CB transformation (17.41) will be substituted into (17.1) presuming equal potential and kinetic energies, hence

$$[\Psi]^T[M][\Psi]\{\ddot{X}\} + [\Psi]^T[K][\Psi]\{X\} = [\Psi]^T\{F(t)\}. \qquad (17.51)$$

Further elaborated we find

$$\begin{bmatrix} \tilde{M}_{jj} & M_{jp} \\ M_{pj} & \langle m_p \rangle \end{bmatrix}\begin{Bmatrix} \ddot{x}_j \\ \ddot{\eta}_p \end{Bmatrix} + \begin{bmatrix} \tilde{K}_{jj} & K_{jp} \\ K_{pj} & \langle k_p \rangle \end{bmatrix}\begin{Bmatrix} x_j \\ \eta_p \end{Bmatrix} = \begin{bmatrix} \Phi_{ij} & \Phi_p \\ I & 0 \end{bmatrix}^T\begin{Bmatrix} F_i \\ F_j \end{Bmatrix}, \qquad (17.52)$$

with

- $[\tilde{M}_{jj}]$ the Guyan reduced mass matrix (j-set)
- $[\tilde{K}_{jj}]$ the Guyan reduced stiffness matrix (j-set)
- $\langle m_p \rangle$ the diagonal matrix of generalised masses, $\langle m_p \rangle = [\Phi_p]^T[M][\Phi_p]$
- $\langle k_p \rangle$ the diagonal matrix of generalised stiffnesses,

  $$\langle k_p \rangle = [\Phi_p]^T K[\Phi_p] = \langle \lambda_p \rangle \langle m_p \rangle$$
- $[K_{ip}] = [\Phi_{ij}]^T[K_{ii}][\Phi_p] + [K_{ji}][\Phi_p] = (-[K_{ij}]^T[K_{ii}]^{-1}[K_{ii}] + [K_{ji}])[\Phi_p] = [0]$ (see (17.45))
- $[K_{pi}] = [K_{ip}]^T = [0]$

Thus (17.52) becomes

$$\begin{bmatrix} \tilde{M}_{jj} & M_{jp} \\ M_{pj} & \langle m_p \rangle \end{bmatrix} \begin{Bmatrix} \ddot{x}_j \\ \ddot{\eta}_p \end{Bmatrix} + \begin{bmatrix} \tilde{K}_{jj} & 0 \\ 0 & \langle k_p \rangle \end{bmatrix} \begin{Bmatrix} x_j \\ \eta_p \end{Bmatrix} = \begin{bmatrix} \Phi_{ij} & \Phi_p \\ I & 0 \end{bmatrix}^T \begin{Bmatrix} F_i \\ F_j \end{Bmatrix}. \tag{17.53}$$

Finally

$$[M_{\text{CB}}]\{\ddot{X}\} + [K_{\text{CB}}]\{X\} = [\Psi]^T\{F\}, \tag{17.54}$$

with

- $[M_{\text{CB}}]$ the CB reduced-mass matrix
- $[K_{\text{CB}}]$ the CB reduced-stiffness matrix

The CB matrices are $j + p$, $j + p$ sized matrices. Equation (17.54) is frequently applied for component-mode synthesis methods (dynamic substructuring).

The accuracy of the CB reduction technique is very satisfactory and was discussed in [Claessens 96].

17.6 Generalised Dynamic Reduction

The generalised dynamic reduction (GDR), [Gockel 83, Wijker 91], is a very elegant method to reduce considerably the number of degrees of freedom of a dynamic finite element model. The GDR method is a mathematical extension of the Guyan static reduction technique [Guyan 68]. A GDR reduced dynamic model will represent, to a certain specified maximum frequency $f_{\max}$ (Hz), the modal properties very well; i.e. the natural frequencies, the mode shapes, the effective masses, etc. with reference to the complete finite element model. The undamped homogeneous equations of motion are (17.1)

$$[M]\{\ddot{x}\} + [K]\{x\} = \{0\}.$$

The boundary and subsidiary conditions are already incorporated into the undamped equations of motion, (17.1). The total number of independent degrees of freedom is indicated with n. The n degrees of freedom are partitioned into two sets of degrees of freedom:

1. the kept a degrees of freedom
2. the eliminated e degrees of freedom

For a reduced dynamic model the number of a degrees of freedom $a \ll n$.

In (17.1) the mass matrix $[M]$, the stiffness matrix $[K]$ and the displacement vector can be partitioned as follows

$$\begin{bmatrix} M_{aa} & M_{ae} \\ M_{ea} & M_{ee} \end{bmatrix} \begin{Bmatrix} \ddot{x}_a \\ \ddot{x}_e \end{Bmatrix} + \begin{bmatrix} K_{aa} & K_{ae} \\ K_{ea} & K_{ee} \end{bmatrix} \begin{Bmatrix} x_a \\ x_e \end{Bmatrix} = \begin{Bmatrix} 0 \\ 0 \end{Bmatrix}. \tag{17.55}$$

The elimination of the e degrees of freedom is done with the static condensation method [Guyan 68].

The reduced-mass matrix $[\overline{M}_{aa}]$ becomes with (17.9)

$$[\overline{M}_{aa}] = [T_{ea}]^T[M][T_{ea}].$$

Analogous to the reduced mass matrix $[\overline{M}_{aa}]$ the reduced-stiffness matrix $[\overline{K}_{aa}]$, using (17.11), becomes

$$[\overline{K}_{aa}] = [T_{ea}]^T[K][T_{ea}]$$

To improve the static condensation [Guyan 68] an additional condensation will be applied on the e dofs. When the a dofs are fixed the following eigenvalue problem is obtained.

$$([K_{ee}] - \lambda_e[M_{ee}])\{\phi_e\} = \{0\}. \tag{17.56}$$

Equation (17.56) can be rearranged

$$([K_{ee}] - \lambda_e[M_{ee}] + \upsilon[M_{ee}] - \upsilon[M_{ee}])\{\phi_e\} = \{0\}, \tag{17.57}$$

or

$$([K_{ee}] + \upsilon[M_{ee}])\{\phi_e\} = (\lambda_e + \upsilon)[M_{ee}]\{\phi_e\}. \tag{17.58}$$

Equation (17.58) can be written as

$$([K_{ee}] + \upsilon[M_{ee}])^{-1}[M_{ee}]\{\phi_e\} = [A_{ee}]\{\phi_e\} = \frac{1}{(\lambda_e + \upsilon)}\{\phi_e\}. \tag{17.59}$$

In the eigenvalue problem (17.59) the eigenvalue λ_e has been shifted by $-\nu$. The eigenvector $\{\phi_e\}$ remains the same.

The eigenvalue can be approximated using the power method. The power method is a matrix iteration method. The power method will converge to the largest eigenvalue $\frac{1}{(\lambda_e + \upsilon)}$ or the smallest value of $(\lambda_e + \upsilon)$.

The iteration scheme after the $n+1$-th iteration is

$$\{\Psi_{e,n+1}\} = [A_{ee}]\{\Psi_{e,n}\}, \tag{17.60}$$

or

$$([K_{ee}] + \upsilon[M_{ee}])\{\Psi_{e,n+1}\} = [M_{ee}]\{\Psi_{e,n}\}. \tag{17.61}$$

Assume the start vector $\{\Psi_{e,1}\}$ can be expressed in the original eigenvectors $[\Phi_e]$

$$\{\Psi_{e,1}\} = \alpha_1\{\phi_{e,1}\} + \alpha_2\{\phi_{e,2}\} + ...+\alpha_e\{\phi_{e,e}\} = \sum_{k=1}^{e} \alpha_k\{\phi_{e,k}\}. \tag{17.62}$$

After n iterations

$$\{\Psi_{e,n}\} = \sum_{k=1}^{e} \alpha_k\left(\frac{1}{\lambda_{e,k} + \upsilon}\right)^n \{\phi_{e,k}\} \tag{17.63}$$

After n iterations a good approximation of the largest eigenvalue will be obtained if

$$\left(\frac{\left\{\frac{1}{\lambda_{e,2} + \upsilon}\right\}}{\left\{\frac{1}{\lambda_{e,1} + \upsilon}\right\}}\right)^n \le \delta \ll 1. \tag{17.64}$$

For a good convergency of the eigenvalue $\lambda_{e,1}$ in a range of

$$0 \le \lambda_{e,1} \le \lambda_{\max} = (2\pi f_{\max})^2 \tag{17.65}$$

with

- $f_{\max}$ the maximum frequency (Hz) of interest

To achieve a good convergency ($\lambda_{e,1} \ll \upsilon$ and $\lambda_{e,2} \approx \lambda_{\max}$)

$$\upsilon \le \lambda_{max}\left(\frac{\varepsilon}{1-\varepsilon}\right), \varepsilon = \sqrt[n]{\delta}. \tag{17.66}$$

With $\delta = 10^{-8}$ and the number of iterations $n = 10$ the shift factor becomes

$$\upsilon = 0.18834\lambda_{\max}. \tag{17.67}$$

The elements of m start vectors $[\Psi_{e,0}]$ are random numbers. The start vectors are independent of each other. After n iterations and using (17.60) we have obtained groups of vectors

$$[\Psi_{e,0}], [\Psi_{e,1}], ..., [\Psi_{e,n-1}], [\Psi_{e,n}]. \tag{17.68}$$

A new set of q vectors $[X_{e,q}]$ will be defined, with $q \ll e$,

$$[X_{e,q}] = ([\Psi_{e,n}], [\Psi_{e,n-1}], [\Psi_{e,n-2}], ...). \tag{17.69}$$

The number q dofs must be taken γ times the number of eigenvalues below λ_{max}. [Gockel 83] suggest that $\gamma \approx 1.5$. The number of eigenvalues below λ_{max} can be obtained with the Sturm sequence [Strang 88] method. The vectors $[X_{e,q}]$ are made orthogonal with respect to the mass matrix $[M_{ee}]$, hence

$$[Y_{e,q}]^T[M_{ee}][Y_{e,q}] = \text{diagonal matrix}. \tag{17.70}$$

This can be done with the QR method [Strang 88] or the Gram–Schmidt method [Strang 88]. For an illustration of this the QR method will be applied. A square matrix $[A]$ may be decomposed with the QR method as follows

$$[A] = [Q][R], \tag{17.71}$$

with
- $[Q]$ the matrix containing a set of orthonormal vectors, and $[Q]^T[Q] = [I]$.

Define the matrix $[Y_{e,q}]$ such that

$$[Y_{e,q}] = [X_{e,q}][\beta_q]. \tag{17.72}$$

Cholesky decomposes [Strang 88] the mass matrix $[M_{ee}]$ as follows

$$[M_{ee}] = [L_{ee}][L_{ee}]^T. \tag{17.73}$$

Equation (17.70) now becomes

$$[\beta_q]^T[X_{e,q}][L_{ee}][L_{ee}]^T[X_{e,q}][\beta_q] = \text{diagonal matrix}. \tag{17.74}$$

This means that vectors in the matrix $[L_{ee}]^T[X_{e,q}]\{\beta_q\}$ form an orthogonal set of vectors. The matrices $[Q]$ and $[R]$ are the result of a QR decomposition on the matrix $[L_{ee}]^T[X_{e,q}]$. Thus $[L_{ee}]^T[X_{e,q}] = [Q][R]$. Assume $[\beta_q] = [R]^{-1}$, which results in

$$[X_{e,q}] = [Y_{e,q}][R], \tag{17.75}$$

and

$$[L_{ee}]^T[X_{e,q}] = [L_{ee}]^T[Y_{e,q}][R] = [Q][R]. \tag{17.76}$$

Finally we obtain

$$[Y_{e,q}] = [G_{eq}] = ([L_{ee}]^T)^{-1}[Q]. \tag{17.77}$$

The displacement $\{x\}$ is expressed as follows

$$\{x\} = \begin{Bmatrix} x_a \\ x_e \end{Bmatrix} = \begin{bmatrix} I \\ G_{ea} \end{bmatrix}\{x_a\} + \begin{bmatrix} 0 \\ G_{eq} \end{bmatrix}\{\eta_q\} = \begin{bmatrix} I & 0 \\ G_{ea} & G_{eq} \end{bmatrix}\begin{Bmatrix} x_a \\ \eta_q \end{Bmatrix} = [T_{nv}]\{x_v\}. \tag{17.78}$$

The reduced-mass matrix $[M_{vv}]$ becomes

$$[M_{vv}] = [T_{nv}]^T[M][T_{nv}]. \tag{17.79}$$

Analogous to the reduced-mass matrix $[M_{vv}]$ the reduced-stiffness matrix $[K_{vv}]$ is

$$[K_{vv}] = [T_{nv}]^T[K][T_{nv}]. \tag{17.80}$$

The GDR reduction method is very similar to the CB method.

17.7 System Equivalent Reduction Expansion Process (SEREP)

The SEREP is proposed by [Kammer 87] and is based upon a partitioning of the calculated mode shapes in combination with pseudo-inversion of matrices.

The displacement vector $x(t)$ is projected on the modal matrix $[\Phi]$. The number of m kept mode shapes is much less than the total number of degrees of freedom n, hence $m \ll n$. The displacement vector can be written as

$$x(t)) = [\Phi]\{\eta(t)\}, \tag{17.81}$$

with

- $\{\eta(t)\}$ the vector of generalised coordinates.

The displacement vector $x(t)$ will be partitioned into two sets; the kept set of degrees of freedom denoted by a and the eliminated set of degrees of freedom denoted by e, thus

$$\begin{Bmatrix} x_a \\ x_e \end{Bmatrix} = \begin{bmatrix} \Phi_a \\ \Phi_e \end{bmatrix} \{\eta\}, \tag{17.82}$$

We will express $\{x\}$ in $\{x_a\}$ as follows

$$\begin{Bmatrix} x_a \\ x_e \end{Bmatrix} = \begin{bmatrix} I \\ T_{ea} \end{bmatrix} \{x_a\} = [T_{\text{Kammer}}]\{x_a\}. \tag{17.83}$$

The a set of the displacement vector $\{x_a\}$ can be written as

$$\{x_a\} = [\Phi_a]\{\eta\} \tag{17.84}$$

We want to express the vector of generalised coordinates $\{\eta\}$ in $\{x_a\}$. However, the inverse of the rectangular matrix $[\Phi_a]$ does not exist. Both sides of (17.84) will be multiplied by $[\Phi_a]^T$, thus

$$[\Phi_a]^T\{x_a\} = [\Phi_a]^T[\Phi_a]\{\eta\}. \tag{17.85}$$

The matrix $[\Phi_a]^T[\Phi_a]$ is a square matrix and in general the inverse of that matrix exists. The generalised coordinates $\{\eta\}$ are expressed in $\{x_a\}$

$$\{\eta\}= ([\Phi_a]^T[\Phi_a])^{-1}[\Phi_a]^T\{x_a\} \tag{17.86}$$

The matrix $([\Phi_a]^T[\Phi_a])^{-1}[\Phi_a]^T$ is called the pseudo-inverse matrix of the modal matrix $[\Phi_a]$, hence

$$[\Phi_a]^{-1} = ([\Phi_a]^T[\Phi_a])^{-1}[\Phi_a]^T. \tag{17.87}$$

The displacement vector of eliminated degrees of freedom $\{x_e\}$ can be expressed in the set of kept degrees of freedom $\{x_a\}$. From (17.82) we can write

$$\{x_e\} = [\Phi_e]\{\eta\}, \tag{17.88}$$

and with (17.86) we obtain

$$\{x_e\} = [\Phi_e]([\Phi_a]^T[\Phi_a])^{-1}[\Phi_a]^T\{x_a\} = [T_{ea}]\{x_a\}. \qquad (17.89)$$

The complete displacement vector $\{x\}$ can be expressed in $\{x_a\}$, see (17.83)

$$\begin{Bmatrix} x_a \\ x_e \end{Bmatrix} = \begin{bmatrix} I \\ [\Phi_e]([\Phi_a]^T[\Phi_a])^{-1}[\Phi_a]^T \end{bmatrix}\{x_a\} = \begin{bmatrix} I \\ T_{ea} \end{bmatrix}\{x_a\} = [T_{\text{Kammer}}]\{x_a\}. \qquad (17.90)$$

The reduced-mass matrix $[M_{\text{SEREP}}]$ becomes

$$[M_{\text{SEREP}}] = [T_{\text{Kammer}}]^T[M][T_{\text{Kammer}}]. \qquad (17.91)$$

Analogous to the reduced-mass matrix $[M_{\text{SEREP}}]$ the reduced-stiffness matrix $[K_{\text{SEREP}}]$ is

$$[K_{\text{SEREP}}] = [T_{\text{Kammer}}]^T[K][T_{\text{Kammer}}]. \qquad (17.92)$$

The SEREP reduction method will provide 'physical' reduced matrices. In general, the kept degrees of freedom $\{x_a\}$ will correspond to measurement locations and directions.

We select the following kept dofs; $\{x_a\} = \lfloor x_1, x_4, x_5, x_7 \rfloor^T$ (Fig. 17.1) and take into account the first and second modes to reduce the model with the SEREP method.

The modes $[\Phi]$ are scaled such that

$$[\Phi]^T[M][\Phi] = [I] \text{ and } [\Phi]^T[K][\Phi] = \langle\omega^2\rangle = \langle(2\pi f)^2\rangle$$

$$[\Phi] = \begin{bmatrix} 0.3064 & -0.3878 \\ 0.2819 & -0.1817 \\ 0.2819 & -0.1817 \\ 0.3064 & -0.3878 \\ 0.3204 & 0.6591 \\ 0.2948 & 0.3088 \\ 0.2583 & -0.0304 \\ 0.1750 & -0.0333 \\ 0.0777 & -0.0185 \\ 0 & 0 \end{bmatrix}.$$

The first two calculated natural frequencies of the reduced system are

$$\{f_a\} = \lfloor 14.25, 36.69 \rfloor^T.$$

Four natural frequencies are calculated but the third and the fourth natural frequencies have no physical meaning. Only two modes of the reference dynamic model are taken into account.

The modes $[\Phi_{\text{SEREP}}]$ are scaled such that $[\Phi_{\text{SEREP}}]^T[M][\Phi_{\text{SEREP}}] = [I]$

$$[\Phi_{\text{SEREP}}] = \begin{bmatrix} -0.3064 & -0.3878 \\ -0.2819 & -0.1817 \\ -0.2819 & -0.1817 \\ -0.3064 & -0.3878 \\ -0.3204 & 0.6591 \\ -0.2948 & 0.3088 \\ -0.2583 & -0.0304 \\ -0.1750 & -0.0333 \\ -0.0777 & -0.0185 \\ 0 & 0 \end{bmatrix}.$$

The MAC and the modified MAC become

$$\text{MAC}= \frac{([\Phi]^T[\Phi_{\text{SEREP}}])^2}{([\Phi]^T[\Phi])([\Phi_{\text{SEREP}}]^T[\Phi_{\text{SEREP}}])} = \begin{bmatrix} 1.0000 & -0.0003 \\ -0.0006 & 1.0000 \end{bmatrix},$$

and

$$\text{NCO}= \frac{([\Phi]^T[M][\Phi_{\text{SEREP}}])^2}{[\Phi]^T[M][\Phi]([\Phi_{\text{SEREP}}]^T[M][\Phi_{\text{SEREP}}])} = \begin{bmatrix} 1.0000 & -0.0004 \\ -0.0004 & 1.0000 \end{bmatrix}.$$

The first two modes of the reduced model, associated with the first two natural frequencies, $\{f_a\} = \lfloor 14.25, 36.69 \rfloor^T$, do correlate very well with the first two modes of the reference model.

17.8 Ritz Vectors

The set of independent base vectors, the so-called Ritz vectors, to reduce the dynamic model mathematically are created using only a static analysis. The dynamic loads will be applied to the model of the structure quasistatically. The undamped equations of motion of the structure, applying dynamic loads, of (17.1), are

$$[M]\{\ddot{x}\} + [K]\{x\} = \{F\}.$$

For a base excitation the quasistatic loads (QL) are defined as:

$$\{F_{\text{QL}}\} = [M][T]\{\ddot{u}\}, \tag{17.93}$$

where:

- $\{F_{QL}\}$ is the vector of quasistatic loads
- $[T]$ is the rigid-body motion vector in a certain direction with respect to the point of base excitation, translation or rotation.
- $\{\ddot{u}\}$ the base acceleration, translation or rotation.
- $[T]\{\ddot{u}\}$ is analogous to the gravitational field.

Proper boundary conditions prevent a singular stiffness matrix $[K]$, so the flexibility matrix $[G]$ can be calculated $[G] = [K]^{-1}$.

We will solve the following static problem

$$\{\hat{y}_1\} = [G]\{F_{QL}\}. \tag{17.94}$$

The displacement vector $\{\hat{y}_1\}$ is normalised with respect to the mass matrix $[M]$,

$$\{y_1\}^T[M]\{y_1\} = 1. \tag{17.95}$$

The new displacement vector $\{y_1\}$ can be obtained with

$$\{\hat{y}_1\}^T[M]\{\hat{y}_1\} = c_1, \{y_1\} = \frac{\{\hat{y}_1\}}{\sqrt{c_1}} \tag{17.96}$$

The other Ritz vectors can be calculated as follows

- Calculate displacement vectors $\{\hat{y}_k\}$;
  $\{\hat{y}_k\} = [G][M]\{y_{k-1}\}, k = 1, 2, .., p$
- Calculate scalars c_k; $\{y_j\}^T[M]\{\hat{y}_k\} = c_j, j = 1, 2, ..., k-1$
- Create a orthogonal set of vector $\{\bar{y}_k\}, k = 1, 2,..,p$ with the Gramm–Schmitt method $\{\bar{y}_k\} = \{\hat{y}_k\} - \sum_{j=1}^{k-1} c_j\{y_j\}$ and normalise the vector $\{\bar{y}_k\}$ such that $\{y_k\}^T[M]\{y_k\} = 1, k = 1, 2, ..,p$. The same procedure as for $\{y_1\}$ is applied.

The displacement vector $\{x(t)\}$ is projected on the Ritz vectors $[Y]$

$$[Y] = [y_1, y_2, ..., y_p], \tag{17.97}$$

hence

$$\{x(t)\} = [Y]\{\rho(t)\}. \tag{17.98}$$

The reduced mass-matrix $[M_{\mathrm{Ritz}}]$ becomes

$$[M_{\mathrm{Ritz}}] = [Y]^T[M][Y]. \tag{17.99}$$

Analogous to the reduced-mass matrix $[M_{\mathrm{Ritz}}]$ the reduced-stiffness matrix is $[K_{\mathrm{Ritz}}]$

$$[K_{\mathrm{Ritz}}] = [Y]^T[K][Y]. \tag{17.100}$$

17.9 Conclusion

Reduced models are also used to support the modal survey, the experimental modal analysis. The dofs in the reduced dynamic model are related to measurements. The reduced dynamic model will be used to calculate the orthogonality relations between measured and analysed modes. This reduced model is called the test analysis model (TAM) [Kammer 87]. Reduction methods that will give a TAM are:

- The static condensation technique [Guyan 68]
- The dynamic reduction method [Miller 80]
- The improved reduced system (IRS) [O'Callahan 89]
- System equivalent reduction expansion proces (SEREP) [Kammer 87]

The other methods:

- Graig–Bampton (CB) reduction method [Craig 68]
- Generalised dynamic reduction (GDR) method [Gockel 83]
- Ritz vectors [Escobedo 93]

will result in a hybrid reduced mathematical dynamic model; dofs related to physical dofs combined with mathematical (generalised) dofs.

18 Component Mode Synthesis

18.1 Introduction

The component mode synthesis (CMS) or component modal synthesis [Hintz 75] or modal coupling technique [Maia 97] is used when components (substructures) are described by the mode displacement method (MDM) and coupled together (synthesis) via the common boundaries $\{x_b\}$ in order to perform a dynamic analysis, e.g. modal analysis, responses, on the complete structure (assembly of substructures). The CMS method can only be applied to linear structures. The component mode synthesis method can also be applied on components for which the modal characteristics were measured in combination with finite element reduced dynamic models. Many papers and reports are available in the open literature; e.g. [Craig 68, Craig 76, Craig 77, Curnier 81, Lacoste 83, MacNeal 71 Stavrinidis 84, Craig 00].

In general, a component or substructure is a recognisable part of the structure, e.g. for a spacecraft; the primary structure, the solar arrays, the antenna, large instruments, etc.

In the past, the CMS method was applied to significantly reduce the number of dofs due to the imposed limitations on computers, however, nowadays, these limitations are more or less removed but still the CMS method is very popular. Subcontractors deliver their reduced FE dynamic models to the prime contractor who will combine (synthesise) all these models to the spacecraft dynamic FE model to perform the dynamic analysis on the complete spacecraft. The same applies to the coupled dynamic load analysis (CDLA) when the reduced FE model of the complete spacecraft is placed on top of the launch vehicle. In general, the dynamic FE modal of the launch vehicle is a reduced dynamic FE model too.

Dynamic properties of substructures may be defined by experiment and may be coupled to other dynamic FE models of other substructures.

Hence, there are many reasons to apply the CMS method.

For dynamic analyses the components may be obtained by reducing the number of dofs by applying the MDM. The physical dofs $\{x\}$ are, in general, depicted on a small number of kept modes (eigenvectors), the modal base,

$$\{x\} = [\Phi]\{\eta\}, \tag{18.1}$$

- $[\Phi]$ the modal base consists of the kept mode
- $\{\eta\}$ the generalised or principal coordinates.

The number of generalised coordinates $\{\eta\}$ is, in general, much less than the number of physical dofs $\{x\}$.

In this chapter an introduction to the CMS method will be given and a number of methods will be discussed. We will assume undamped components, however, in a later stage during the synthesis the modal damping ratios can be introduced.

18.2 The Unified CMS Method

There are many CMS techniques described in the literature. The modal description of the components strongly depends on the boundary conditions applied by building the reduced FE model of the component. The discussion of the CMS method will be focused on:

- Components with fixed-interface dofs $\{x_b\}$
- Components with free interfaces
- Components with loaded interfaces

The component or substructure can be considered as a linear mdof dynamic system. The equations of motion of a damped component are written as:

$$\begin{bmatrix} M_{ii} & M_{ib} \\ M_{bi} & M_{bb} \end{bmatrix} \begin{Bmatrix} \ddot{x}_i \\ \ddot{x}_b \end{Bmatrix} + \begin{bmatrix} C_{ii} & C_{ib} \\ C_{bi} & C_{bb} \end{bmatrix} \begin{Bmatrix} \dot{x}_i \\ \dot{x}_b \end{Bmatrix} + \begin{bmatrix} K_{ii} & K_{ib} \\ K_{bi} & K_{bb} \end{bmatrix} \begin{Bmatrix} x_i \\ x_b \end{Bmatrix} = \begin{Bmatrix} F_i \\ F_b \end{Bmatrix}. \tag{18.2}$$

However, for the time being we will consider only undamped components, hence

$$\begin{bmatrix} M_{ii} & M_{ib} \\ M_{bi} & M_{bb} \end{bmatrix} \begin{Bmatrix} \ddot{x}_i \\ \ddot{x}_b \end{Bmatrix} + \begin{bmatrix} K_{ii} & K_{ib} \\ K_{bi} & K_{bb} \end{bmatrix} \begin{Bmatrix} x_i \\ x_b \end{Bmatrix} = \begin{Bmatrix} F_i \\ F_b \end{Bmatrix}, \tag{18.3}$$

with:

- $\{x_i\}$ the internal dofs
- $\{x_b\}$ the boundary or external dofs
- $[M]$ the mass matrix
- $[K]$ the stiffness matrix
- $\{F\}$ the force vector

18.2.1 Modal Truncation

A general unified component mode synthesis method has been developed [Curnier 81, Lacoste 83]. We start with equations of motion of two uncoupled components (substructures) A and B, based on (18.3).

$$\begin{bmatrix} M^A{}_{ii} & M^A{}_{ib} & 0 & 0 \\ M^A{}_{bi} & M^A{}_{bb} + m^A{}_{bb} & 0 & 0 \\ 0 & 0 & M^B{}_{ii} & M^B{}_{ib} \\ 0 & 0 & M^B{}_{bi} & M^B{}_{bb} + m^B{}_{bb} \end{bmatrix} \begin{Bmatrix} \ddot{x}^A{}_i \\ \ddot{x}^A{}_b \\ \ddot{x}^B{}_i \\ \ddot{x}^B{}_b \end{Bmatrix}$$

$$+ \begin{bmatrix} K^A{}_{ii} & K^A{}_{ib} & 0 & 0 \\ K^A{}_{bi} & K^A{}_{bb} + k^A{}_{bb} & 0 & 0 \\ 0 & 0 & K^B{}_{ii} & K^B{}_{ib} \\ 0 & 0 & K^B{}_{bi} & K^B{}_{bb} + k^B{}_{bb} \end{bmatrix} \begin{Bmatrix} x^A{}_i \\ x^A{}_b \\ x^B{}_i \\ x^B{}_b \end{Bmatrix} = \begin{Bmatrix} F^A{}_i \\ F^A{}_b \\ F^B{}_i \\ F^B{}_b \end{Bmatrix}, \tag{18.4}$$

with

- $m^A{}_{bb}$ the added mass to component A from adjacent component B
- $m^B{}_{bb}$ the added mass to component B from adjacent component A
- $k^A{}_{bb}$ the added stiffness to component A from adjacent component B
- $k^B{}_{bb}$ the added stiffness to component B from adjacent component A

The added matrices at the boundaries of the components can be obtained with, for example, the static condensation method [Guyan 68]. The boundary dofs $\{x_b\}$ of the adjacent component are considered the $\{x_a\}$ dofs during the reduction process. We distinguish components with:

- fixed-interface dofs: $m^A{}_{bb} \to \infty$ and $k^A{}_{bb} \to 0$
- free interfaces $m^A{}_{bb} \to 0$ and $k^A{}_{bb} \to 0$
- loaded interfaces $0 < m^A{}_{bb} \le \infty$ and $0 < k^A{}_{bb} \le \infty$. If the interface is determinate $k^A{}_{bb} = 0$.

The reduced-stiffness matrix $[k_{bb}]$ and reduced-mass matrix $[m_{bb}]$ of the adjacent component can be obtained using the static condensation method (GR) as proposed by [Guyan 68]. The GR reduced-mass matrices are

$$[m^A_{bb}] = [M^B_{bb}] + [K^B_{ib}][K^B_{ii}]^{-1}[M^B_{ii}][K^B_{ii}]^{-1}[K^B_{ib}] \tag{18.5}$$

$$[m^B_{bb}] = [M^A_{bb}] + [K^A_{ib}][K^A_{ii}]^{-1}[M^A_{ii}][K^A_{ii}]^{-1}[K^A_{ib}], \tag{18.6}$$

and the reduced-stiffness matrices are.

$$[k^A_{bb}] = [K^B_{bb}] - [K^B_{ib}][K^B_{ii}]^{-1}[K^B_{ib}] \tag{18.7}$$

$$[k^B_{bb}] = [K^A_{bb}] - [K^A_{ib}][K^A_{ii}]^{-1}[K^A_{ib}]. \tag{18.8}$$

When the interface between two components is determinate the reduced-stiffness matrices are $[k^A_{bb}] = [k^B_{bb}] = [0]$.

18.2.2 General Synthesis of Two Components

If both components A and B the physical dofs are depicted on a reduced number of modes $[\Phi_A]$ of component A and $[\Phi_B]$ of component B we may write

$$\begin{Bmatrix} x^A_i \\ x^A_b \\ x^B_b \\ x^B_i \end{Bmatrix} = \begin{bmatrix} \Phi^A_{ii} & \Phi^A_{ib} & 0 & 0 \\ \Phi^A_{bi} & \Phi^A_{bb} & 0 & 0 \\ 0 & 0 & \Phi^B_{bb} & \Phi^B_{bi} \\ 0 & 0 & \Phi^B_{ib} & \Phi^B_{ii} \end{bmatrix} \begin{Bmatrix} \eta^A_i \\ \eta^A_b \\ \eta^B_b \\ \eta^B_i \end{Bmatrix}. \tag{18.9}$$

The eigenvectors of a 3-dofs system are

$$\begin{Bmatrix} x_{b1} \\ x_i \\ x_{b2} \end{Bmatrix} = \begin{bmatrix} -0.5774 & -0.7071 & 0.4082 \\ -0.5774 & 0.0000 & -0.8165 \\ -0.5774 & 0.7071 & 0.4082 \end{bmatrix} \cdot$$

The modal matrix

$$\begin{bmatrix} \Phi_{ii}^A & \Phi_{ib}^A \\ \Phi_{bi}^A & \Phi_{bb}^A \end{bmatrix} = \begin{Bmatrix} x_i \\ x_{b1} \\ x_{b2} \end{Bmatrix} = \begin{bmatrix} 0.0000 & -0.5774 & -0.8165 \\ -0.7071 & -0.5774 & 0.4082 \\ 0.7071 & -0.5774 & 0.4082 \end{bmatrix} \cdot$$

Synthesis is the assembly of the uncoupled equations in (18.9) of the two components (substructures) *A* and *B*. The substructures will be coupled via their common interfaces

$$\{x_b^A\} = \{x_b^B\} . \tag{18.10}$$

This means

$$\{x_b^A\} = [\Phi_{bi}^A]\{\eta_i^A\} + [\Phi_{bb}^A]\{\eta_b^A\} = [\Phi_{bi}^B]\{\eta_i^B\} + [\Phi_{bb}^B]\{\eta_b^B\} = \{x_b^B\} . \tag{18.11}$$

This will result in the following equation

$$[\Phi_{bb}^A]\{\eta_b^A\} - [\Phi_{bb}^B]\{\eta_b^B\} = [\Phi_{bi}^B]\{\eta_i^B\} - [\Phi_{bi}^A]\{\eta_i^A\} \tag{18.12}$$

If we introduce new generalised co-ordinates

$$\{\eta_b^{AB}\} = \frac{1}{2}([\Phi_{bb}^A]\{\eta_b^A\} + [\Phi_{bb}^B]\{\eta_b^B\}) . \tag{18.13}$$

Adding and subtracting (18.12) and the (18.13) will make $\{\eta_b^A\}$ and $\{\eta_b^B\}$ explicit

$$\{\eta_b^A\} = [\Phi_{bb}^A]^{-1}\left(\{\eta_b^{AB}\} + \frac{1}{2}[\Phi_{bi}^B]\{\eta_i^B\} - \frac{1}{2}[\Phi_{bi}^A]\{\eta_i^A\}\right) \tag{18.14}$$

$$\{\eta_b^B\} = [\Phi_{bb}^B]^{-1}\left(\{\eta_b^{AB}\} - \frac{1}{2}[\Phi_{bi}^B]\{\eta_i^B\} + \frac{1}{2}[\Phi_{bi}^A]\{\eta_i^A\}\right) . \tag{18.15}$$

We obtain the following relation

$$\begin{Bmatrix} \eta_i^A \\ \eta_b^A \\ \eta_b^B \\ \eta_i^B \end{Bmatrix} = \begin{bmatrix} I & 0 & 0 \\ -\frac{1}{2}[\Phi_{bb}^A]^{-1}[\Phi_{bi}^A] & [\Phi_{bb}^A]^{-1} & \frac{1}{2}[\Phi_{bb}^A]^{-1}[\Phi_{bi}^B] \\ \frac{1}{2}[\Phi_{bb}^B]^{-1}[\Phi_{bi}^A] & [\Phi_{bb}^B]^{-1} & -\frac{1}{2}[\Phi_{bb}^B]^{-1}[\Phi_{bi}^B] \\ 0 & 0 & I \end{bmatrix} \begin{Bmatrix} \eta_i^A \\ \eta_b^{AB} \\ \eta_i^B \end{Bmatrix}, \tag{18.16}$$

and finally the relation between the physical dofs and the reduced set of generalised dofs becomes

$$\begin{Bmatrix} x_i^A \\ x_b^A \\ x_b^B \\ x_i^B \end{Bmatrix} = \begin{bmatrix} \Phi_{ii}^A & \Phi_{ib}^A & 0 & 0 \\ \Phi_{bi}^A & \Phi_{bb}^A & 0 & 0 \\ 0 & 0 & \Phi_{bb}^B & \Phi_{bi}^B \\ 0 & 0 & \Phi_{ib}^B & \Phi_{ii}^B \end{bmatrix} \begin{bmatrix} I & 0 & 0 \\ -\frac{1}{2}[\Phi_{bb}^A]^{-1}[\Phi_{bi}^A] & [\Phi_{bb}^A]^{-1} & \frac{1}{2}[\Phi_{bb}^A]^{-1}[\Phi_{bi}^B] \\ \frac{1}{2}[\Phi_{bb}^B]^{-1}[\Phi_{bi}^A] & [\Phi_{bb}^B]^{-1} & -\frac{1}{2}[\Phi_{bb}^B]^{-1}[\Phi_{bi}^B] \\ 0 & 0 & I \end{bmatrix} \begin{Bmatrix} \eta_i^A \\ \eta_b^{AB} \\ \eta_i^B \end{Bmatrix}. \tag{18.17}$$

18.2.3 General Example

This general example is based on the work of [Curnier 81] and is an application of the three modal synthesis variants to 3 dofs launcher–payload model. This model is illustrated in Fig. 18.1.

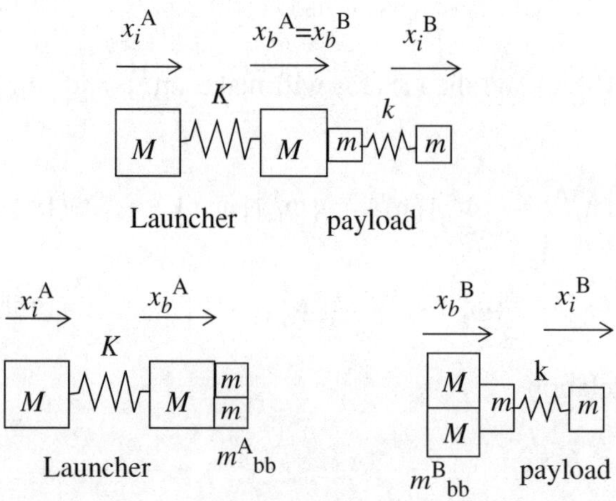

Fig. 18.1. 3-dofs launcher–payload model

The undamped equations of motion of the two uncoupled components are

$$\begin{bmatrix} M & 0 & 0 & 0 \\ 0 & M & 0 & 0 \\ 0 & 0 & m & 0 \\ 0 & 0 & 0 & m \end{bmatrix} \begin{Bmatrix} \ddot{x}_i^A \\ \ddot{x}_b^A \\ \ddot{x}_b^B \\ \ddot{x}_i^B \end{Bmatrix} + \begin{bmatrix} K & -K & 0 & 0 \\ -K & K & 0 & 0 \\ 0 & 0 & k & -k \\ 0 & 0 & -k & k \end{bmatrix} \begin{Bmatrix} x_i^A \\ x_b^A \\ x_b^B \\ x_i^B \end{Bmatrix} = \begin{Bmatrix} 0 \\ 0 \\ 0 \\ 0 \end{Bmatrix}. \tag{18.18}$$

The coupled equations of motion become

$$\begin{bmatrix} M & 0 & 0 \\ 0 & M+m & 0 \\ 0 & 0 & m \end{bmatrix} \begin{Bmatrix} \ddot{x}_i^A \\ \ddot{x}_b^A \\ \ddot{x}_i^B \end{Bmatrix} + \begin{bmatrix} K & -K & 0 \\ -K & K+k & -k \\ 0 & -k & k \end{bmatrix} \begin{Bmatrix} x_i^A \\ x_b^A \\ x_i^B \end{Bmatrix} = \begin{Bmatrix} 0 \\ 0 \\ 0 \end{Bmatrix}, \tag{18.19}$$

with $S = \frac{K}{M} + \frac{k}{m} + \frac{K+k}{M+m}$ and $P = \frac{2Kk}{Mm}$ the eigenvalues of the dynamic system described with (18.19) are

$$\{\lambda\} = \begin{bmatrix} 0 \\ \frac{1}{2}\left\{S - \sqrt{S^2 - 4P}\right\} \\ \frac{1}{2}\left\{S + \sqrt{S^2 - 4P}\right\} \end{bmatrix}, \tag{18.20}$$

and the associated modal matrix $[\Phi]$

$$[\Phi] = \begin{bmatrix} 1 & 1 & \frac{K}{K-2M} \\ 1 & 0 & 1 \\ 1 & -\frac{k}{K} & \frac{k}{k-2m} \end{bmatrix}. \tag{18.21}$$

The equations of motion after the interface treatment m_{bb}^A, m_{bb}^B, k_{bb}^A and k_{bb}^B become

$$\begin{bmatrix} M & 0 & 0 & 0 \\ 0 & M+m_{bb}^A & 0 & 0 \\ 0 & 0 & m+m_{bb}^B & 0 \\ 0 & 0 & 0 & m \end{bmatrix} \begin{Bmatrix} \ddot{x}_i^A \\ \ddot{x}_b^A \\ \ddot{x}_b^B \\ \ddot{x}_i^B \end{Bmatrix} + \begin{bmatrix} K & -K & 0 & 0 \\ -K & K+k_{bb}^A & 0 & 0 \\ 0 & 0 & k+k_{bb}^B & -k \\ 0 & 0 & -k & k \end{bmatrix} \begin{Bmatrix} x_i^A \\ x_b^A \\ x_b^B \\ x_i^B \end{Bmatrix} = \begin{Bmatrix} 0 \\ 0 \\ 0 \\ 0 \end{Bmatrix}. \tag{18.22}$$

The following options can be taken into account

- free-interface $m_{bb}^A = m_{bb}^B = k_{bb}^A = k_{bb}^B = 0$
- fixed-interface $m_{bb}^A = m_{bb}^B = \infty$ and $k_{bb}^A = k_{bb}^B = 0$
- loaded-interface $m_{bb}^A = m + k\frac{1}{k}m\frac{1}{k}k = 2m$, $m_{bb}^B = M + K\frac{1}{K}M\frac{1}{K}K = 2M$,

  $k_{bb}^A = k - k\frac{1}{k}k = 0$ and $k_{bb}^B = K - K\frac{1}{K}K = 0$

The characteristic equations are

- launcher $\begin{vmatrix} K-\lambda M & -K \\ -K & K-\lambda(M+m_{bb}^A) \end{vmatrix} = 0$
- payload $\begin{vmatrix} k-\lambda m & -k \\ -k & K-\lambda(m+m_{bb}^{\boldsymbol{B}}) \end{vmatrix} = 0$

With reference to (18.9) we find

$$\begin{Bmatrix} x_i^A \\ x_b^A \\ x_b^B \\ x_i^B \end{Bmatrix} = \begin{bmatrix} \Phi_{ii}^A & \Phi_{ib}^A & 0 & 0 \\ \Phi_{bi}^A & \Phi_{bb}^A & 0 & 0 \\ 0 & 0 & \Phi_{bb}^B & \Phi_{bi}^B \\ 0 & 0 & \Phi_{ib}^B & \Phi_{ii}^B \end{bmatrix} \begin{Bmatrix} \eta_i^A \\ \eta_b^A \\ \eta_b^B \\ \eta_i^B \end{Bmatrix} = \begin{bmatrix} 1 & 1 & 0 & 0 \\ \alpha = \dfrac{-M}{M+m_{bb}^A} & 1 & 0 & 0 \\ 0 & 0 & 1 & \beta = \dfrac{-m}{m+m_{bb}^B} \\ 0 & 0 & 1 & 1 \end{bmatrix} \begin{Bmatrix} \eta_i^A \\ \eta_b^A \\ \eta_b^B \\ \eta_i^B \end{Bmatrix}. \tag{18.23}$$

Table 18.1. Characteristics general example

Rigid-body natural frequency $\left(\frac{\text{rad}}{\text{s}}\right)^2$	Elastic natural frequency $\left(\frac{\text{rad}}{\text{s}}\right)^2$	Interface	Added mass
λ_0	λ_1		$m_{bb}^A = 2m$
0	$\frac{K}{M}\frac{2M+m_{bb}^A}{M+m_{bb}^A}$		
0	$\frac{2K}{M}$	free	0
0	$\frac{K}{M}\left(\frac{2M+2m}{M+2m}\right)$	loaded	$2m$
0	$\frac{K}{M}$	fixed	∞
Payload			
0	$\frac{k}{m}\frac{2m+m_{bb}^B}{m+m_{bb}^B}$		$m_{bb}^B = 2M$
0	$\frac{2k}{m}$	free	0
0	$\frac{k}{m}\left(\frac{2m+2M}{m+2M}\right)$	loaded	$2M$
0	$\frac{k}{m}$	fixed	∞

Equation (18.17) now becomes

$$\begin{Bmatrix} \eta_i^A \\ \eta_b^A \\ \eta_i^B \\ \eta_b^B \end{Bmatrix} = \begin{bmatrix} I & 0 & 0 \\ -\frac{1}{2}[\Phi_{bb}^A]^{-1}[\Phi_{bi}^A] & [\Phi_{bb}^A]^{-1} & \frac{1}{2}[\Phi_{bb}^A]^{-1}[\Phi_{bi}^B] \\ \frac{1}{2}[\Phi_{bb}^B]^{-1}[\Phi_{bi}^A] & [\Phi_{bb}^B]^{-1} & -\frac{1}{2}[\Phi_{bb}^B]^{-1}[\Phi_{bi}^B] \\ 0 & 0 & I \end{bmatrix} \begin{Bmatrix} \eta_i^A \\ \eta_b^{AB} \\ \eta_i^B \end{Bmatrix} = \begin{bmatrix} 1 & 0 & 0 \\ \frac{-\alpha}{2} & 1 & \frac{\beta}{2} \\ \frac{\alpha}{2} & 1 & \frac{-\beta}{2} \\ 0 & 0 & 1 \end{bmatrix} \begin{Bmatrix} \eta_i^A \\ \eta_b^{AB} \\ \eta_i^B \end{Bmatrix}. \tag{18.24}$$

Equation (18.17) becomes

$$\begin{Bmatrix} x_i^A \\ x_b^A \\ x_b^B \\ x_i^B \end{Bmatrix} = \begin{bmatrix} 1-\frac{\alpha}{2} & 1 & \frac{\beta}{2} \\ \frac{\alpha}{2} & 1 & \frac{\beta}{2} \\ \frac{\alpha}{2} & 1 & \frac{\beta}{2} \\ \frac{\alpha}{2} & 1 & 1-\frac{\beta}{2} \end{bmatrix} \begin{Bmatrix} \eta_i^A \\ \eta_b^{AB} \\ \eta_i^B \end{Bmatrix} = [T]\{\Pi\}. \tag{18.25}$$

The synthesised mass matrix $[M_{\text{syn}}] = [T]^T[M_{tot}][T]$ becomes

$$[M_{\text{syn}}] = \begin{bmatrix} (1-\alpha)M + \frac{\alpha^2}{2}(M+m) & M+\alpha m & \frac{\beta M + \alpha m}{2} \\ M + \alpha m & 2(M+m) & m + \beta M \\ \frac{\beta M + \alpha m}{2} & m + \beta M & (1-\beta)m + \frac{\beta^2}{2}(M+m) \end{bmatrix}, \tag{18.26}$$

and the synthesised stiffness matrix $[K_{\text{syn}}] = [T]^T[K_{\text{tot}}][T]$ becomes

$$[K_{\text{syn}}] = \begin{bmatrix} (1-\alpha)^2 K & 0 & 0 \\ 0 & 0 & 0 \\ 0 & 0 & (1-\beta)^2 k \end{bmatrix}. \tag{18.27}$$

The evaluation of the parameters is shown in Table 18.2.

Table 18.2. Evaluation parameters

Parameters	Free	Loaded	Fixed
$\alpha = \frac{-M}{M + m_{bb}^A}$	-1	$\frac{-M}{M+2m}$	0
$\beta = \frac{-m}{m + m_{bb}^B}$	-1	$\frac{-m}{m+2M}$	0

The following properties are assumed:

- Launcher: $K = 75$ N/m, $M = 200$ kg
- Payload: $k = 10$ N/m, $m = 5$ kg

The following eigenvalues for the two synthesised components (launcher, payload) can be calculated

- free: 0.0000, 0.7273, 2.0623
- loaded: 0.0000, 0.7273, 2.0623
- fixed: 0.0000, 0.7273, 2.0623

which corresponds with the theoretical solution.

18.3 Special CMS Methods

In this section three special CMS methods will be presented:

- The Craig–Bampton fixed-interface method.
- The free-interface method with improved accuracy
- The general CMS method, which combines the fixed- and free-interface CMS methods.

18.3.1 Craig–Bampton Fixed-Interface Method

The fixed-interface method (Craig–Bampton method) is discussed in several publications [Craig 68, Craig 77, Craig 81, Craig 00, Gordon 99] and is one of the favourite methods used in the CMS.

We denote the external or boundary degrees of freedom with the index *b* and the internal degrees of freedom with the index *i*. The matrix equations (18.3) may be partitioned as follows

$$\begin{bmatrix} M_{ii} & M_{ib} \\ M_{bi} & M_{bb} \end{bmatrix} \begin{Bmatrix} \ddot{x}_i \\ \ddot{x}_b \end{Bmatrix} + \begin{bmatrix} K_{ii} & K_{ib} \\ K_{bi} & K_{bb} \end{bmatrix} \begin{Bmatrix} x_i \\ x_b \end{Bmatrix} = \begin{Bmatrix} F_i \\ F_b \end{Bmatrix}.$$

In [Craig 68] it is proposed to depict the displacement vector $\{x(t)\}$ on a basis of static or constraint modes $[\Phi_c]$ with $\{x_b\} = [I]$ and elastic mode shapes $[\Phi_i]$ with fixed external degrees of freedom $\{x_b\} = \{0\}$ and the eigenvalue problem $[[K_{ii}] - \langle\lambda_p\rangle[M_{ii}]][\Phi_{ii}] = [0]$. We can express $\{x\}$ as

$$\{x\} = [\Phi_c]\{x_b\} + [\Phi_i]\{\eta_i\} = [\Phi_c, \Phi_i]\begin{Bmatrix} x_b \\ \eta_i \end{Bmatrix}. \tag{18.28}$$

The static modes can be obtained, assuming zero inertia effects, $\{F_i\} = \{0\}$, and successively prescribe a unit displacement for the boundary degrees of freedom, thus $\{x_b\} = [I]$. So we may write (18.3) as follows

$$\begin{bmatrix} K_{ii} & K_{ib} \\ K_{bi} & K_b \end{bmatrix} \begin{Bmatrix} x_i \\ x_b \end{Bmatrix} = \begin{Bmatrix} 0 \\ R_b \end{Bmatrix}. \tag{18.29}$$

From the first equation of (18.29) we find for $\{x_i\}$

$$[K_{ii}]\{x_i\} + [K_{ib}][x_b] = 0, \tag{18.30}$$

hence

$$\{x_i\} = -[K_{ii}]^{-1}[K_{ib}]\{x_b\}, \tag{18.31}$$

and therefore

$$[\Phi_{ib}] = -[K_{ii}]^{-1}[K_{ib}][I] = -[K_{ii}]^{-1}[K_{ib}]. \tag{18.32}$$

The static transformation now becomes

$$\{x\} = \begin{Bmatrix} x_i \\ x_b \end{Bmatrix} = \begin{bmatrix} \Phi_{ib} \\ I \end{bmatrix} \{x_b\} = [\Phi_c]\{x_b\}. \tag{18.33}$$

Assuming fixed external degrees of freedom $\{x_b\} = \{0\}$ and also assuming harmonic motions $x(t) = X(\omega)e^{j\omega t}$ the eigenvalue problem can be stated as

$$([K_{ii}] - \omega_i^2[M_{ii}])\{\Phi_{ii}\} = \{0\}. \tag{18.34}$$

The internal degrees of freedom $\{x_i\}$ will be projected on the set of orthogonal mode shapes (modal matrix) $[\Phi_{ip}]$, thus

$$\{x_i\} = [\Phi_{ii}]\{\eta_i\}. \tag{18.35}$$

The modal transformation becomes

$$\{x\} = \begin{Bmatrix} x_i \\ x_b \end{Bmatrix} = \begin{bmatrix} \Phi_{ii} \\ 0 \end{bmatrix} \{\eta_i\} = [\Phi_i]\{\eta_i\}. \tag{18.36}$$

The Craig–Bampton (CB) transformation matrix is (18.28)

$$[\Phi_c, \Phi_i] = \begin{bmatrix} -[K_{ii}]^{-1}[K_{ib}] & \Phi_{ii} \\ I & 0 \end{bmatrix}.$$

The Craig–Bampton (CB) transformation matrix is (18.28)

$$\{x\} = [\Phi_c, \Phi_i]\begin{Bmatrix} x_b \\ \eta_i \end{Bmatrix} = [\Psi]\{X\}.$$

with

- $[\Phi_c]$ the static or constraint modes
- $[\Phi_i]$ the modal matrix
- $\{x_b\}$ the external or boundary degrees of freedom
- $\{\eta_i\}$ the generalised coordinates. In general, the number of generalised coordinates i is much less than the total number of degrees of freedom $n = b + i$, $i \ll n$.

The CB transformation (18.28) will be substituted into (18.3) presuming equal potential and kinetic energies, hence

$$[\Psi]^T[M][\Psi]\{\ddot{X}\} + [\Psi]^T[K][\Psi]\{X\} = [\Psi]^T\{F(t)\}. \tag{18.37}$$

On further elaboration we find

$$\begin{bmatrix} \tilde{M}_{bb} & M_{bi} \\ M_{ib} & \langle m_i \rangle \end{bmatrix}\begin{Bmatrix} \ddot{x}_b \\ \ddot{\eta}_i \end{Bmatrix} + \begin{bmatrix} \tilde{K}_{bb} & K_{bi} \\ K_{ib} & \langle k_i \rangle \end{bmatrix}\begin{Bmatrix} x_b \\ \eta_i \end{Bmatrix} = \begin{bmatrix} \Phi_{ib} & \Phi_i \\ I & 0 \end{bmatrix}^T\begin{Bmatrix} F_i \\ F_b \end{Bmatrix}, \tag{18.38}$$

with

- $[\tilde{M}_{bb}]$ the Guyan reduced-mass matrix (b-set)
- $[\tilde{K}_{bb}]$ the Guyan reduced-stiffness matrix (b-set)
- $\langle m_i \rangle$ the diagonal matrix of generalised masses, $\langle m_i \rangle = \Phi_i^T[M][\Phi_i]$
- $\langle k_i \rangle$ the diagonal matrix of generalised stiffnesses,

$$\langle k_i \rangle = [\Phi_i]^T[K][\Phi_i] = \langle \lambda_i \rangle \langle m_i \rangle = \langle \omega_i^2 \rangle \langle m_i \rangle$$

- $[K_{ib}] = [\Phi_{ib}]^T[K_{ii}][\Phi_i] + [K_{bi}][\Phi_i] = (-[K_{ib}]^T[K_{ii}]^{-1}[K_{ii}] + [K_{bi}])[\Phi_i] = [0]$
- $[K_{ib}] = K_{bi}^T = [0]$

Thus (18.38) becomes

$$\begin{bmatrix} \tilde{M}_{bb} & M_{bi} \\ M_{ib} & \langle m_i \rangle \end{bmatrix}\begin{Bmatrix} \ddot{x}_b \\ \ddot{\eta}_i \end{Bmatrix} + \begin{bmatrix} \tilde{K}_{bb} & 0 \\ 0 & \langle k_i \rangle \end{bmatrix}\begin{Bmatrix} x_b \\ \eta_i \end{Bmatrix} = \begin{bmatrix} \Phi_{ib} & \Phi_i \\ I & 0 \end{bmatrix}^T\begin{Bmatrix} F_i \\ F_b \end{Bmatrix}. \tag{18.39}$$

Finally

$$[M_{CB}]\{\ddot{X}\} + [K_{CB}]\{X\} = [\Psi]^T\{F\} = \{F_{CB}\}, \quad (18.40)$$

with

- $[M_{\mathrm{CB}}]$ the CB reduced-mass matrix
- $[K_{\mathrm{CB}}]$ the CB reduced-stiffness matrix

The CB matrices are $b+i\,,b+i$ sized matrices.

If we look at the reduced-mass matrix $[M_{CB}]$ and the reduced-stiffness matrix $[K_{\mathrm{CB}}]$ in more detail we observe only a mass coupling between the internal dofs $\{x_i\}$ and the external dofs $\{x_b\}$ in the reduced-mass matrix $[M_{\mathrm{CB}}]$ via the sub- matrices $[M_{bi}]$ and $[M_{ib}]$ consisting of the modal participation factors. We now write (18.4) as follows

$$\begin{bmatrix} M^A_{\mathrm{CB}} & 0 \\ 0 & M^B_{\mathrm{CB}} \end{bmatrix} \begin{Bmatrix} \ddot{X}^A \\ \ddot{X}^B \end{Bmatrix} + \begin{bmatrix} K^A_{\mathrm{CB}} & 0 \\ 0 & K^B_{\mathrm{CB}} \end{bmatrix} \begin{Bmatrix} X^A \\ X^B \end{Bmatrix} = \begin{Bmatrix} F^A_{\mathrm{CB}} \\ F^B_{\mathrm{CB}} \end{Bmatrix} \quad (18.41)$$

$$[M_{\mathrm{tot}}]\{\ddot{Q}_{\mathrm{tot}}\} + [K_{\mathrm{tot}}]\{Q_{\mathrm{tot}}\} = \{F_{\mathrm{tot}}\}\ , \quad (18.42)$$

or

$$\begin{bmatrix} \langle m_i\rangle^A & M^A_{ib} & 0 & 0 \\ M^A_{bi} & \tilde{M}^A_{bb} & 0 & 0 \\ 0 & 0 & \langle m_i\rangle^B & M^B_{ib} \\ 0 & 0 & M^B_{bi} & \tilde{M}^B_{bb} \end{bmatrix} \begin{Bmatrix} \ddot{\eta}^A{}_i \\ \ddot{x}^A{}_b \\ \ddot{\eta}^B{}_i \\ \ddot{x}^B{}_b \end{Bmatrix} + \begin{bmatrix} \langle k_i\rangle^A & 0 & 0 & 0 \\ 0 & \tilde{K}^A_{bb} & 0 & 0 \\ 0 & 0 & \langle k_i\rangle^B & 0 \\ 0 & 0 & 0 & \tilde{K}^B_{bb} \end{bmatrix} \begin{Bmatrix} \eta^A{}_i \\ x^A{}_b \\ \eta^B{}_i \\ x^B{}_b \end{Bmatrix} = \begin{Bmatrix} F^A{}_{CB,\,i} \\ F^A{}_{CB,\,b} \\ F^B{}_{CB,\,i} \\ F^B{}_{CB,\,b} \end{Bmatrix} . \quad (18.43)$$

For coupling the substructures A and B we assume equal displacement and acceleration of the external dofs $x^A{}_b = x^B{}_b$ and $\ddot{x}^A{}_b = \ddot{x}^B{}_b$. Therefore, the total displacement vector can be written as

$$\{Q_{\mathrm{tot}}\} = \begin{Bmatrix} \eta^A{}_i \\ x^A{}_b \\ \eta^B{}_i \\ x^B{}_b \end{Bmatrix} = \begin{bmatrix} I & 0 & 0 \\ 0 & 0 & I \\ 0 & I & 0 \\ 0 & 0 & I \end{bmatrix} \begin{Bmatrix} \eta^A{}_i \\ \eta^B{}_i \\ x^A{}_b = x^B{}_b \end{Bmatrix} = [L]\{Q_{\mathrm{red}}\} \quad (18.44)$$

The transformation matrix $[L]$ also applies to the acceleration. Using the equations of Lagrange or assuming equal potential and kinetic energies (18.44) can be incorporated into (18.43).

$$[L]^T[M_{\text{tot}}][L]\{\ddot{Q}_{\text{red}}\} + [L]^T[K_{\text{tot}}][L]\{Q_{\text{red}}\} = [L]^T\{F_{\text{tot}}\} = \{F_{\text{red}}\} \quad (18.45)$$

$$\begin{bmatrix} \langle m_i \rangle^A & 0 & M^A_{ib} \\ 0 & \langle m_i \rangle^B & M^B_{ib} \\ M^A_{bi} & M^B_{bi} & \tilde{M}^A_{bb} + \tilde{M}^B_{bb} \end{bmatrix} \begin{Bmatrix} \ddot{\eta}^A_i \\ \ddot{\eta}^B_i \\ \ddot{x}^A_b \end{Bmatrix} + \begin{bmatrix} \langle k_i \rangle^A & 0 & 0 \\ 0 & \langle k_i \rangle^B & 0 \\ 0 & 0 & \tilde{K}^A_{bb} + \tilde{K}^B_{bb} \end{bmatrix} \begin{Bmatrix} \ddot{\eta}^A_i \\ \ddot{\eta}^B_i \\ \ddot{x}^A_b \end{Bmatrix}$$

$$= \begin{Bmatrix} F^A_{\text{CB},i} \\ F^B_{\text{CB},i} \\ F^A_{\text{CB},b} + F^B_{\text{CB},b} \end{Bmatrix}. \quad (18.46)$$

The reduced-mass matrices $[\tilde{M}^A_{bb}]$, $[\tilde{M}^B_{bb}]$, the reduced_stiffness matrices $[\tilde{K}^A_{bib}]$ and $[\tilde{K}^B_{bb}]$, related to the common boundary dofs $\{\ddot{x}^A_b\} = \{\ddot{x}^B_b\}$ and $\{x^A_b\} = \{x^B_b\}$ are added. The generalised masses $\langle m_i \rangle^A$ and $\langle m_i \rangle^B$ are coupled via the modal participation factors (matrices $[M_{bi}]$ and $[M_{ib}]$) to the reduced- mass matrix $[\tilde{M}^A_{bb} + \tilde{M}^B_{bb}]$. The generalised stiffnesses are not coupled with the reduced-stiffness matrix $[\tilde{K}^A_{bb} + \tilde{K}^B_{bb}]$.

The Craig–Bampton method is widely applied in the cases when the component dynamic properties are described by their mass and stiffness matrices.

A linear free-free dynamic system consists of 19 dofs; 1 to 19. The lumped masses at dof 1 and dof 19 are $m_1 = m_{19} = 0.5$ kg The masses lumped to the other dofs, 2 to 18, are $m_2 = m_3 = ... = m_{17} = m_{18} = 1$ kg. The 18 springs between the dofs 1 to 19 are equal, $k_{12} = k_{23} = ... = k_{1718} = k_{1819} = 10000$ N/m. The free-free dynamic system is illustrated in Fig. 18.2.

Fig. 18.2. Free dynamic system with 19 dofs

A linear free-free substructure consists of 7 dofs; 1 to 7. The lumped masses at dof 1 and dof 7 are $m_1 = m_7 = 0.5$ kg The other masses lumped to the other dofs, 2 to 6, are $m_2 = ... = m_6 = 1$ kg. The 6 springs between the dofs 1 to 19 are equal, $k_{12} = k_{23} = ... = k_{67} = 10000$ N/m. The substructure is shown in Fig. 18.3.

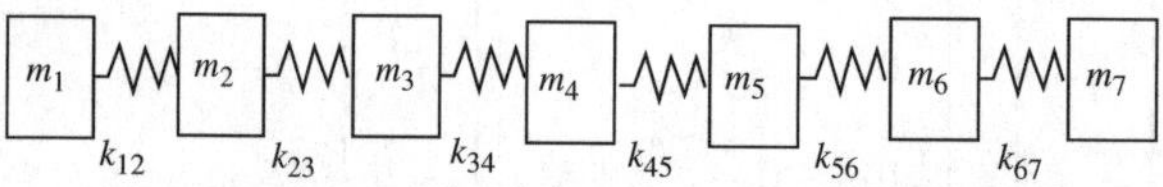

Fig. 18.3. Free-Free substructure with 7 dofs

Three substructures will build up the total structure as illustrated in Fig. 18.2. The results of the analyses are shown in Table 18.3.

Table 18.3. Results of CMS natural-frequency calculations

#	Complete Model (Hz)	Model A 1 mode per substructure (7 dofs) (Hz)	Model B 2 modes per substructure (10 dofs) (Hz)
1	0.0000	0.0000	0.0000
2	2.7743	2.7771 (0.1%)	2.7752 (0.0%)
3	5.5274	5.5803 (1.5%)	5.5316 (0.1%)
4	8.2385	8.9437 (9.5%)	8.2535 (0.2%)
5	10.8868	11.9825 (10.0%)	10.9253 (0.4%)
6	13.4524	15.5038	13.6279
7	15.9155	17.3217	17.3217
8	18.2575		19.7890
9	20.4606		22.5478
10	22.5079		23.7726

18.3.2 Free-Interface Method

The principle of CMS with the free-interface method (unconstrained boundaries) is discussed by Graig, [Craig 76, Craig 77, Craig 00]. The basic free-free undamped equations of motion are taken from (18.3), simply written as

$$[M]\{\ddot{x}\} + [K]\{x\} = \{F\}. \tag{18.47}$$

The eigenvalue problem is

$$\{[K] - \omega_i^2[M]\}\{\phi_i\} = \{0\}. \tag{18.48}$$

The n physical dofs $\{x\}$ are projected on the linear independent set of eigenvectors, the so-called modal matrix

$$[\Phi] = [\phi_1, \phi_2, \ldots\ldots, \phi_n], \tag{18.49}$$

hence

$$\{x\} = [\Phi]\{\eta\}. \tag{18.50}$$

The modal matrix $[\Phi]$ is orthogonal with respect to the mass matrix $[M]$, thus

$$[\Phi]^T[M][\Phi] = \langle m \rangle, \tag{18.51}$$

and orthogonal with respect to the stiffness matrix $[K]$

$$[\Phi]^T[K][\Phi] = \langle m\omega_i^2 \rangle. \tag{18.52}$$

Equation (18.3) can be transformed (coordinate transformation) into a set of decoupled n sdof equations of the generalised coordinates $\{\eta\}$

$$m_k\ddot{\eta}_k + m_k\omega_k^2\eta_k = \{\phi_k\}\{F\},\ k = 1, 2, \ldots, n, \tag{18.53}$$

with

- m_k the generalised or modal mass
- $m_k\omega_k^2$ the generalised or modal stiffness

In the frequency domain with

- $\{x(t)\} = \{X(\omega)\}j^{\omega t}$
- $\{\eta(t)\} = \{\Pi(\omega)\}j^{\omega t}$
- $\{F(t)\} = \{F(\omega)\}j^{\omega t}$

The solution of $\Pi_k(\omega)$ is

$$\Pi_k(\omega) = \frac{\{\phi_k\}^T\{F(\omega)\}}{m_k(\omega_k^2 - \omega^2)}. \tag{18.54}$$

The solution for the vector of generalised coordinates $\{\Pi(\omega)\}$ in the frequency domain becomes

$$\{\Pi(\omega)\} = \langle\frac{1}{m_k(\omega_k^2 - \omega^2)}\rangle[\Phi]^T\{F(\omega)\}. \tag{18.55}$$

The physical displacement vector $\{X(\omega)\}$ is obtained from

$$\{X\{\omega\}\} = [\Phi]\{\Pi(\omega)\} = [\Phi]\langle\frac{1}{m_k(\omega_k^2 - \omega^2)}\rangle[\Phi]^T\{F(\omega)\}. \tag{18.56}$$

The modal matrix $[\Phi]$ may be partitioned in the kept modes and the deleted modes:

$$[\Phi] = [\Phi_k, \Phi_d]. \tag{18.57}$$

Reconstructing the flexibility matrix $[G]$, with $\omega \to 0$

$$[G] = [\Phi_k]\langle m_k\omega_k^2\rangle^{-1}[\Phi_k]^T + [\Phi_d]\langle m_d\omega_d^2\rangle^{-1}[\Phi_d]^T = [G_k] + [G_r], \tag{18.58}$$

with

- $[G_r]$ the residual flexibility matrix, $[G_r] = [G] - [\Phi_k]\langle m_k\omega_k^2\rangle^{-1}[\Phi_k]^T$
- $[G] = [K]^{-1}$ the flexibility matrix (the inverse of the stiffness matrix $[K]$ is only allowed if the structure is constrained such that rigid body motions are eliminated)

If the rigid-body modes are eliminated we can express $\{X(\omega)\}$, assuming for the modes $k > m$, $\omega_k^2 \gg \omega^2$

$$\{X(\omega)\} = \sum_{k=1}^{m}\{\phi_k\}\left(\frac{\{\phi_k\}\{F(\omega)\}}{m_k[\omega_k^2 - \omega^2]}\right) + \sum_{k=m+1}^{n}\{\phi_k\}\left(\frac{\{\phi_k\}\{F(\omega)\}}{m_k\omega_k^2}\right). \tag{18.59}$$

Equation can be transformed back in the time domain, such that

$$\{x(t)\} = [\Phi_k]\{\eta_k(t)\} + [G_r]\{F(t)\} \tag{18.60}$$

and this is done in combination with (18.53).

If a substructure has rigid-body modes, the flexibility matrix $[G]$ does not exist, however, an alternative formulation can be derived. We write the displacement vector $\{x\}$ as follows

$$\{x\} = [\Phi_r]\{\eta_r\} + [\Phi_e]\{\eta_e\}, \tag{18.61}$$

with

- $[\Phi_r]$ the rigid-body modes ($\omega_r = 0$)
- $[\Phi_e]$ the elastic modes of a free-free component ($\omega_e \neq 0$)

- $\{\eta_r\}$ generalised coordinates associated with the rigid body motions
- $\{\eta_e\}$ generalised coordinates associated with the elastic motions

The modal matrices $[\Phi_r]$ and $[\Phi_e]$ are orthonormal with respect to the mass matrix $[M]$, thus

$$[\Phi_r]^T[M][\Phi_r] = \langle m_r \rangle,\ [\Phi_e]^T[M][\Phi_e] = \langle m_e \rangle,\ [\Phi_r]^T[M][\Phi_e] = [0], \tag{18.62}$$

and orthogonal with respect to the stiffness matrix $[K]$

$$[\Phi_r]^T[K][\Phi_r] = [0],\ [\Phi_e]^T[K][\Phi_e] = \langle m_e\omega_e^2 \rangle,\ [\Phi_r]^T[K][\Phi_e] = [0]. \tag{18.63}$$

Equation (18.47) can be written

$$[M][\Phi_r]\{\ddot{\eta}_r\} + [M][\Phi_e]\{\ddot{\eta}_e\} + [K][\Phi_r]\{\eta_r\} + [K][\Phi_e]\{\eta_e\} = \{F\}. \tag{18.64}$$

Taking into account that $[K][\Phi_r] = [0]$ (18.64) becomes

$$[M][\Phi_e]\{\ddot{\eta}_e\} + [K][\Phi_e]\{\eta_e\} = \{F\} - [M][\Phi_r]\{\ddot{\eta}_r\}. \tag{18.65}$$

Using (18.53), and referring to (18.62), it can be easily proved that

$$\{\ddot{\eta}_r\} = \langle m_r \rangle^{-1}[\Phi_r]^T\{F\}. \tag{18.66}$$

Equation (18.65) now becomes

$$[M][\Phi_e]\{\ddot{\eta}_e\} + [K][\Phi_e]\{\eta_e\} = \{F\} - [M][\Phi_r]\langle m_r \rangle^{-1}[\Phi_r]^T\{F\} = [A]\{F\} \tag{18.67}$$

$$\langle m_e \rangle\{\ddot{\eta}_e\} + \langle m_e\omega_e^2 \rangle\{\eta_e\} = [\Phi_e]^T[A]\{F\} = [\Phi_e]^T\{F\}, \tag{18.68}$$

with

- $[A]$ the inertia-relief filter matrix with the following properties: $[\Phi_r]^T[A]\{F\} = [0]$ and $[\Phi_e]^T[A] = [\Phi_e]^T$. The first equation means that $[A]\{F\}$ is an equilibrium force system.

Because the force system $[A]\{F\}$ is in equilibrium the free-free substructure may be constrained in an arbitrarily point "*B*", which will take out the rigid-body motions. This has no influence on the elastic deformation in the substructure. The elastic deformation, with respect to "*B*", is $\{x_{B,e}\}$ and can be calculated with

$$\{x_{B,e}\} = [G_{B,e}][A]\{F\}. \tag{18.69}$$

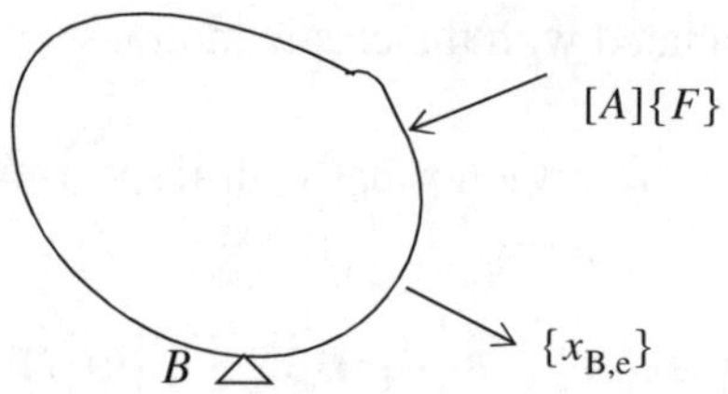

Fig. 18.4. Constrained substructure in point B

The stiffness matrix $[K]$ can be partitioned, giving

$$[K] = \begin{bmatrix} K_{ee} & K_{eB} \\ K_{Be} & K_{BB} \end{bmatrix}. \tag{18.70}$$

The dofs in point B are constrained, so the rows and columns in the matrix $[K]$ with respect to dofs associated with point B are removed. The stiffness matrix $[K_{ee}]$ is regular. To calculate the complete vector of deformation (inclusive the dofs associated with point B) we can define the constrained flexibility matrix $[G_{B,e}]$ as follows

$$[G_{B,e}] = \begin{bmatrix} K_{ee}^{-1} & 0 \\ 0 & 0 \end{bmatrix}. \tag{18.71}$$

The total displacement of the free-free substructure can be written as

$$\{x_{\text{rel}}\} = \{x_{B,e}\} + [\Phi_r]\{\theta_r\}. \tag{18.72}$$

We force the displacement $\{x_{\text{rel}}\}$ to be mass orthogonal with the rigid-body modes $[\Phi_r]$, thus

$$[\Phi_r]^T[M]\{x_{\text{rel}}\} = \{0\}. \tag{18.73}$$

This will result in

$$\{\theta_r\} = -\langle m_r \rangle^{-1}[\Phi_r]^T[M]\{x_{B,e}\}. \tag{18.74}$$

Thus, the free-free displacement $\{x_{rel}\}$ becomes

$$\{x_{\text{rel}}\} = ([I] - [\Phi_r]\langle m_r \rangle^{-1}[\Phi_r]^T[M])\{x_{B,e}\} = [A]^T[G_{B,e}][A]\{F\} = [G]\{F\}. \tag{18.75}$$

The matrix $[G]$ in (18.75) is called the elastic flexibility matrix in inertia-relief format. (18.75) must be used when the substructure is unconstrained (free-free). (18.61) can now be written as

$$\{x\} = [\Phi_r]\{\eta_r\} + [\Phi_{e,k}]\{\eta_{e,k}\} + \{x_{\mathrm{rel}}\}, \tag{18.76}$$

or the displacement vector $\{x\}$ becomes

$$\{x\} = [\Phi_r]\{\eta_r\} + [\Phi_{e,k}]\{\eta_{e,k}\} + [G]\{F\} = [\Phi_{\mathrm{k}}]\{\eta_k\} + [G]\{F\}, \tag{18.77}$$

with

- $[\Phi_{\mathrm{k}}]$ the kept elastic modes (inclusive rigid-body modes)

Equation (18.77) will be partitioned in internal dofs $\{x_i\}$ and external or boundary dofs $\{x_b\}$.

$$\begin{Bmatrix} x_i \\ x_b \end{Bmatrix} = \begin{bmatrix} \Phi_{k,i} \\ \Phi_{k,b} \end{bmatrix}\{\eta_k\} + \begin{bmatrix} G_{ii} & G_{ib} \\ G_{bi} & G_{bb} \end{bmatrix}\begin{Bmatrix} F_i \\ F_b \end{Bmatrix}, \tag{18.78}$$

and the associated undamped equations of motion expressed in the generalised coordinates $\{\eta_k\}$ (including the rigid-body modes, $\omega_k^2 = 0$)

$$\langle m_k\rangle\{\ddot{\eta}_k\} + \langle m_k\omega_k^2\rangle\{\eta_k\} = [\Phi_{\mathrm{k}}]^T\{F\} = \begin{bmatrix} \Phi_{k,i} \\ \Phi_{k,b} \end{bmatrix}^T\begin{Bmatrix} F_i \\ F_b \end{Bmatrix}. \tag{18.79}$$

Coupling of Two Substructures *A* and *B*

To couple the two substructures A and B we must force continuity with respect to the external or boundary displacement $\{x_b^A\}$ and $\{x_b^A\}$, hence

$$\{x_b^A\} = \{x_b^B\}, \tag{18.80}$$

and at the boundaries the external forces of substructure A $\{F_b^A\}$ and substructure B $\{F_b^B\}$ are at equilibrium, hence

$$\{F_b^A\} + \{F_b^B\} = \{0\}. \tag{18.81}$$

If we substitute the second part of (18.79) into (18.80) the following equation is found

$$\Phi_{k,b}^A\{\eta_k^A\} + [G_{bi}^A]\{F_i^A\} + [G_{bb}^A]\{F_b^A\} = \Phi_{k,b}^B\{\eta_k^B\} + [G_{bi}^B]\{F_i^B\} + [G_{bb}^B]\{F_b^B\}, \tag{18.82}$$

With the introduction of equilibrated forces at the boundaries in (18.81), (18.82) can be rewritten as

$$\{F_b^A\}\{[G_{bb}^A]+[G_{bb}^B]\} = \Phi_{k,b}^B\{\eta_k^B\}-\Phi_{k,b}^A\{\eta_k^A\}+[G_{bi}^B]\{F_i^B\}-[G_{bi}^A]\{F_i^A\}, \quad (18.83)$$

or

$$\{F_b^A\} = \{[G_{bb}^A]+[G_{bb}^B]\}^{-1}(\Phi_{k,b}^B\{\eta_k^B\}-\Phi_{k,b}^A\{\eta_k^A\}+[G_{bi}^B]\{F_i^B\}-[G_{bi}^A]\{F_i^A\}) . \quad (18.84)$$

and

$$\{F_b^B\} = -\{[G_{bb}^A]+[G_{bb}^B]\}^{-1}(\Phi_{k,b}^B\{\eta_k^B\}-\Phi_{k,b}^A\{\eta_k^A\}+[G_{bi}^B]\{F_i^B\}-[G_{bi}^A]\{F_i^A\}) \quad (18.85)$$

Substitution of the last two equations (18.84) and (18.85) into (18.79) in respectively substructures A and B, with $[K_{bb}^{AB}] = \{[G_{bb}^A]+[G_{bb}^B]\}^{-1}$, we get

$$\langle m_k^A\rangle\{\ddot{\eta}_k^A\}+\langle m_k^A\omega_k^{2A}\rangle\{\eta_k^A\} = \begin{bmatrix}\Phi_{k,i}^A\\ \Phi_{k,b}^A\end{bmatrix}^T\begin{Bmatrix}F_i^A\\ F_b^A\end{Bmatrix}$$

$$= \begin{bmatrix}\Phi_{k,i}^A\\ \Phi_{k,b}^A\end{bmatrix}^T\begin{bmatrix}F_i^A\\ [K_{bb}^{AB}]([\Phi_{k,b}^B]\{\eta_k^B\}-[\Phi_{k,b}^A]\{\eta_k^A\}+[G_{bi}^B]\{F_i^B\}-[G_{bi}^A]\{F_i^A\})\end{bmatrix}, \quad (18.86)$$

and

$$\langle m_k^B\rangle\{\ddot{\eta}_k^B\}+\langle m_k^B\omega_k^{2B}\rangle\{\eta_k^B\} = \begin{bmatrix}\Phi_{k,i}^B\\ \Phi_{k,b}^B\end{bmatrix}^T\begin{Bmatrix}F_i^B\\ F_b^B\end{Bmatrix}$$

$$= \begin{bmatrix}\Phi_{k,i}^B\\ \Phi_{k,b}^B\end{bmatrix}^T\begin{bmatrix}F_i^B\\ (-[K_{bb}^{AB}])([\Phi_{k,b}^B]\{\eta_k^B\}-[\Phi_{k,b}^A]\{\eta_k^A\}+[G_{bi}^B]\{F_i^B\}-[G_{bi}^A]\{F_i^A\})\end{bmatrix} . \quad (18.87)$$

Rewriting (18.86) and (18.87) we obtain

$$\begin{bmatrix}\langle m_k^A\rangle & 0\\ 0 & \langle m_k^B\rangle\end{bmatrix}\begin{Bmatrix}\ddot{\eta}_k^A\\ \ddot{\eta}_k^B\end{Bmatrix}$$

$$+\begin{bmatrix}\langle k_k^A\rangle+[\Phi_{k,b}^A]^T[K_{bb}^{AB}][\Phi_{k,b}^A] & -[\Phi_{k,b}^A]^T[K_{bb}^{AB}][\Phi_{k,b}^B]\\ -[\Phi_{k,b}^{\boldsymbol{B}}]^T[K_{bb}^{AB}][\Phi_{k,b}^A] & \langle k_k^A\rangle+[\Phi_{k,b}^A]^T[K_{bb}^{AB}][\Phi_{k,b}^A]\end{bmatrix}\begin{Bmatrix}\eta_k^A\\ \eta_k^B\end{Bmatrix}$$

$$= \begin{bmatrix} \left[[\Phi^A_{k,i}]^T - [\Phi^A_{k,b}]^T [G^A_{bi}]\, [\Phi^A_{k,b}]^T \right] & [\Phi^A_{k,i}]^T [K^{AB}_{bb}][G^B_{bi}] \\ [\Phi^B_{k,i}]^T [K^{AB}_{bb}][G^A_{bi}] & \left[[\Phi^B_{k,i}]^T - [\Phi^B_{k,b}]^T [G^A_{bi}]\, [\Phi^B_{k,b}]^T \right] \end{bmatrix} \begin{Bmatrix} F^A_i \\ F^B_i \end{Bmatrix} \quad (18.88)$$

In the final synthesised dynamic system (substructures *A*, *B*, etc.) only the generalised coordinates are left, while the interface dofs have been cancelled out. The coupling of the substructures is done via the stiffness matrix. The mass matrix is a diagonal matrix with the generalised masses on the main diagonal.

18.3.3 General-Purpose CMS Method

The general-purpose CMS method has been addressed by [Herting 79, UAI 93]. Both constrained and unconstrained substructures are covered by this CMS method. We assume undamped substructures. The undamped equations of motion for a substructure or component can be written using, (18.3)

$$[M]\{\ddot{x}\} + [K]\{x\} = \{F\}.$$

In the previous section we have derived the solution for the physical displacement vector in the frequency domain, (18.56)

$$\{X\{\omega\}\} = [\Phi]\langle\frac{1}{m_k(\omega_k^2 - \omega^2)}\rangle[\Phi]^T\{F(\omega)\}.$$

Three groups of responses can be considered:

1. The rigid-body modes; $\omega_k^2 = 0$, $[\Phi] = [\Phi_0]$ and $m_k = m_0$, $k = 1, 2, ..., 6$
2. The kept elastic modes, the natural frequencies which are in the frequency range of interest; $\omega_k^2 \approx O(\omega^2)$, $[\Phi] = [\Phi_k]$ and m_k, $k = 6, 7, ..., m$
3. The deleted elastic modes: $\omega_k^2 \gg \omega^2$, $[\Phi] = [\Phi_d]$ and m_d, $d = m + 1, ...$

Equation (18.56) can be written

$$\{X(\omega)\} = \left[[\Phi_0]\langle\frac{1}{-m_0\omega^2}\rangle[\Phi_0]^T + [\Phi_k]\langle\frac{1}{m_k(\omega_k^2 - \omega^2)}\rangle[\Phi_k]^T\right]\{F(\omega)\}$$

$$\left[+[\Phi_d]\langle\frac{1}{m_d(\omega_d^2)}\rangle[\Phi_d]^T\right]\{F(\omega)\}. \tag{18.89}$$

The constant acceleration, when $\omega \to 0$, becomes

$$\ddot{x}_c = \lim_{\omega \to 0}\{\ddot{X}(\omega)\} = -\omega^2\{X(\omega)\} = [\Phi_0]\langle\frac{1}{m_0}\rangle[\Phi_0]^T\{F(t)\}, \tag{18.90}$$

and the static displacement vector ($\omega \to 0$), premultiplied by the stiffness matrix $[K]$, is given by

$$[K]\{x_{\text{stat}}\} = \lim_{\omega \to 0}[K]\{X(\omega)\} = [K]\left([\Phi_k]\langle\frac{1}{m_k\omega_k^2}\rangle[\Phi_k]^T\right)\{F(t)\}$$

$$+[K]\left([\Phi_d]\langle\frac{1}{m_d\omega_d^2}\rangle[\Phi_d]^T\right)\{F(t)\}. \tag{18.91}$$

If (18.90) and (18.91) are substituted into (18.3), with $\omega \to 0$, we can express $[K]\left([\Phi_d]\langle\frac{1}{m_d\omega_d^2}\rangle[\Phi_d]^T\right)\{F(t)\}$ as follows

$$\left([\Phi_d]\langle\frac{1}{m_d\omega_d^2}\rangle[\Phi_d]^T\right)\{F(t)\} = \{F(t)\} - [M][\Phi_0]\langle\frac{1}{m_0}\rangle[\Phi_0]^T\{F(t)\}$$

$$-[K]\left([\Phi_k]\langle\frac{1}{m_k\omega_k^2}\rangle[\Phi_k]^T + \ldots\right)\{F(t)\}. \tag{18.92}$$

Equation (18.89) is transferred in the time domain giving:

$$\{x(t)\} = [\Phi_0]\{\eta_0(t)\} + [\Phi_k]\{\eta_k(t)\} + [\Phi_d]\langle\frac{1}{m_d\omega_d^2}\rangle[\Phi_d]^T\{F(t)\}. \tag{18.93}$$

If the result of (18.92) is substituted into (18.89) and if (18.93) is premultiplied by the stiffness matrix $[K]$ we get

$$[K]\{x(t)\} = [K][\Phi_0]\{\eta_0(t)\} + [K][\Phi_k]\{\eta_k(t)\} + \{F(t)\}$$

$$-[M][\Phi_0]\langle\frac{1}{m_0}\rangle[\Phi_0]^T\{F(t)\} - [K][\Phi_k]\langle\frac{1}{m_k\omega_k^2}\rangle[\Phi_k]^T\{F(t)\}. \tag{18.94}$$

Making use of $[K][\Phi_0] = \{0\}$, (18.94) can be written as

$$[K]\{x(t)\} = [K][\Phi_k](\{\eta_k(t)\} - \{\eta_{\text{stat}}(t)\}) - [M][\Phi_0]\langle\frac{1}{m_0}\rangle[\Phi_0]^T\{F(t)\} + \{F(t)\}, \tag{18.95}$$

or

$$[K]\{x(t)\} = [K][\Phi_k]\{\delta_k(t)\} - [M][\Phi_0]\{\delta_0(t)\} + \{F(t)\}, \tag{18.96}$$

with

- $\{\delta_k(t)\} = (\{\eta_k(t)\} - \{\eta_{\text{stat}}(t)\})$ the normal mode generalised coordinates.
- $\{\delta_0(t)\} = \langle\frac{1}{m_0}\rangle[\Phi_0]^T\{F(t)\}$ the inertia-relief coordinates ($n_0 \le 6$), [UAI 93], and have units of acceleration.

We return to (18.3)

$$\begin{bmatrix} M_{ii} & M_{ib} \\ M_{bi} & M_{bb} \end{bmatrix}\begin{Bmatrix} \ddot{x}_i \\ \ddot{x}_b \end{Bmatrix} + \begin{bmatrix} K_{ii} & K_{ib} \\ K_{bi} & K_{bb} \end{bmatrix}\begin{Bmatrix} x_i \\ x_b \end{Bmatrix} = \begin{Bmatrix} F_i \\ F_b \end{Bmatrix}.$$

With use of (18.96) we can express $[K_{ii}]\{x_i\}$ as follows

$$\begin{aligned}[K_{ii}]\{x_i\} &= \{F_i\} + ([K_{ii}][\Phi_{k,i}] + [K_{ib}][\Phi_{k,b}]\{\delta_k(t)\}) \\ &\quad -([M_{ii}][\Phi_{0,i}] + [M_{ib}][\Phi_{0,b}])\{\delta_0(t)\} - [K_{ib}]\{x_b\},\end{aligned} \tag{18.97}$$

and furthermore,

$$\begin{aligned}\{x_i\} &= [K_{ii}]^{-1}\{F_i\} + ([\Phi_{k,i}] - [G_{ib}][\Phi_{k,b}])\{\delta_k(t)\} \\ &\quad -[K_{ii}]^{-1}([M_{ii}][G_{ib}] + [M_{ib}])[\Phi_{0,b}]\{\delta_0(t)\} + [G_{ib}]\{x_b\},\end{aligned} \tag{18.98}$$

with

- $[G_{ib}] = -[K_{ii}]^{-1}[K_{ib}]$
- $[G_{ib}][\Phi_{0,b}] = [\Phi_{0,i}]$

The displacement vector $\{x\}$, using (18.98), can be written as

$$\{x\} = \begin{Bmatrix} \bar{x}_i \\ x_b \end{Bmatrix} = [\Psi]\begin{Bmatrix} \delta_0 \\ \delta_k \\ x_b \end{Bmatrix} = [\Psi]\{\vartheta\}, \tag{18.99}$$

with

$$\{x_i\} = \{\bar{x}_i\} + [K_{ii}]^{-1}\{F_i\}. \tag{18.100}$$

Thus we can write for the transformation matrix $[\Psi]$

$$[\Psi] = \begin{bmatrix} -[K_{ii}]^{-1}([M_{ii}][G_{ib}] + [M_{ib}])[\Phi_{0,b}] & ([\Phi_{k,i}]-[G_{ib}][\Phi_{k,b}]) & G_{ib} \\ 0 & 0 & I \end{bmatrix}. \quad (18.101)$$

Some remarks can be made;

- When the number of modes is zero and the inertia-relief effects are ignored, the transformation matrix $[\Psi]$ in (18.101) is the same as the Guyan reduction of matrix condensation transformation.
- Modes provide dynamic motion relative to the static deformation.
- Rigid-body motion and redundant constraint information are contained in the $[G_{ib}]$ transformation
- Inertia-relief deformation shapes are contained in the $-[K_{ii}]^{-1}([M_{ii}][G_{ib}] + [M_{ib}])[\Phi_{0,b}]$ matrix.
- The sum of rigid-body dofs $\{\delta_0\}$ and elastic generalised dofs $\{\delta_k\}$ shall be less than or equal to the number of internal dofs $\{x_i\}$.

The general undamped equations of motion are

$$[M]\{\ddot{x}\} + [K]\{x\} = \{F(t)\} .$$

If we apply the transformation (18.99) the following undamped equation of motion are obtained

$$[\Psi]^T[M][\Psi]\{\ddot{\vartheta}\} + [\Psi]^T[K][\Psi]\{\vartheta\} = [\Psi]^T\{F(t)\}, \quad (18.102)$$

or

$$[M_{\vartheta\vartheta}]\{\ddot{\vartheta}\} + [K_{\vartheta\vartheta}]\{\vartheta\} = \{F_{\vartheta}(t)\} \quad (18.103)$$

The data recovery of the physical dofs, displacement $\{x\}$, velocities $\{\dot{x}\}$ and acceleration $\{\ddot{x}\}$, can be obtained as follows. The displacements $\{x\}$ become

$$\{x\}= \begin{Bmatrix} x_i \\ x_b \end{Bmatrix} = \begin{Bmatrix} \bar{x}_i \\ x_b \end{Bmatrix} + \begin{Bmatrix} x_{i,stat} \\ 0 \end{Bmatrix} = [\Psi]\begin{Bmatrix} \delta_0 \\ \delta_k \\ x_b \end{Bmatrix} + \begin{bmatrix} [K_{ii}]^{-1}\{F_i\} \\ 0 \end{bmatrix}, \quad (18.104)$$

and the velocities $\{\dot{x}\}$

$$\{\dot{x}\} = \left\{ \begin{matrix} \dot{x}_i \\ \dot{x}_b \end{matrix} \right\} = \left\{ \begin{matrix} \dot{x}_i \\ \dot{x}_b \end{matrix} \right\} = [\Psi] \left\{ \begin{matrix} \dot{\delta}_0 \\ \dot{\delta}_k \\ \dot{x}_b \end{matrix} \right\}, \tag{18.105}$$

and the accelerations $\{\ddot{x}\}$ are

$$\{\ddot{x}\} = \left\{ \begin{matrix} \ddot{x}_i \\ \ddot{x}_b \end{matrix} \right\} = \left\{ \begin{matrix} \ddot{x}_i \\ \ddot{x}_b \end{matrix} \right\} = [\Psi] \left\{ \begin{matrix} \ddot{\delta}_0 \\ \ddot{\delta}_k \\ \ddot{x}_b \end{matrix} \right\}. \tag{18.106}$$

We may improve the solution of the displacements $\{x\}$ by using the mode acceleration method [MAM]

$$\{x_{\text{MAM}}\} = \left\{ \begin{matrix} x_{i,\,\text{MAM}} \\ x_b \end{matrix} \right\} = \begin{bmatrix} G_{ib} \\ I \end{bmatrix} \{x_b\} + \begin{bmatrix} [K_{ii}]^{-1}(\{F_i\} - [M_{ib}]\{\ddot{x}_b\} - [M_{ii}]\{\ddot{x}_i\}) \\ 0 \end{bmatrix}. \tag{18.107}$$

The damping effects are ignored here.

A free-free dynamic system consists of 20 discrete masses, each $m = 1$ kg, connected with springs, each $k = 10000$ N/m. The total mass matrix and stiffness matrix are

$$[M] = \begin{bmatrix} 1 & 0 & \dots & 0 & 0 \\ 0 & 1 & \dots & 0 & 0 \\ \dots & \dots & \dots & \dots & \dots \\ 0 & 0 & \dots & 1 & 0 \\ 0 & 0 & \dots & 0 & 1 \end{bmatrix}, \ [K] = 10000 \begin{bmatrix} 1 & -1 & \dots & 0 & 0 \\ -1 & 2 & \dots & 0 & 0 \\ \dots & \dots & \dots & \dots & \dots \\ 0 & 0 & \dots & 2 & -1 \\ 0 & 0 & \dots & -1 & 1 \end{bmatrix}.$$

In this example only the natural frequencies of the complete and reduced models, using the general-purpose CMS method, will be given. The results of the reduction process are shown in Table 18.4. The number of elastic modes taken into account are reflected in the accuracy of the natural frequencies. The accuracy of the modes is not considered in this example.

Table 18.4. Results of reduction process, natural frequencies

#	Natural frequency (Hz)	Reduced model natural frequency (Hz)	Reduced model natural frequency (Hz)
	Complete model	$nb^{a}=2$, $nr^{b}=1$, $ne^{c}=5$	$nb=2$, $nr=1$, $ne=5$
1	0.0000	0.0000	0.0000
2	2.4974	2.4974	2.4974
3	4.9795	4.9795	4.9795
4	7.4308	8.8043	7.4308
5	9.8363	11.7091	9.8363
6	12.1812		12.1812
7	14.4510		16.8204
8	16.6316		19.0208
9	18.7098		
10	20.6726		

a. number of boundary dofs, constarint modes
b. number of rigid-body modes
c. number of elastic modes

18.4 Problems

18.4.1 Problem 1

A linear free-free substructure consists of 7 dofs; 1 to 7. The lumped masses at dof 1 and dof 7 is $m_1 = m_7 = 0.5$ kg The other masses lumped to the other dofs, 2 to 6, are $m_2 = \ldots = m_6 = 1$ kg. The 6 springs between the dofs 1 to 19 are equal, $k_{12} = k_{23} = \ldots = k_{67} = 10000$ N/m. The substructure is shown in Fig. 18.3.

Couple two substructures with each other, substructure 1 node 7 with node 1 of substructure 2, and calculate the natural frequencies and associated modes. There are no other boundary conditions (free-free structure). Use the following CMS methods:

1. Craig–Bampton method
2. Craig–Chang method
3. Herting method

18.4.2 Problem 2

A structure may be identified with two components or substructures; component 1 and component 2, as illustrated in Fig. 18.5.

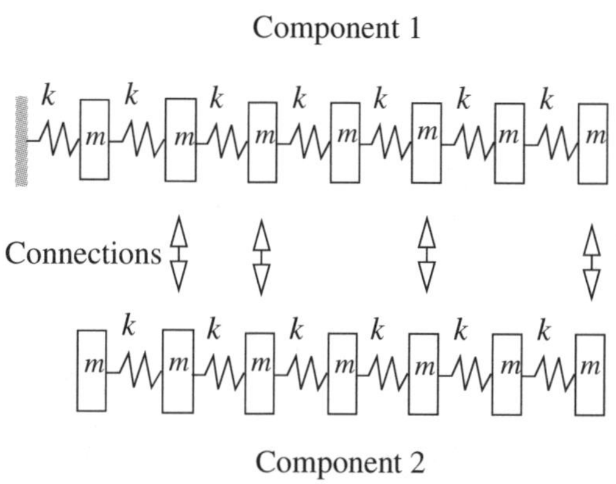

Fig. 18.5. Component 1 and component 2

Calculate the modal characteristics (natural frequencies, mode shapes and effective masses). All masses have a mass $m = 1$ kg and all springs have a spring stiffness $k = 100000$ N/m

Calculate all elastic modes per component (except for the Herting method).

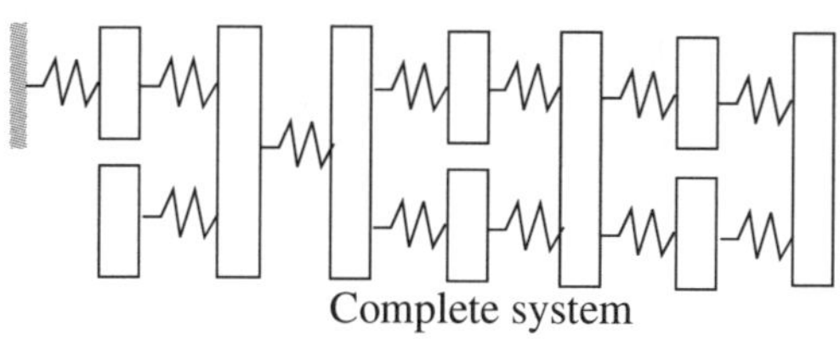

Fig. 18.6. Synthesised components

- Component 1
    1. Craig–Bampton method (using constraint modes)
    2. Herting method
- Component 2
    1. Craig–Bampton method (using constraint modes)
    2. Craig–Bampton method (free-free)
    3. Herting method
- Synthesis
    1. Component 1 - method 1, component 2 method 1
    2. Component 1 - method 1, component 2 method 2
    3. Component 1 - method 1, component 2 method 3
    4. Component 1 - method 2, component 2 method 3

19 Load Transformation Matrices

19.1 Introduction

The mathematical reduced (condensed) dynamic model consists of the reduced- mass and reduced-stiffness matrices. The damping matrix is, in general, not delivered in a reduced form because the damping characteristics will be introduced later in the dynamic response analyses.

Because the reduced dynamic model only consists of reduced matrices during the dynamic response analyses no direct information about physical responses; e.g. forces, stresses, can be made available. The reduced dynamic model will only produce response characteristics of physical (e.g. I/F dofs) and generalised degrees of freedom; displacements, velocities and accelerations.

To be able to produce responses, stresses and forces, in selected structural elements during the dynamic response analyses using (coupled) reduced dynamic models the so-called load transformation matrix (LTM) can be used. The LTM defines a relation between forces and stresses in certain structural elements and the degrees of freedom of the reduced dynamic model. In general the transformation matrix is called the output transformation matrix (OTM) [Chung 98, Fransen 02]. Besides LTMs displacement transformation matrices (DTM), acceleration transformation matrices (ATM) can also be defined [Bray 91], however, in this chapter only LTMs will be discussed. The creation of DTMs and ATMs is quite the same as the generation of LTMs.

In the following sections two methods of obtaining LTMs will be discussed. Both methods are based upon the mode displacement method (MDM) and the mode acceleration method (MAM) [Craig 81]. The methods described are:

- The LTMs of a reduced dynamic model with boundary conditions, such that rigid-body motions are prevented, thus $\{x_j\} = \{0\}$,

- The LTMs of a free-free reduced dynamic model. In general, six rigid-body motions will exist.

19.2 Reduced Model with Boundary Conditions

The reduced dynamic model has been created from a reference finite element model mass and stiffness matrix for which the boundary conditions prevent at least the six motions as a rigid-body. The stiffness matrix is not singular in that case.
Using the MAM the displacement vector $\{x\}$ can be written as:

$$\{x\} = ([K]^{-1} - [\Phi_k]\langle\lambda_k\rangle^{-1}[\Phi_k]^T)\{F(t)\} + [\Phi_k]\{\eta_k\}, \tag{19.1}$$

with:

- $[K]$ the stiffness matrix
- $[\Phi_k]\{\eta_k\}$ the MDM solution
- $[\Phi_k]$ the modal matrix of the kept modes
- $\{\eta_k\}$ the vector of generalised coordinates
- $\langle\lambda_k\rangle = \langle\omega_k^2\rangle$ the diagonal matrix of eigenvalues (natural frequencies)
- $\{F(t)\}$ the external force vector

The flexibility matrix $[G] = [K]^{-1}$ exists and can be expressed as:

$$[G] = [\Phi]\langle\omega\rangle^{-1}[\Phi]^T = [\Phi_k]\langle\lambda_k\rangle^{-1}[\Phi_k]^T + [\Phi_d]\langle\lambda_d\rangle^{-1}[\Phi_d]^T \tag{19.2}$$

$$[G] = [\Phi]\langle\lambda\rangle^{-1}[\Phi]^T = [G_k] + [G_r]. \tag{19.3}$$

Finally the displacement vector becomes:

$$\{x\} = [\psi_{MAM}] + [\Phi_k]\{\eta_k\}, \tag{19.4}$$

with:

- $[G_r]$ the residual flexibility matrix
- $[G_k]$ the residual flexibility matrix based upon the kept modes
- $[\Phi]$ the modal matrix of all modes
- $[\Phi_d]$ the modal matrix of the deleted modes
- $[\psi_{MAM}] = [G_r]\{F(t)\}$ the residual attachment modes, in fact the MAM correction on the MDM

In [Chung 98] an efficient manner to create the residual attachment modes has been suggested. The displacement vector $\{x\}$ becomes using [Chung 98]:

$$\{x\} = [\hat{\psi}_{\mathrm{MAM}}]\{F(t)\} + [\Phi_{\mathrm{k}}]\{\eta_k\}, \tag{19.5}$$

with

$$[\hat{\Psi}_{\mathrm{MAM}}] = [G_r][T], \tag{19.6}$$

and with:

- $[\hat{\Psi}_{\mathrm{MAM}}]$ the "Chung" residual attachment modes
- $[T]$ the square load distribution matrix, with unit loads in the columns at locations where the loads $\{F(t)\}$ are applied.

The internal generalised forces $\{\sigma\}$; forces and stresses, in certain structural elements are proportional to the displacement vector $\{x\}$:

$$\{\sigma\} = [D_\sigma]\{x\} , \tag{19.7}$$

with:

- $[D_\sigma]$ the output transformation matrix, also denoted with [OTM]

Rewriting (19.7) we obtain the following expression for $\{\sigma\}$:

$$\{\sigma\} = [D_\sigma][\hat{\psi}_{\mathrm{MAM}}]\{F(t)\} + [D_\sigma][\Phi_k]\{\eta_k\}, \tag{19.8}$$

or

$$\{\sigma\} = [\mathrm{LTM}_F]\{F(t)\} + [\mathrm{LTM}_\eta]\{\eta_k\}. \tag{19.9}$$

Quit often, (19.9) is written as

$$\{\sigma_{\mathrm{MAM}}\} = [\mathrm{LTM}_F]\{F(t)\} + \{\sigma_{\mathrm{MDM}}\}. \tag{19.10}$$

The undamped equations of motion of the 3 mass–spring dynamic system (Fig. 19.1) are:

$$m\begin{bmatrix}1&0&0\\0&1&0\\0&0&1\end{bmatrix}\begin{Bmatrix}\ddot{x}_1\\\ddot{x}_2\\\ddot{x}_3\end{Bmatrix}+k\begin{bmatrix}1{,}5&-1&-0{,}5\\-1&3&-1\\-0{,}5&-1&2{,}5\end{bmatrix}\begin{Bmatrix}x_1\\x_2\\x_3\end{Bmatrix} = -m\ddot{u}\begin{bmatrix}1&0&0\\0&1&0\\0&0&1\end{bmatrix}\begin{Bmatrix}1\\1\\1\end{Bmatrix}.$$

The displacement vector $\{x\}$ is with respect to the base (relative displacements). The right side of the equations of motion form the inertia loads due to the base acceleration $\ddot{u}$.

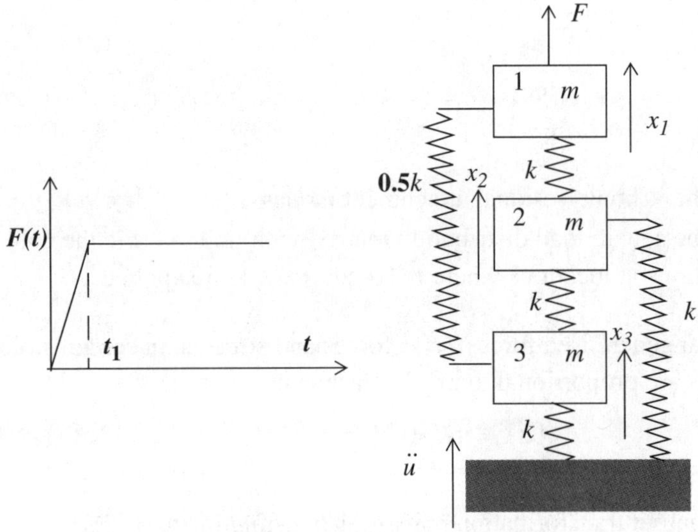

Fig. 19.1. 3 mass–spring dynamic system (constrained)

The eigenvalue problem of the dynamic system is defined as:

$$\left(-\lambda\begin{bmatrix}1&0&0\\0&1&0\\0&0&1\end{bmatrix}+\frac{k}{m}\begin{bmatrix}1.5&-1&-0.5\\-1&3&-1\\-0.5&-1&2.5\end{bmatrix}\right)\begin{Bmatrix}x_1\\x_2\\x_3\end{Bmatrix}=\begin{Bmatrix}0\\0\\0\end{Bmatrix}.$$

The first eigenvalue λ_1 and the associated mode shape $\{\phi_1\}$ are:

$$\lambda_1 = 0.5539\frac{k}{m}\text{, and } \{\phi_1\} = \frac{1}{\sqrt{m}}\begin{Bmatrix}0.7511\\0.4886\\0.4440\end{Bmatrix} \text{ with } \{\phi_1\}^T[M]\{\phi_1\} = [I] = 1\,.$$

The LTMs will be based upon the first natural frequency and associated mode shape. Thus the reduced model will only consist of the first mode shape.

The forces in the springs can be calculated with:

$$\{\sigma\} = [D_\sigma]\{x\},$$

with a stress matrix $[D_\sigma]$ defined as follows:

$$[D_\sigma] = k\begin{bmatrix} 1 & -1 & 0 \\ 0 & 1 & -1 \\ 0 & 0 & 1 \\ 0.5 & 0 & -0.5 \\ 0 & 1 & 0 \end{bmatrix}.$$

The flexibility matrix $[G]$ is:

$$[G] = [K]^{-1} = \frac{1}{k}\begin{bmatrix} 1.1818 & 0.5455 & 0.4545 \\ 0.5455 & 0.6364 & 0.3636 \\ 0.4545 & 0.3636 & 0.6364 \end{bmatrix}.$$

The flexibility matrix $[G_k]$, which is associated with the first natural frequency λ_1 and mode shape $\{\phi_1\}$, is:

$$[G_k] = \frac{\{\varphi_1\}\{\varphi_1\}^T}{\lambda_1} = \frac{1}{k}\begin{bmatrix} 1.0185 & 0.6625 & 0.6021 \\ 0.6625 & 0.4310 & 0.3917 \\ 0.6021 & 0.3917 & 0.3559 \end{bmatrix}.$$

The residual flexibility matrix $[G_r]$ is

$$[G_r] = [G] - [G_k] = \frac{1}{k}\begin{bmatrix} 0.1634 & -0.1170 & -0.1475 \\ -0.1170 & 0.2054 & -0.0280 \\ -0.1475 & -0.0280 & 0.2804 \end{bmatrix}.$$

The modified residual attachment mode matrix $[\hat{\psi}_{MAM}]$ becomes:

$$[\hat{\psi}_{MAM}] = [G_r][T] = [G_r][I] = [G_r].$$

The load transformation matrices were defined as:

$$\{\sigma\} = [\mathrm{LTM}_F]\{P(t)\} + [\mathrm{LTM}_\eta]\{\eta_1(t)\}.$$

This gives:

$$[\mathrm{LTM}_F] = [D_\sigma][\hat{\psi}_{MAM}] = \begin{bmatrix} 0.2804 & -0.3225 & -0.1195 \\ 0.0305 & 0.2334 & -0.3084 \\ -0.1475 & -0.0280 & 0.2804 \\ 0.1555 & -0.0445 & -0.2140 \\ -0.1170 & 0.2054 & -0.0280 \end{bmatrix},$$

and

$$[\mathrm{LTM}_\eta] = [D_\sigma][\phi_1] = \begin{Bmatrix} 0.2625 \\ 0.0446 \\ 0.4440 \\ 0.1535 \\ 0.4886 \end{Bmatrix}.$$

19.3 Reduced Free-Free Dynamic Model

The calculation of the stresses and forces that are related to the degrees of freedom at the unconstrained boundary $\{x_j\}$ and the generalised coordinates $\{\eta_p\}$ may be inaccurate when the contribution of the high natural frequency modes is neglected. With the aid of the MAM the stresses and forces in the structural elements become more accurate.

The equations of motion for a component or a substructure are:

$$\begin{bmatrix} M_{ii} & M_{ij} \\ M_{ji} & M_{jj} \end{bmatrix} \begin{Bmatrix} \ddot{x}_i \\ \ddot{x}_j \end{Bmatrix} + \begin{bmatrix} K_{ii} & K_{ij} \\ K_{ji} & K_{jj} \end{bmatrix} \begin{Bmatrix} x_i \\ x_j \end{Bmatrix} = \begin{Bmatrix} 0 \\ F_j \end{Bmatrix}. \tag{19.11}$$

The force vector $\{F_j\}$ represents the interface forces between components, and

- $\{x_i\}$ the internal degrees of freedom
- $\{x_j\}$ the external degrees of freedom (in general at the boundary)

The internal degrees of freedom $\{x_i\}$ may be written as [Klein 88]:

$$\{x_i^*\} = -[K_{ii}]^{-1} \left\{ \begin{bmatrix} M_{ii} & M_{ij} \end{bmatrix} \begin{Bmatrix} \ddot{x}_i \\ \ddot{x}_j \end{Bmatrix} + [K_{ij}]\{x_j\} \right\}. \tag{19.12}$$

For a Craig–Bampton model [Craig 68], [Craig 81]

$$\begin{Bmatrix} \ddot{x}_i \\ \ddot{x}_j \end{Bmatrix} = \begin{bmatrix} \phi_p & \phi_{ij} \\ 0 & I \end{bmatrix} \begin{Bmatrix} \ddot{\eta}_p \\ \ddot{x}_j \end{Bmatrix}, \tag{19.13}$$

- $[\phi_{ij}] = -[K_{ii}]^{-1}[K_{ij}]$ the constrained modes
- $([K_{ii}] - \lambda_i[M_{ii}])\{\phi_{pi}\} = \{0\}$ the eigenvalues problem of the internal degrees of freedom
- $\{\eta_p\}$ the generalised coordinates (modal amplitude coefficients)

The displacement vector $\{x_i^*\}$ can be written as:

$$\{x_i^*\} = -[K_{ii}]^{-1}\left\{\begin{bmatrix} M_{ii} & M_{ij} \end{bmatrix}\begin{bmatrix} \phi_p & \phi_{ij} \\ 0 & I \end{bmatrix}\begin{Bmatrix} \ddot{\eta}_p \\ \ddot{x}_j \end{Bmatrix} + [K_{ij}]\{x_j\}\right\}. \tag{19.14}$$

The complete displacement vector $\{x\}$ becomes:

$$\{x\} = \begin{bmatrix} x_i^* \\ x_j \end{bmatrix} = \begin{bmatrix} -[K_{ii}]^{-1}\begin{bmatrix} M_{ii} & M_{ij} \end{bmatrix}\begin{bmatrix} \phi_p & \phi_{ij} \\ 0 & I \end{bmatrix}\begin{Bmatrix} \ddot{\eta}_p \\ \ddot{x}_j \end{Bmatrix} \\ 0 \end{bmatrix} + \begin{bmatrix} -[K_{ii}]^{-1}[K_{ij}] \\ I \end{bmatrix}\{x_j\}. \tag{19.15}$$

The stresses or forces in particular structural elements of the component can be expressed as:

$$\{\sigma\} = [D_\sigma]\{x\} = \begin{bmatrix} D_{\sigma i} & D_{\sigma j} \end{bmatrix}\begin{Bmatrix} x_i^* \\ x_j \end{Bmatrix}, \tag{19.16}$$

or

$$\{\sigma\} = \begin{bmatrix} D_{\sigma i} & D_{\sigma j} \end{bmatrix}\left\{\begin{bmatrix} -[K_{ii}]^{-1}\begin{bmatrix} M_{ii} & M_{ij} \end{bmatrix}\begin{bmatrix} \phi_p & \phi_{ij} \\ 0 & I \end{bmatrix}\begin{Bmatrix} \ddot{\eta}_p \\ \ddot{x}_j \end{Bmatrix} \\ 0 \end{bmatrix} + \begin{bmatrix} -[K_{ii}]^{-1}[K_{ij}] \\ I \end{bmatrix}\{x_j\}\right\}, \tag{19.17}$$

or

$$\{\sigma\} = [\mathrm{LMT}_1]\begin{Bmatrix} \ddot{\eta}_p \\ \ddot{x}_j \end{Bmatrix} + [\mathrm{LMT}_2]\{x_j\}, \tag{19.18}$$

with:

$$[\mathrm{LMT}_1] = \begin{bmatrix} D_{\sigma i} & D_{\sigma j} \end{bmatrix} \left(-[K_{ii}]^{-1} \begin{bmatrix} M_{ii} & M_{ij} \end{bmatrix} \begin{bmatrix} \phi_p & \phi_{ij} \\ 0 & I \end{bmatrix} \right), \tag{19.19}$$

and

$$[\mathrm{LMT}_2] = \begin{bmatrix} D_{\sigma i} & D_{\sigma j} \end{bmatrix} \begin{bmatrix} -[K_{ii}]^{-1}[K_{ij}] \\ I \end{bmatrix}, \tag{19.20}$$

The load transformation matrix $[\mathrm{LMT}_1]$ can be defined by setting

$\{x_j\} = \{0\}$ and $\begin{Bmatrix} \ddot{\eta}_p \\ \ddot{x}_j \end{Bmatrix} = [I]$, and the load transformation matrix $[\mathrm{LMT}_2]$

by setting $\{x_j\} = \{I\}$ and $\begin{Bmatrix} \ddot{\eta}_p \\ \ddot{x}_j \end{Bmatrix} = \{0\}$.

For $\{x_j\} = \{0\}$ the external degrees of freedom have been fixed and if

$\begin{Bmatrix} \ddot{\eta}_p \\ \ddot{x}_j \end{Bmatrix} = \{0\}$ no inertia forces are active.

The delivery of a reduced dynamic model is frequently accompanied by the load transformation matrices. During the coupled dynamic loads analysis the stresses and forces may be calculated in selected structural elements.

The undamped equations of motion of the “free-free” 4 mass–spring dynamic system are (shown in Fig. 19.2):

$$m \begin{bmatrix} 1 & 0 & 0 & 0 \\ 0 & 1 & 0 & 0 \\ 0 & 0 & 1 & 0 \\ 0 & 0 & 0 & 1 \end{bmatrix} \begin{Bmatrix} \ddot{x}_1 \\ \ddot{x}_2 \\ \ddot{x}_3 \\ \ddot{x}_4 \end{Bmatrix} + k \begin{bmatrix} 1.5 & -1 & -0.5 & 0 \\ -1 & 3 & -1 & -1 \\ -0.5 & -1 & 2.5 & -1 \\ 0 & -1 & -1 & 2 \end{bmatrix} \begin{Bmatrix} x_1 \\ x_2 \\ x_3 \\ x_4 \end{Bmatrix} = \begin{Bmatrix} 0 \\ 0 \\ 0 \\ F_4 \end{Bmatrix},$$

with:

- $\{x_i\} = \begin{Bmatrix} x_1 \\ x_2 \\ x_3 \end{Bmatrix}$ the internal degrees of freedom
- $x_j = x_4$ the external degree of freedom
- F_4 the interface force

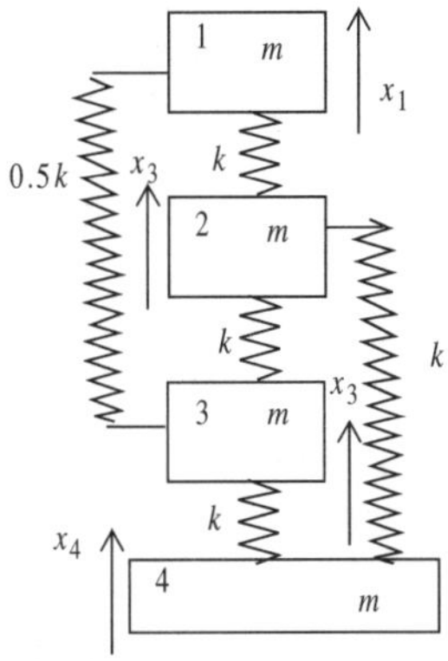

Fig. 19.2. 4 mass–spring dynamic system (free-free)

The partitioned mass matrices become:

$$[M_{ii}] = m\begin{bmatrix} 1 & 0 & 0 \\ 0 & 1 & 0 \\ 0 & 0 & 1 \end{bmatrix}, \ [M_{jj}] = m \text{ and } [M_{ij}] = [M_{ji}]^T = m\begin{Bmatrix} 0 \\ 0 \\ 0 \end{Bmatrix},$$

and the partitioned stiffness matrices:

$$[K_{ii}] = k\begin{bmatrix} 1.5 & -1 & -0.5 \\ -1 & 3 & 0-1 \\ -0.5 & -1 & 2.5 \end{bmatrix}, \ [K_{jj}] = 2k \text{ and } [K_{ij}] = [K_{ji}]^T = k\begin{Bmatrix} 0 \\ -1 \\ -1 \end{Bmatrix}.$$

The following eigenvalue problem for the $\{x_i\}$ degrees of freedom must now be solved. This eigenvalue problem of the internal degrees of freedom is defined as:

$$([K_{ii}] - \lambda[M_{ii}])\{x_i\} = \{0\}.$$

The eigenvalue λ_1 and the associated mode shape $\{\phi_1\}$ are:

$$\lambda_1 = 0.5539\frac{k}{m}\text{, and } \{\phi_1\} = \{\phi_p\} = \frac{1}{\sqrt{m}}\begin{Bmatrix} 0.7511 \\ 0.4886 \\ 0.4440 \end{Bmatrix}\text{, with}$$

$$\{\phi_1\}^T[M]\{\phi_1\} = [I].$$

The LTMs will be based upon the first natural frequency and associated mode shape. Thus the reduced model will only consist of the first mode shape.

The constrained mode $[\phi_{ij}]$ is:

$$[\phi_{ij}] = -[K_{ii}]^{-1}[K_{ij}] = \begin{Bmatrix} 1 \\ 1 \\ 1 \end{Bmatrix}.$$

The forces in the springs can be calculated with:

$$\{\sigma\} = [D_\sigma]\{x\},$$

with a stress matrix $[D_\sigma]$ defined as follows:

$$[D_\sigma] = k\begin{bmatrix} 1 & -1 & 0 & 0 \\ 0 & 1 & -1 & 0 \\ 0 & 0 & 1 & -1 \\ 0.5 & 0 & -0.5 & 0 \\ 0 & 1 & 0 & -1 \end{bmatrix}, [D_{\sigma i}] = k\begin{bmatrix} 1 & -1 & 0 \\ 0 & 1 & -1 \\ 0 & 0 & 1 \\ 0.5 & 0 & -0.5 \\ 0 & 1 & 0 \end{bmatrix} \text{ and } [D_{\sigma j}] = k\begin{Bmatrix} 0 \\ 0 \\ -1 \\ 0 \\ -1 \end{Bmatrix}.$$

The load transformation matrices were defined as:

$$\{\sigma\} = [\mathrm{LTM}_1]\begin{Bmatrix} \ddot{\eta}_p \\ \ddot{x}_j \end{Bmatrix} + [\mathrm{LTM}_2]\{x_j\},$$

and are as follows:

$$[\mathrm{LTM}_1] = [D_{\sigma i}]\left[-[K_{ii}]^{-1}\left\{\begin{bmatrix} M_{ii} & M_{ij} \end{bmatrix}\begin{bmatrix} \phi_p & \phi_{ij} \\ 0 & I \end{bmatrix}\right\}\right] = k\begin{bmatrix} -0.4739 & -0.6364 \\ -0.0804 & -0.0909 \\ -0.8016 & -1.4545 \\ -0.2772 & -0.3636 \\ -0.8821 & -1.5455 \end{bmatrix},$$

and

$$[\text{LTM}_2] = \begin{bmatrix} D_{\sigma i} & D_{\sigma j} \end{bmatrix} \begin{bmatrix} -K_{ii}^{-1} K_{ij} \\ I \end{bmatrix} = \begin{Bmatrix} 0 \\ 0 \\ 0 \\ 0 \\ 0 \end{Bmatrix}.$$

$[\text{LTM}_2] = \{0\}$ means that the structure has a determinate interface.

19.4 Continuous Dynamic Systems

The LTMs are generally applied to discrete dynamic systems (finite element model), however, for continuous dynamic systems the same procedures can be used. The continuous dynamic system must be discretised using assumed modes within the framework of the Rayleigh–Ritz method. The real discretization in generalised coordinates can be done with the aid of the Lagrange equations. Once the continuous dynamic system has been discretised and transformed into a set of generalised coordinates, associated with the assumed mode approach, the same procedure may be followed as described in previous sections.

The procedure to generate the LTMs for continuous dynamic systems will be illustrated with two examples, a massless "clamped" beam, with a discrete mass at the tip with constrained (Fig. 19.3) and free-free boundary conditions (Fig. 19.4). The bending moment and shear force at the "clamped" positions are to be calculated.

Fig. 19.3. Clamped beam

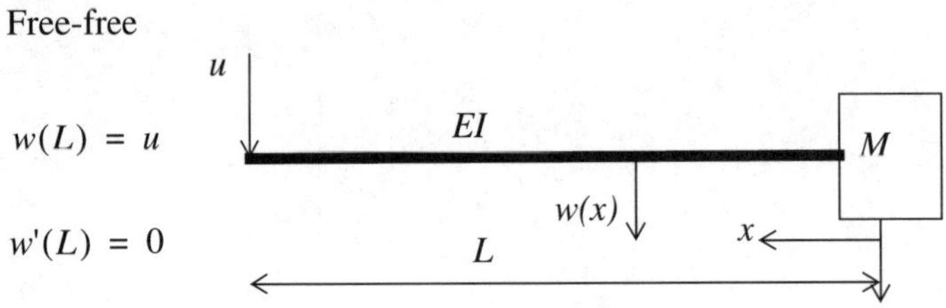

Fig. 19.4. Free-free beam

The displacement function $w(x)$ due to the unit load $F = 1$ N at the tip of the clamped beam is [Prescott 24] (Fig. 19.5):

$$EIw(x) = \frac{1}{6}x^3 - \frac{1}{2}xL^2 + \frac{1}{3}L^3. \tag{19.21}$$

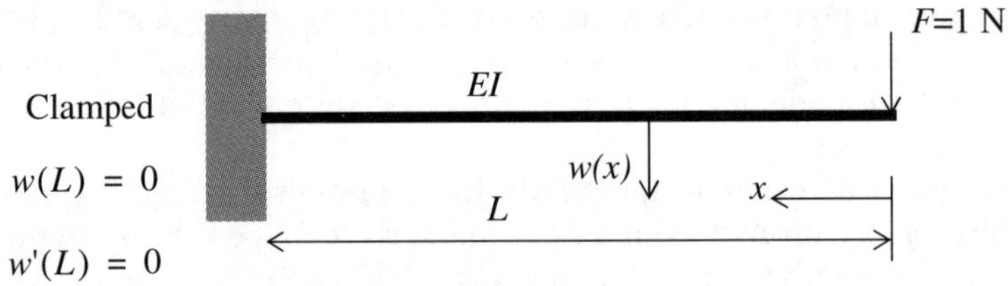

Fig. 19.5. Clamped beam

The displacement $w(x, t)$ is expressed as an assumed mode $\phi(x)$ multiplied by a generalised coordinate $\eta(t)$:

$$w(x, t) = \Phi(x)\eta(t) \tag{19.22}$$

with:

$$\phi(x) = \left(\frac{x}{L}\right)^3 - 3\left(\frac{x}{L}\right) + 2 \tag{19.23}$$

The potential (strain) energy of the clamped beam is (Fig. 19.4):

$$U = \frac{1}{2}\int_0^L EI(w'')^2 dx = \frac{1}{2}EI\eta^2\int_0^L (\phi''(x))^2 dx = \frac{1}{2}\frac{12\mathrm{EI}}{L^3}\eta^2. \tag{19.24}$$

The kinetic energy of the discrete mass is:

$$T = \frac{1}{2}M\phi^2(0)\dot{\eta}^2 = \frac{1}{2}4M\dot{\eta}^2. \tag{19.25}$$

The virtual work applied by the external force is:

$$\delta W = -ML\ddot{u}\delta w(0) = -2ML\ddot{u}\delta\eta = Q\delta\eta. \tag{19.26}$$

With the Lagrange equations an equivalent mass–spring system is established:

$$\frac{\partial}{\partial t}\frac{T}{\partial\dot{\eta}} - \frac{\partial T}{\partial\eta} + \frac{\partial U}{\partial\eta} = Q \tag{19.27}$$

This results in the equation of motion of the equivalent mass–spring system:

$$\ddot{\eta} + \frac{3\mathrm{EI}}{ML^3}\eta = -\frac{\ddot{u}}{2}. \tag{19.28}$$

The bending moment M_{bend} and the shear force D may be calculated:

$$M_{\text{bend}}(x) = -EIw''(x, t) \text{ and } D(x) = -EIw'''(x, t). \tag{19.29}$$

Hence

$$\begin{Bmatrix} M_{\text{bend}} \\ D \end{Bmatrix} = -\frac{6EI}{L^3}\begin{Bmatrix} x \\ 1 \end{Bmatrix}\eta(t) \tag{19.30}$$

$$\begin{Bmatrix} M_{\text{bend}} \\ D \end{Bmatrix} = [\mathrm{LTM}_F]\{F(t)\} + [\mathrm{LTM}_\eta]\{\eta_1(t)\}. \tag{19.31}$$

The load transformation matrix $[\mathrm{LTM}_F] = [0]$ and

$$[\mathrm{LTM}_\eta] = \frac{-6\mathrm{EI}}{L^3}\begin{Bmatrix} x \\ 1 \end{Bmatrix}, \tag{19.32}$$

with:

$$\ddot{\eta} + \frac{3\mathrm{EI}}{ML^3}\eta = -\frac{\ddot{u}}{2}.$$

This example (Fig. 19.4) is the free-free beam with $x_i = \eta$ and $x_j = u$ and

$$w(x, t) = u(t) + \phi(x)\eta(t) \tag{19.33}$$

The potential (strain) energy of the clamped beam is:

$$U = \frac{1}{2}\int_0^L EI(w'')^2 dx = \frac{1}{2}EI\eta^2\int_0^L (\phi''(x))^2 dx = \frac{1}{2}\frac{12\text{EI}}{L^3}\eta^2. \tag{19.34}$$

The kinetic energy of the discrete mass is:

$$\frac{1}{2}U = \frac{1}{2}M\{\phi(0)\dot{\eta} + \dot{u}\}^2 = \frac{1}{2}M(2\dot{\eta} + \dot{u})^2. \tag{19.35}$$

With the Lagrange equations an equivalent mass–spring system is established:

$$\frac{\partial}{\partial t}\frac{T}{\partial\dot{\eta}} - \frac{\partial T}{\partial\eta} + \frac{\partial U}{\partial\eta} = 0 \text{ and } \frac{\partial}{\partial t}\frac{T}{\partial\dot{u}} - \frac{\partial T}{\partial u} + \frac{\partial U}{\partial u} = R. \tag{19.36}$$

The reaction force R is due to the base excitation.
The equations of motions of the $2x2$ dynamic system are:

$$M\begin{bmatrix} 4 & 2 \\ 2 & 1 \end{bmatrix}\begin{Bmatrix} \ddot{\eta} \\ \ddot{u} \end{Bmatrix} + \begin{bmatrix} \frac{12\text{EI}}{L^3} & 0 \\ 0 & 0 \end{bmatrix}\begin{Bmatrix} \eta \\ u \end{Bmatrix} = \begin{Bmatrix} 0 \\ F \end{Bmatrix}. \tag{19.37}$$

Applying the MAM, the solution of the generalised coordinate η can be written as:

$$\eta^* = \frac{-ML^3}{12\text{EI}}\begin{bmatrix} 4 & 2 \end{bmatrix}\begin{Bmatrix} \ddot{\eta} \\ \ddot{u} \end{Bmatrix} + 0u, \tag{19.38}$$

with

$$\begin{bmatrix} M_{\text{bend}}(x) \\ D(x) \end{bmatrix} = -\frac{6EI}{L^3}\begin{Bmatrix} x \\ 1 \end{Bmatrix}\eta^*(t).$$

The load transformation matrices were defined as:

$$\{\sigma\} = [\text{LTM}_1]\begin{Bmatrix} \ddot{\eta}_p \\ \ddot{x}_j \end{Bmatrix} + [\text{LTM}_2]\{x_j\}$$

$$[\text{LTM}_1] = M\begin{bmatrix} 2x & x \\ 2 & 1 \end{bmatrix} \tag{19.39}$$

$$[\mathrm{LTM}_2] = \begin{Bmatrix} 0 \\ 0 \end{Bmatrix}. \qquad (19.40)$$

19.5 Problems

19.5.1 Problem 1

A clamped beam with a bending stiffness EI, a length L and a mass per unit of length m has been connected at the tip with a simple mass–spring system with mass M and stiffness k (Fig. 19.6).

The displacement $w(x, t)$ is expressed as an assumed mode $\phi(x)$ multiplied by a generalised coordinate $\eta(t)$: $w(x, t) = \Phi(x)\eta_1(t)$with: $\phi(x) = \left(\frac{x}{L}\right)^3 - 3\left(\frac{x}{L}\right) + 2$.

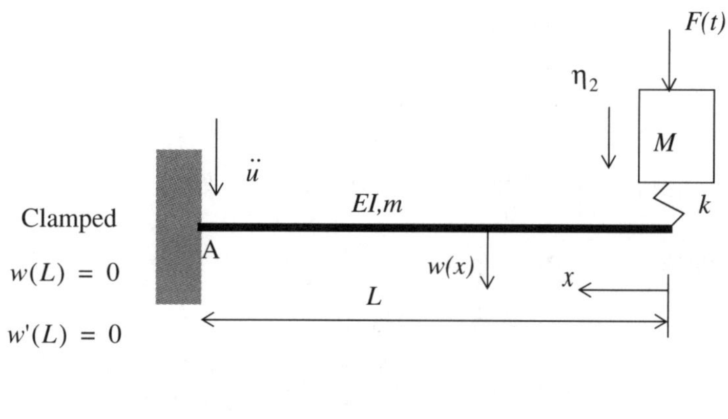

Fig. 19.6. Beam mass-spring system

Perform the following steps:

- Set up the equations of motion for clamped beam and mass–spring system with a base excitation $\ddot{u}$.

- Set up the 'free-free" load transformation matrices

$$\{\sigma\} = [\mathrm{LTM}_1]\begin{Bmatrix} \ddot{\eta}_p \\ \ddot{x}_j \end{Bmatrix} + [\mathrm{LTM}_2]\{x_j\}$$ for the bending moment $M(L)$, the shear force $D(L)$ and the force in the spring with spring stiffness k expressed in $\ddot{u}$, $\ddot{\eta}$ and u.

- What is the physical meaning of the mass matrices $[M_{ij}] = [M_{ji}]^T$ and what are they called?

Answers: The undamped equations of motion:

$$\begin{bmatrix} 0.94286mL & 0 & 1.5mL \\ 0 & M & 0 \\ 1.5mL & 0 & mL \end{bmatrix}\begin{Bmatrix} \ddot{\eta}_1 \\ \ddot{\eta}_2 \\ \ddot{u} \end{Bmatrix} + \begin{bmatrix} \frac{12\mathrm{EI}}{L^3} + 4k & -2k & 0 \\ -2k & k & 0 \\ 0 & 0 & 0 \end{bmatrix}\begin{Bmatrix} \eta_1 \\ \eta_2 \\ u \end{Bmatrix} = \begin{Bmatrix} 0 \\ F \\ R \end{Bmatrix}.$$

The submatrices:

$$[K_{ii}] = \begin{bmatrix} \frac{12\mathrm{EI}}{L^3} + 4k & -2k \\ -2k & k \end{bmatrix}, \; [K_{ij}] = \begin{Bmatrix} 0 \\ 0 \end{Bmatrix}$$

$$[M_{ii}] = \begin{bmatrix} 0.94286mL & 0 \\ 0 & M \end{bmatrix}, \; [M_{ij}] = \begin{Bmatrix} 1.5mL \\ 0 \end{Bmatrix}.$$

The force matrix:

$$[D_\sigma] = \begin{bmatrix} -\frac{6EI}{L^3}x & 0 & 0 \\ -\frac{6EI}{L^3} & 0 & 0 \\ 2k & -k & 0 \end{bmatrix}.$$

19.5.2 Problem 2

A bending beam with a length L, a bending stiffness EI, and a mass per unit length m, is simply supported at both ends. Both supports are enforced with the acceleration $\ddot{u}$. The following questions are to be answered:

- Derive the equation of motion expressed in a relative deflection w(x,t).
- Derive the analytical solutions for the bending eigenfunctions $\phi(x)$.
- Derive the equations of motion using the generalised coordinates $\eta(t)$. The physical deflection can be expressed as: $w(x, t) = \phi(x)\eta(t)$.
- Derive the ATM in the middle of the supported beam at $x = \frac{1}{2}L$.
- Derive the LTM in the middle of the beam at $x = \frac{1}{2}L$, with $M = -EIw'$ and $D = -EIw''$

19.5.3 Problem 3

The dynamic system is illustrated in Fig. 19.2 with the degrees of freedom $x_1 = x_4 = 0$. A load F is applied in degree of freedom x_2, however, in the opposite direction. The stress matrix $[D_\sigma]$ is defined as follows:

$$[D_\sigma] = k\begin{bmatrix} -1 & 0 \\ 1 & -1 \\ 0 & 1 \\ 0 & -0{,}5 \\ 1 & 0 \end{bmatrix}.$$

Calculate $[\mathrm{LTM}_F]$ and $[\mathrm{LTM}_\eta]$.

19.5.4 Problem 4

The dynamic system is illustrated in Fig. 19.2. The degrees of freedom are $\{x_j\} = \begin{Bmatrix} x_1 \\ x_4 \end{Bmatrix}$ and $\{x_i\} = \begin{Bmatrix} x_2 \\ x_3 \end{Bmatrix}$. The stress matrix $[D_\sigma]$ is defined as follows

$$[D_\sigma] = k\begin{bmatrix} 1 & -1 & 0 & 0 \\ 0 & 1 & -1 & 0 \\ 0 & 0 & 1 & -1 \\ 0.5 & 0 & -0.5 & 0 \\ 0 & 1 & 0 & -1 \end{bmatrix}.$$

Calculate $[\mathrm{LTM}_1]$ and $[\mathrm{LTM}_2]$.

References

Abramowitz, M., Stegun, I.A., *Handbook of Mathematical Functions With Formulas*, Graphs, and Mathematical Tables, National Bureau of Standards, Applied Mathematics Series. 55, 1970.

Allen T., *Automatic ASET Selection for Dynamics Analysis*, NASA CP 2303, Twenty-first NASTRAN User' s conference, 2630 April, 1993, pages 175–181.

Allen T., Cook, G., Walls, W., *Improved Omit Aset Displacement Recovery in Dynamics Analysis*, NASA CP 2303, Twenty-first NASTRAN User' s Conference, 2630 April, 1993, pages 8–16.

Appel, S., *Calculation of Modal participation Factors and Effective Mass with the Large Mass Approach*, Fokker Space report FSS-R-92-0027, 1992.

Assink, F.C.J., *Guidelines for the calculation of Shock Response Spectra*, Memo Environmental Test Laboratory, Signaal, RDT/950828/02, The Netherlands, 1995 [in Dutch].

Babuska, I., Prager, M., Vistasek, E., *Numerical Processes in Differential Equations*, Interscience Publishers, John Wiley, 1966.

Barnoski, R.L., Piersol, A.G., Van der Laan, W.F., White, P.H., Winter, E.F., *Summary of Random Vibration Prediction Procedures*, NASA CR- 1302, 1969.

Beards, C.F., *Structural Vibration: Analysis and Damping*, Arnold, 1996, ISBN 0 340 64580 6.

Belytscho, T., Liu, W.K., Moran, B., *Nonlinear Finite Elements for Continua and Structures*, Wiley, ISBN 471-98774-3, 2000.

Benoroya, H., *Mechanical Vibration, Analysis, Uncertainties, and Control*, Prentice Hall, ISBN 0-13-948373, 1998.

Beranek, L.L. (ed.), *Noise and Vibration Control*, McGraw-Hill, ISBN 07-004841-X, 1971.

Bismark-Nasr, M.N., *Structural Dynamics in Aeronautical Engineering*, ISBN 1-56347-323-2, AIAA Education Series, 1999.

Bolotin, Vladimir, V., *Wahrscheinlichkeitsmethoden zur Berechnung von Konstructionen*, VEB Verlag fuer Bauwesen, GDR, 1981, translation from Russian.

Bray, E.L., *Specification for Dynamic Models of Polar Platform*, British Aerospace Ltd., Earth Observation and Science Division, SPE-1211368-003, Issue 3, May 1991.

Brock, J.E., *Dunkerly-Miklin Estimates of Gravest Frequency of a Vibrating System*, Journal of Applied Mechanics, June 1976, pages 345–348.

Bucher, I. Braun, S.G., *Left Eigenvectors: Extraction From Measurements and Physical Interpretation*, Journal of Applied Mechanics, Vol. 64, March 1997, pages 97-105.

Carrington, H.G., *A Survey of Data on Damping in Spacecraft Structures*, Final report, ESRO CR-539, 1975.

Ceasar, B., Vogel, F., *The Introduction of Clampbands in Spacecraft Structural Mathematical Models*, ESA study 5099/82/NL/PB(SC), EMSB-12/83, 1983.

Chopra, A.K., Dynamics of Structures, *Theory and Applications to Earthquake Engineering*, ISBN 0-13-855214-2, Prentice Hall, 1995.

Chung, J, Hulbert, G.M., *A Time Integration Algoritm for Structural Dynamics With Improved Numerical Dissipation: The Generalized-Method*, Journal of Applied Mechanics, June 1993, pages 371–375.

Chung, Y.T., *Dynamic Loads Recovery using Alternative Mode Acceleration Approach*, AIAA-98-1719, 39th AIAA/ASME/ASCE/AHS/ ASC/ Structures, Structural Dynamics, and Materials Conference and Exhibit and AIAA/ ASME/AHS Adaptive Structures Forum, Long Beach, California, April 2023, 1998.

Claessens, G.J.T.J. Wijker, J.J., *The Accuracy of Reduced Component Models*, Proc. Int. Conf.: "Spacecraft Structures and Mechanical testing", Noordwijk, The Netherlands, 2729 March 1996, pages 533547, ESA SP-386.

Cook, R. D., Malkus, D. S., Plesha, M. E., *Concepts and Applications of Finite Element Analysis*, Third edition, John Wiley and Sons, ISBN 0-471-84788-7, 1989.

Craig, R.R, Jr. Bampton, M.C.C., *Coupling of Substructures for Dynamic Analysis*, AIAA Journal, Vol. 6 No. 7, July 1968, pages 1313–1319.

Craig, R.R, Jr. Chang, Ching-Jone, *Free-Interface Methods of Substructure Coupling for Dynamic Analysis*, AIAA Journal, Vol. 14, No. 11, pages 1633–1635, 1976.

Craig, R.R, Jr. Chang, Ching-Jone, *Substructure Coupling for Dynamic Analysis and Testing*, NASA CR-2781, 1977.

Craig, R.R, Jr., Structural Dynamics, *An Introduction to Computer Methods*, John Wiley & Sons, 1981, ISBN 0-471-04499-7.

Craig, R.R, Jr., *Coupling of Substructures for Dynamic Analysis: An Overview*, AIAA-2000-1573, 41st AIAA/ASME/ASCE/AHS/ASC Structures, Structural Dynamics, and Materials Conference and Exhibit, 38 April 2000, Atlanta, GA.

Crandall S.H. (ed.), *Random Vibration Volume 2*, M.I.T. Press, 1963.

Crandall S.H., Mark W.D., *Random Vibration in Mechanical Systems,* Academic Press, 1973.

Cremer, L., Heckl, M., Ungar, E.E., *Structure-Borne Sound, Structural Vibrations and Sound Radiation at Audio Frequencies*, Springer-Verlag, ISBN 3-540-06002-2, 1973.

Curnier, A., *Application of Flight Loads in the Structural Dynamic Design of Spacecraft Deploying Non Transient Analysis Methods*, ESI report ED 80-307/RD, 1981.

Dickens J.M. Stroeve A., *Modal Truncation Vectors for Reduced Dynamic Substructures*, AIAA-2000-1578, 41st AIAA/ASME/ASCE/ AHS/ASC Structures, Structural Dynamics, and Materials Conference and Exhibit, 38 April 2000, Atlanta, GA.

Dokainish, M.A., Subbraraj, K, *A Survey of Direct Time- Integration Methods in Computational Structural Dynamics-I, Explicit Methods*, Computer & Structures, Vol. 32, No. 6, 1989, pages 1371–1386.

D' Souza, A.F., Garg, V.K.*Advanced Dynamics Modelling and Analysis*, Prentice-Hall, 1984, ISBN 0-13-011312-3.

Eaton, D., *Structural Acoustics Design Manual*, ESA PSS03-204, Issue 1, March 1996, European Space Agency, ESTEC, Noordwijk.

Ebeling, R.M., Green, R.A., French, S.E., *Accuracy of Response of Single-Degree-of-Freedom Systems to Ground Motion*, Technical Report ITL-97-7, US Army Corps of Engineers, WES, 1997.

Elishakoff I., *Probalistic Methods in the Theory of Structures*, ISBN 0-471-87572-4, John Wiley and Sons, 1983.

Elishakoff I., Lin, Y.K., Zhu, L.P., *Probabilistic and Convex Modelling of Acoustically Excited Structures*, ISBN 0-444-81624-0, Elsevier, 1994.

Escobedo-Torres, J. Ricles, J.M., *Improved Dynamic Analysis method using Load-Dependent Ritz Vectors*, AIAA-93-1489-CP, 34th AIAA/ASME/ ASCE/AHS/ASC Structures, Structural Dynamics and Materials Conference AIAA/ASME Adaptive Structures Forum, April 1922, 1993/La Jolla, CA, USA.

Ewins, D.J., *Modal Testing: Theory, Practice and Application*, Second edition, Research Studies Press, ISBN 086380 218 4, 2000.

Francesconi, D., *Statistical Energy Analysis Application on Pressurised Modules: Test and Analysis of Typical Structures*, Conference on Spacecraft Structures, Materials & Mechanical testing, ESA SP-386, Noordwijk, The Netherlands, 27-29 March, 1996.

Fransen, S.H.J.A., *An Overview and Comparison of OTM Formulations on the Basis of the Mode Displacement Method and the Mode Acceleration Method*, paper 17, Proceedings of the Worldwide Areospace Conference and Technology Showcase 2002, 810 April, 2002, Toulouse.

Friswell, M.I., Mottershead, J.E., *Finite Element Updating in Structural Dynamics*, Kluwer, 1995.

Friswell, M.I., Lees, A.,W., *Resonance Frequencies of Viscously Damped Structures*, Journal of Sound and Vibration (1998) 217(5), pages 950-959.

Gatti, P. L., Ferrari V., *Applied Structural and Mechanical Vibrations, Theory, Methods and Measuring Instrumentation*, E&FN SPON, 1999, ISBN 0-419-22710-5.

Gecko, M.A., *MSC/NASTRAN Handbook for Dynamic Analysis V63*, MSR-64, The MacNeal-Schwendler Corporation, 1983.

Gordon, S., *The Craig-Bampton Method*, FEMCI Presentation, NASA Goddard Space Flight Center, May 6, 1999, http://analyst.gsfc.nasa.gov/FEMCI/craig_bampton/.

Grygier, M.S., *Payload Loads Design Guide*, Lyndon B. Johnson Center, NASA, 1997.

Gupta, A.K., *Response Spectrum Method in Seismic Analysis and Design of New Structures*, ISBN 0-8493-8628-4, CRC Press, 1992.

Guyan, R.J., *Reduction of Stiffness and Mass Matrices*, AIAA Journal, Vol. 3, No. 2, 1968, page 380.

Healsig, R.T., *DYNRE4 Technical Discussion*, STARDYNE Finite Element Program, 1972.

Hairer, E., Norsett, S.P., Wanner, G., *Solving Ordinary Differential Equations I, Nonstiff Problems*, ISBN 0-540-56670-8, Springer, 1992.

Hairer, E., Wanner, G., *Analysis by Its History*, ISBN 0-387- 94551-2, Springer, 1996.

Haris, R.W., Ledwidge, T.J., *Introduction to Noise Analysis*, Pion Limited, ISBN 0 85086 041 5, 1974.

Heinrich W., Hennig K., *Zufallsschwingungen Mechanischer Systemen*, Vieweg, ISBN 3 528 06822 1, 1978.

Herting, D.N., Morgan, M.J., *A General Purpose, Multi-stage Component Modal Synthesis Method*, AIAA/ASME/ASCE/AHS, 20th Structures,

Structural Dynamics, and Materials Conference, April 6, 1979, presented by Universal Analytics, Inc.

Hintz, R.M., *Analytical Methods in Component Modal Synthesis*, AIAA Journal, Vol. 13, No. 8, 1975, pages 1007–1016.

Houbolt, J.C., *A Recurrence Matrix Solution for the Dynamic Response of Elastic Aircraft*, Journal Aeronautical Sciences, September 1950, pages 540–550.

Hughes, T.J.R. Belytschko, T., *A Precis of Developments in Computational Methods for Transient Analysis*, Journal of Applied Mechanics, December 1983, pages 1033–1041.

James, Lyn, *Advanced Modern Engineering Mathematics*, Addison-Wesley, 1993.

Journal of Acoustical Society of America (JASA), *Correlation Coefficients in Reverberant Sound*, Vol. 27, 1955, page 1073.

Kammer, D.C., *Test-Analysis-Model Development using an Exact Modal Reduction*, Journal of Modal Analysis, October 1987, pages 174–179.

Keane, A.J., Price, W.G., *Statistical Energy Analysis, An Overview, with Applications in Structural Dynamics*, ISBN, 0 512 55175 7, Cambridge University Press, 1994.

Kelly, R.D., Richman, G., *Principles and Techniques of Shock Data Analysis*, SVM-5, The Shock and Vibration Information Centre, US DoD, 1969.

Keltie, R., *Applications of Statistical Energy Analysis*, Lecture notes, North Carolina State University, Mechanical & Aerospace Engineering, Box 7910, Raleigh, N.C. 27695, USA, 2001.

Kenny, A., *Statistical Energy Analysis (SEA), A Vital CAE solution for Noise and Vibration: Part I-What Can it Do For Me*, Benchmark, Nafems, April 2002 a.

Kenny, A., *Statistical Energy Analysis (SEA), A Vital CAE solution for Noise and Vibration: Part II-SAE Theory & Relationship to FEM*, Benchmark, Nafems, July 2002 b.

Klein, M., Reynolds, J., Ricks, E., *Derivation of improved load transformation matrices for launchers-spacecraft coupled analysis, and direct computation of margins of safety*, Proc. Int. Conf.: "Spacecraft Structures and Mechanical testing", Noordwijk, The Netherlands, 1921 October 1988, pages 703–719.

Kreyszig, E., *Advanced Engineering Mathematics*, Seventh edition, John Wiley, 1993.

Lanczos, C., *The Variational Principles of Mechanics*, Fourth Edition, University of Toronto Press, ISBN 0-8020-1743-6, 1972.

Lacoste, P., *Modal deformations and Stresses*, ESI report ED 81-343/ R&D, ESA CR(P) 1777, 1983.

Lin, Y.K., *Probabilistic Theory of Structural Dynamics*, R.E. Krieger, ISBN 0-88275-377-0, 1976.

Lutes, L.D., Sarkani Shahram, *Stochastic Analysis of Structural and Mechanical Vibrations*, Prentice Hall, 1996.

Lyon, R.H., Maidanik, G., *Power Flow between Linearly Coupled Oscillators*, The Journal of Acoustical Society of America, Vol. 34, Nr. 5, May, 1962, pages 623–639.

Lyon, R.H., Maidanik, G., *Statistical Methods in Vibration Analysis*, AIAA Journal, Vol. 2, No. 6, June 1964, pages 1015–1024.

Lyon, R.H., *Statistical Energy Analysis of Dynamical Systems, Theory and Applications*, MIT Press, ISBN 0-262-12071-2, 1975.

Lyon, R.H., Dejong, R.G., *Theory and Application of the Statistical Energy Analysis*, Second edition, Butterworth-Heinemann, ISBN 0-7506-9111-5, 1995.

MacNeal, R.H., *A Hybrid Method of Component Mode Synthesis*, Computer & Structures, Vol. 1, pp. 581601, 1971.

Madayag, A.F., *Metal Fatigue: Theory and Design*, Chapter 7 Response to Random Loadings; Sonic Fatigue, Smith, H.P., Wiley, 1969, SBN 471 56315 3.

Maia, N.M.M., Silva, J.M.M., *Theoretical and Experimental Modal Analysis*, John Wiley, ISBN 0 471 97067 0, 1997.

Maymon G., *Some Engineering Applications in Random Vibrations and Random Structures*, Volume 178, Progress in Astronautics and Aeronautics, AIAA, ISBN 1-56347-258-9, 1998.

McConnel, K.G., *Vibration Testing, Theory and Practice*, John Wiley, ISBN 0-471-30435-2, 1995.

McGowan, D.M., Bosic S.W., *Comparison of Advanced Reduced-Basis Methods for Transient Structural Analysis*, AIAA Journal, Vol. 31, No. 9, September 1993, pages 1712–1719.

Meirovitch, L., *Methods of Analytical Dynamics*, MacGraw-Hill, 1970.

Meirovitch, L., *Elements of Vibration Analysis*, McGraw-Hill, ISBN 0-07-041340-1, 1975.

Meirovtch, L., *Principles and Techniques of Vibrations*, Prentice Hall, ISBN 0-02-380141-7, 1997.

Michlin, S.C., *Variationsmethoden der Mathematischen Physik*, Akademie-Verlag, Berlin, 1962.

Miles, J.W., *On Structural Fatigue Under Random Loading*, Journal of the Aeronautical Sciences, November 1954, pages 753–762.

Miller, C.A., *Dynamic Reduction of Structural Models*, Proceedings of the American Society of Civil Engineers, Vol. 16, ST10, October 1980, pages 2097–2108.

Moretti, P.M., *Modern Vibrations Primer*, CRC Press, ISBN 0- 8493-2038-0, 2000.

Mulville, D.R., *Pyroshock Test Criteria*, NASA Technical Standard, NASA-STD-7003, May 18, 1999, http://standards.nasa.gov.

Nahin, P.J., *An Imaginary Tale, The Story of* $\sqrt{-1}$, Princeton University Press, ISBN 0-691-02795-1, 1998.

Nelson, D.B., Prasthofer, P.H., *A Case for Damped Oscillatory Excitation as a Natural Pyrotechnic Shock Simulation*, Shock and Vibration Bulletin, No 44, Part 3, pages 67–71, 1974.

Newland, D.E., *Mechanical Vibration Analysis and Computation*, Longman Scientific & Technical, ISBN 0-470-21388 (Wiley) or ISBN 0-582-02744, 1989.

Newland, D. E., *An Introduction to Random Vibrations, Spectral & Wavelet Analysis*, Third edition, Longman, ISBN 0582 21584 6, 1994.

Nigam, N.C., Jennings, P.C., *Digital Calculation of Response from Strong-Motion Earthquake Records*, California Institute of Technology, 1968, printed in STARDYNE Theoretical Manual, III. DYNRE5 Program Analysis.

Nigam, N.C., Narayanan, S., *Applications of Random Vibrations*, ISBN 3-540-9861-X, Springer Verlag, 1994.

Norton, M.P., *Fundamentals of noise and vibration analysis for engineers*, Cambridge University Press, ISBN 0 521 34148 5, 1989.

O' Callahan, J.,*A Procedure for an Improved Reduced System (IRS) Model*, Proceedings of the 7th International Modal Analysis Conference, Las Vegas, pages 17-21.

Papoulis, A., *Probability, Random Variables and Stochastic Processes*, McGraw-Hill, International Student Edition, 1965.

Park, K.C., *An Improved Stiffly Stable Method for Direct Integration of Nonlinear Structural Dynamic Equations*, Journal of Applied Mechanics, June 1975, pages 464–470.

Petyt, M., *Introduction to finite element vibration analysis*, Cambridge University Press, ISBN 0-521-26607-6, 1990.

Pinnington, R.J., *Approximate Mobilities of Built up Structures*, ISVR contract report 86/16, University of Southampton, 1986.

Pilkey, W.D., Wunderlich, W., *Mechanics of Structures, Variational and Computational Methods*, CRC Press, ISBN 0-8493-4435-2, 1994.

Piszczek K., Niziol J., *Random Vibration of Mechanical Systems*, John Wiley, ISBN 0-85312-347-0, 1986, Ellis Horwood Limited.

Prescott, J., *Applied Elasticity*, Dover Publications, Inc., 1924.

Preumont, A., *Vibration Control of Active Structures*, Kluwer, ISBN 0-7923-4392-1, 1997.

Przemieniecki, J.S., *Theory of Matrix Structural Analysis*, Dover, 1985, ISBN 0-486-64948-2.

Ricks, E.G., *Guidelines for loads analyses and dynamic verification of Shuttle cargo elements*, NASA MSFC-HDBK-1974, 1991.

Robson J.D., Dodds, C.J., Macvean, D.B., Paling, V.R., *Random Vibrations*, Springer Verlag, ISBN 3-211-81223-7, 1971.

Rose, T., *Using Residual Vectors in MSC/NASTRAN Dynamic Analysis to Improve Accuracy*, 1991 MSC World Users' Conference.

Schueller, G.J., Shinozuka, M., *Stochastic Methods in Structural Dynamics*, Nijhoff, ISBN 90-247-3611-0, 1987.

Schwarz, H.R., *Numerical Analysis, A Comprehensive Introduction*, Wiley, ISBN 0 471 92064 9, Third edition, 1989.

Seide P., *Influence Coefficients for End-Loaded Conical Shells*, AIAA Journal, Vol. 10, No. 12, 1972, pages 1717–1718.

Shunmugavel, P., *Modal Effective Masses for Space Vehicles*, Rockwell Space Systems Division, Downey, California, AIAA-95-1252, 1995.

Skudrzyk, E., *Simple and Complex Vibratory Systems*, The Pennsylvania State University Press, 1968.

Smallwood, D.A., *Time History Synthesis for Shock Testing on Shakers*, Shock and Vibration Bulletin, No 44, Part 3, pages 23–41, 1974 a.

Smallwood, D.A., Nord, A.R., *Matching Shock Spectra with Sums of Decaying Sinusoids Compensated for Shaker Velocity and Displacement Limitations*, Shock and Vibration Bulletin, No 44, Part 3, pages 43–56, 1974 b.

Smith, W.P., Lyon, R.H., *Sound and Structural Vibration*, NASA CR-160, 1965.

Spiegel, M.R., *Theory and Problems of Complex Variables with an Introduction to Conformal Mapping and its Applications*, Schaums' s Outline Series, McGraw-Hill, 1964.

Stephenson, G., *An Introduction to Partial Differential Equations for Science Students*, Longman, 1970, ISBN 0582 44430 6.

Strang, G., *Linear Algebra and its Applications*, Third edition, Harcourt Brace Javanovich Inc., ISBN 0-15551005-3, 1988.

Stravrinidis, C., *Dynamic synthesis and evaluation of spacecraft structures*, ESA STR-208, March 1984.

Subbraraj, K., Dokainish, M.A., *A Survey of Direct Time- Integration Methods in Computational Structural Dynamics-II, Implicit Methods*, Computer and Structures, Vol. 32, No. 6, 1989, pages 1387–1401.
Temple, G., Bickley, W.G., Rayleigh' s principle and Its Applications to Engineering, Dover edition, 1956.
Thomson, W.T., Dahleh, M.D., *Theory of Vibration with Applications*, 5th edition, Prentice Hall, ISBN 0-13-651068-X, 1998.
Thonon, C., Rixen, D., Geradin, M., *Unification of Impedance/ Admittance and Component Mode Formulations for the Assembling of Flexible Structures*, Workshop Proceedings "Advanced Mathematical in the Dynamics of Flexible Bodies, ESA WPP-113, January 1998, ISSN 1022-6656.
Trudell, R.W., Yano, L.I., *Statistical Energy Analysis of complex structures Phase II*, Final report, MDC G9203, NASA CR-161576, 1980.
Universal Analytics Inc. (UAI), *UAI/NASTRAN Seminar on Substructuring and Modal Synthesis*, November 1993, Universal Analytics, Inc, 3625 Del Amo Blvd, Suite 370, Torrance, CA 90503, USA.
Ungar, E.E., *Fundamentals of Statistical Energy Analysis of Vibration Structures*, AFFDL-TR-66-52, AF Flight Dynamics Laboratory, Wright-Patterson Air Force Base, Ohio 45433, 1966.
Scheidt vom, J., Fellenberg, B., Woehrl, U., *Analyse und Simulation stochastischer Schwingungssysteme*, B.G. Teubner Stuttgart, ISBN 3- 519-02376-8, 1994.
Wax N. (ed.), *Selected Papers on Noise and Stochastic Processes*, Dover Publications, 1954.
Wepner G., *Mechanics of Solids*, PWS Publishing Company, 1995, ISBN 0-534-92739-4.
Wijker, J.J., *Reduction of Finite Element System Matrices using GDR*, Fokker Space report, FSS-R-91-0023, 1991.
Wirsching, P.H., Paez, T.L., Ortiz, K., *Random Vibrations, Theory and Practice*, John Wiley, ISBN 0-471-58579-3, 1995.
Witting M., Klein M., *Modal Selection by Means of Effective Masses and Effective Modal Forces an Application Example*, Proc. Conference on Spacecraft Structures, Materials & Mechanical Testing, 2729 March, 1996 (ESA SP-386, June 1996).
Wood, W.L., *Practical Time-stepping Schemes*, Oxford Applied Mathematics and Computing Science Series, 1990, ISBN 0-19-853208-3.
Woodhouse, J., *An approach to the theoretical background of statistical energy analysis applied to structural vibration*, J. Acoustical Soc. Am., 69(6), June 1981, pages 1695–1709.

Zurmuehl, R., *Matrizen und ihre technische Anwendungen*, 4th edition, Springer-Verlag, 1964.

Author Index

Subject Index

W

Z

Druck: Strauss Offsetdruck, Mörlenbach
Verarbeitung: Schäffer, Grünstadt